REEDS MARINE ENGINEERING AND TECHNOLOGY

BASIC ELECTROTECHNOLOGY
FOR MARINE ENGINEERS

Reeds Marine Engineering and Technology Series

Vol. 1 Mathematics for Marine Engineers

Vol. 2 Applied Mechanics for Marine Engineers

Vol. 3 Applied Thermodynamics for Marine Engineers

Vol. 4 Naval Architecture

Vol. 5 Ship Construction

Vol. 6 Basic Electrotechnology for Marine Engineers

Vol. 7 Advanced Electrotechnology for Marine Engineers

Vol. 8 General Engineering Knowledge for Marine Engineers

Vol. 9 Steam Engineering Knowledge for Marine Engineers

Vol. 10 Instrumentation and Control Systems

Vol. 11 Engineering Drawings for Marine Engineers

Vol. 12 Motor Engineering Knowledge for Marine Engineers

Vol. 13 Ship Stability, Resistance and Powering

Vol. 14 Stealth Warship Technology

Vol. 15 Electronics, Navigational Aids and Radio Theory for Electrotechnology Officers

6

REEDS MARINE ENGINEERING AND TECHNOLOGY

BASIC ELECTROTECHNOLOGY
FOR MARINE ENGINEERS

Revised by Christopher Lavers
Edmund G R Kraal
Stanley Buyers

R E E D S
LONDON · OXFORD · NEW YORK · NEW DELHI · SYDNEY

REEDS
Bloomsbury Publishing Plc
50 Bedford Square, London, WC1B 3DP, UK

BLOOMSBURY, REEDS, and the Reeds logo are trademarks of Bloomsbury Publishing Plc

First published by Thomas Reed Publications 1965
Second edition 1973
Reprinted 1977, 1980
Third edition 1985
Reprinted 1994, 1995, 1996, 2000. 2002
Reprinted by Adlard Coles Nautical 2003, 2006, 2008 and 2010
Fourth edition 2013
Fifth edition 2020

A catalogue record for this book is available from the British Library

PB: 978-1-4729-6383-3
ePDF: 978-1-4729-6384-0
ePub: 978-1-4729-6382-6

2 4 6 8 10 9 7 5 3 1

Typeset in Myriad Pro 10/14 by Newgen Imaging Systems Pvt. Ltd, Chennai, India
Printed and bound in Great Britain by CPI Group (UK) Ltd, Croydon CR0 4YY

Bloomsbury Publishing Plc makes every effort to ensure that the papers used in the manufacture
of our books are natural, recyclable products made from wood grown in well-managed forests.
Our manufacturing processes conform to the environmental regulations of the country of origin.

CONTENTS

PREFACE

This book is intended to cover the basic theory underpinning applied work detailed in various syllabi for Electrotechnology written examinations, including the Maritime Coastguard Agency (www.dft.gov.uk/mca) Department of Transport Examinations for Class 1 and Class 2 Engineers (GENERAL ENGINEERING SCIENCE II, STCW III/2 Second Engineer, and STCW III/2 Chief Engineer) but is designed to be much more than an examination focused book. It is also suitable for Marine Engineering students completing Electrical and Electronic Engineering Principles, Electrical Technology and Further Electrical Principles units of the Level 3 Business and Technician Education Council (BTEC) Nationals specifications in Engineering, as well as various International Naval, Coastguard and Marine Science Degree students, such as those taught at Britannia Royal Naval College, Dartmouth, Devon.

As third editor and author of the latest edition of the Basic Electrotechnology Volume 6, and the accompanying Advanced Electrotechnology Volume 7, now at their important half-century waymarks, I am conscious of inheriting a valuable global resource for engineers and marine engineers, specifically building upon the splendid work of both EDMUND G R KRAAL CEng, DFH (Hons), MIEE, MIMarE and STANLEY BUYERS BEd, TEng, MIELecIE.

Sir Isaac Newton stated: 'If I have seen further, it is by standing on the shoulders of giants', and echoing this sentiment I have been careful not to delete many vital principles – even if they have temporarily fallen out of favour with current examinations' syllabi, where appropriate I have included some detail and quotes from several pioneers of Electrotechnology. This marine engineering series was created to emphasise the engineering and applied physics ideas starting from first principles, and with reference to numerous illustrations, to provide worked examples within the text, supplemented with many problems for students to work through on their own to develop a deep understanding of the practical outworking of these key engineering concepts. Typical examination questions at the end of each chapter, and at the end of the book, provide students with further opportunities to test themselves thoroughly against a high standard of questions before attempting the examinations themselves. Fully worked out step-by-step solutions are given to each and every problem, which is especially useful to a marine engineer at sea without recourse to a college tutor. The methodical approach taken throughout the book, with questions followed by detailed logical answers, lends itself to the student generating their own questions and using the laid out established methods to test how well they have understood the process. The student thereby begins to take ownership of his own learning – the experiential doing – partly, I anticipate, by trial and error! And it is the author's hope that among those reading and studying these subjects there will perhaps be a new Tesla, Faraday or Edison capable of lighting up the world with a new generation of electrical devices.

The revolution in global communications within the past two decades, especially for those at sea, and the massive increase in digital storage capacity provide additional tools to this generation of engineers which were not possible in the past, yet these do not provide substitute for *understanding* and applying these key principles. It is also a sobering note that this generation is both more *reliant* than ever on the use of electromagnetic devices and systems and at the same time is most *vulnerable* to their failure. This has always been well understood by the marine engineer, often far from land with the potential failure of vital on-board systems, but is one that has only recently begun to be fully appreciated more widely. We now rely on a multitude of global systems – not just those on land, but sea, air and space – that are especially vulnerable to EMP or the caprices of solar weather, which has the real capability, in its most severe form, to remove, if unprepared, much of the electrical infrastructure that has been so painstakingly developed from the early work by Faraday and Tesla in only a few hours. Marine and terrestrial engineers alike would be well advised to consider the implications of another such 'Carrington Event' solar storm of the magnitude experienced in 1859, which resulted in disruption at the infancy of electrical devices, and even the electrocution of telegraphy operators – that first Victorian Internet.

I, like my predecessors, would like to acknowledge the constructive comments made over the past 25 years by colleagues, officers and students of various nationalities educated and trained at the Britannia Royal Naval College, Dartmouth, Devon, UK and of course the editorial services team at Newgen.

CHRISTOPHER LAVERS Ph.D. (Exon), B.Sc. (Hons), M.Inst.P., C. Phys, PGCE (LTHE), Subject Matter Expert (Radar and Telecommunications), Britannia Royal Naval College 2019

THE S.I. SYSTEM

Prefixes, Symbols, Multiples and Submultiples

Prefix	Symbol	Units multiplying factor
tera	—	$\times 10^{12}$
giga	G	$\times 10^{9}$
mega	M	$\times 10^{6}$
kilo	k	$\times 10^{3}$
milli	m	$\times 10^{-3}$
micro	μ	$\times 10^{-6}$
nano	n	$\times 10^{-9}$
pico	p	$\times 10^{-12}$

Examples

1 megawatt (MW) $\qquad = 1 \times 10^{3}$ kilowatts (kW)

$\qquad\qquad\qquad\qquad = 1 \times 10^{6}$ watts (W)

1 kilovolt (kV) $\qquad = 1 \times 10^{3}$ volts (V)

1 milliampere (mA) $\qquad = 1 \times 10^{-3}$ ampere (A)

1 microfarad (μF) $\qquad = 1 \times 10^{-6}$ farad (F)

Physical Quantities (Electrical), Symbols and Units

The table has been compiled from recommendations in B.S. 1991 and the List of Symbols and Abbreviations issued by the I.E.E.

Quantity	Symbol	Unit	Abbreviation of unit after numerical value
Force	F	newton	N
Work	W	joule	J
or		or	
Energy		newton metre	Nm
Torque	T	newton metre	Nm
Power	P	watt	W
Time	t	second	s
Angular velocity	ω (omega)	radians per second	rad/s
Speed	N	revolutions per minute	rev/min
	n	revolutions per second	rev/s
Electric charge	Q	coulomb	C
Potential difference (P.D.)	V	volt	V
Electromotive force (e.m.f.)	E	volt	V
Current	I	ampere	A
Resistance	R	ohm	Ω (omega)
Resistivity (specific resistance)	ρ (rho)	Ohm-metre	Ωm
Conductance	G	siemens	S
Magnetomotive force (m.m.f.)	F	ampere-turn	At
Magnetic field strength	H	ampere-turn per metre or	At/m
		ampere per metre	A/m

Quantity	Symbol	Unit	Abbreviation of unit after numerical value
Magnetic flux	Φ (phi)	weber	Wb
Magnetic flux density	B	tesla	T
Reluctance	S	ampere-turn or ampere per weber	At/Wb or A/Wb
Absolute permeability of free space	μ_0	henry per metre	H/m
Absolute permeability	μ (mu)	henry per metre	H/m
Relative permeability	μ_r	–	–
Self-inductance	L	henry	H
Mutual inductance	M	henry	H
Reactance	X	ohm	Ω
Impedance	Z	ohm	Ω
Frequency	f	hertz	Hz
Capacitance	C	farad	F
Absolute permittivity of free space	ϵ_0 (epsilon)	farad per metre	F/m
Absolute permittivity	ϵ	farad per metre	F/m
Relative permittivity (dielectric constant, specific inductive capacity)	ϵ_r	–	–
Electric field strength, electric force	E	volt per metre	V/m
Electric flux	ψ (psi)	coulomb	C
Electric flux density, electric displacement	D	coulomb per square metre	C/m^2
Active power	P	watt	W
Reactive power	Q	volt amperes reactive	VAr
Apparent power	S	volt ampere	VA
Phase difference	ϕ (phi)	degree	°
Power factor (p.f.)	$\cos \phi$	–	–

1

FUNDAMENTAL ELECTRICAL THEORY TERMS AND LAWS

I shall make electricity so cheap that only the rich can afford to burn candles.

Thomas Edison

Electron Theory

The nature of electricity

To enable the student engineer to achieve practice with relevant problems and to appreciate fundamental concepts, a start is made in this chapter with an introduction to the subatomic nature of electricity before considering basic circuit theory and relevant calculations. A more detailed explanation is developed as needed in later chapters but it is hoped that the student will understand from the start that electronics and electrical engineering are *related* and that the nature of electricity and many electrical phenomena all have their origin in atomic structure.

The structure of the atom

It is accepted that electrical transmission is due to a flow of *electrons*. However, as there is no observable indication of such a flow in a conductor, we must accept the classical atomic theory on the structure of matter and its effects upon electron movement and their rearrangement. Matter is defined as anything that occupies space; it may be in solid, liquid or gas form, but basically consists of *molecules* of a substance. A molecule is the smallest neutral particle of a substance that exists by itself. Molecules have the properties of the substance that they form but are themselves made of groups of *atoms*. For example, a molecule of water, written H_2O, consists of 2 atoms of hydrogen and one of oxygen. The atom is defined as the smallest particle that can enter into chemical action, but is itself a complex structure consisting of subatomic particles. A substance containing only atoms with the same identical properties is called an *element*, but one containing atoms of different properties is called a *compound*. All atoms of a given element are identical. Atoms of different elements differ only in the number and arrangement of the subatomic particles contained in them. Subatomic particles can be *charged* or *uncharged*. Reference to charge will be made throughout this volume, but at this stage, it is stated that electricity consists of charges and these are of 2 kinds only, positive (+ve) and negative (−ve). Like charges *repel* each other and unlike charges attract each other. Generally the space in which a physical force exists between charges is referred to as an *Electric Field*. A more detailed consideration will, however, be made in Chapter 8 when dealing with electrostatics.

According to the theory proposed by scientists like Rutherford and Bohr (1911), and tested by experiment (1909), each atom has a core or *nucleus* surrounded by *orbital electrons*. The nucleus consists of tiny masses of positively charged subatomic particles or *protons*, and *neutrons*, which have no charge. The main purpose of the neutrons is to 'fix' or 'cement' the positively charged protons together within the nucleus. In a normal stable atom the number of protons is equal to the number of orbiting electrons. The number of protons in an atom determines the *atomic number* (Z) of the element and thus the number of negatively charged orbital electrons. An electron has a mass of 9.04×10^{-28} g and a charge of 1.6×10^{-19} coulomb. A proton has a mass some 1850 times greater than that of an electron, while a neutron has a mass slightly greater than a proton. The concept of the atom is shown in figure 1.1.

The negatively charged electrons are considered to spin about an axis and to revolve around the nucleus similar to the structure of a miniature 'solar system'. The nucleus represents the 'sun' and the electrons represent the 'planets'. Under normal conditions an atom is said to be stable or unexcited. The planetary electrons together neutralise

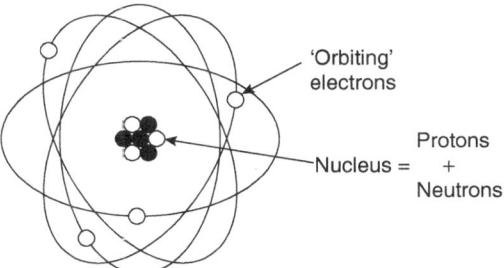

▲ **Figure 1.1** *The atomic concept*

the positively charged protons in the nucleus, so a complete atom itself has no electrical charge. Figure 1.2 shows examples of atomic structure for different elements but these illustrations are drawn in 1 plane only. The simplest atom is that of the element hydrogen, consisting of a nucleus with 1 proton (having a +ve charge) around which orbits 1 electron. The electron with its −ve charge neutralises that of the proton. In the diagrams, electrons are denoted by circles, with their charges shown, and are considered to move on dotted orbits. The nucleus is shown with a full circle, has a net positive charge attributed to the protons and these are shown by + marked circles. Neutrons are shown by small circles with no charge sign.

The next element considered is helium, which has 2 planetary electrons and a nucleus consisting of 2 protons and 2 neutrons. The electrons of most atoms are associated with the nucleus in a definite manner, i.e. the electrons are in groups or *shells*, such that the planetary path of each shell is different. This is shown if an oxygen atom is considered, oxygen has a nucleus of 8 protons and 8 neutrons. The planetary electrons are 8 in 2 orbits or shells – 6 in the outer shell and 2 in the inner shell. For any one atom, the electrons in the first shell may be less than, but never more than 2 electrons and no more than 8 in the second shell.

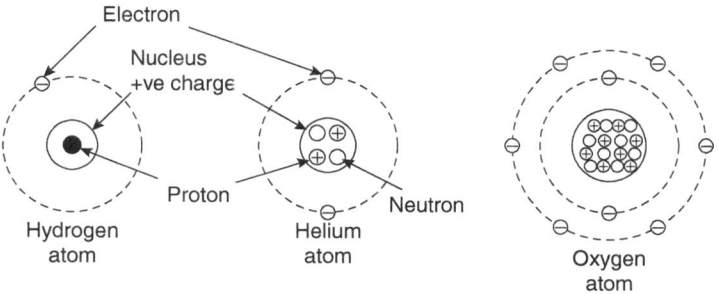

▲ **Figure 1.2** *Atomic structure for hydrogen, helium and oxygen*

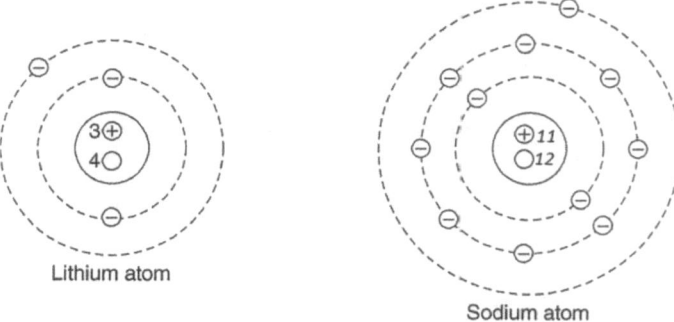

▲ **Figure 1.3** *Atomic structure for lithium and sodium*

Figure 1.3 represents the atomic structure of 2 metals: lithium and sodium. In each case, and if other metals are considered, it will be seen that all metals have 1 or 2 electrons in the outermost shell, which is considered the reason metals have good electrical conducting properties. It is suggested that for metals in their normal crystalline state, the atoms are arranged so their outermost electrons are partially screened from the +ve attractive force of the nucleus and are not so strongly bound. Thus electrons can move relatively freely between one atom and its neighbours. Such outer orbital electrons, or *mobile* or *valence electrons*, move randomly from one atom to another atom and constitute a 'pool' of moving negative charges, which helps to explain the transmission of electricity or current in a circuit. Note that 'valency' is a chemical term of which mention will be made later.

Current as electron movement

Current, according to electron theory, is due to movement of electrons from one atom to the next, each electron carrying a −ve charge. As mobile electrons move randomly between atoms, transfer of charge, and thus electricity in a particular direction, fails to occur and no current flows. If an electrical force, in the form of an electromotive force (e.m.f.) or potential difference (P.D.) is applied across a good conductor then mobile electrons are forced to move towards the higher potential or +ve terminal. The required electrical force produced by a battery or generator can be thought of as a pump moving electrons round a circuit. A 'stream' or movement of electrons is said to constitute an electric current but there is a key difference between the direction of conventional current flow and actual electron flow. If a length of wire is connected to 2 terminals, between which an e.m.f. or P.D. exists, a current will flow from the +ve terminal through the wire to the −ve terminal. However, electron flow will be from the −ve terminal to the +ve terminal. This fundamental difference between conventional current and electron flow must always be remembered and is illustrated in figures 1.4a and 1.4b. This distinction arises from a practical lack of understanding in the earliest days that electrons were the mobile charge carriers. It is noted that the electrical generator or battery, which maintains the e.m.f. or P.D. between the ends of a conductor, does not itself *make* electricity but merely causes movement of electrons that are present in the circuit.

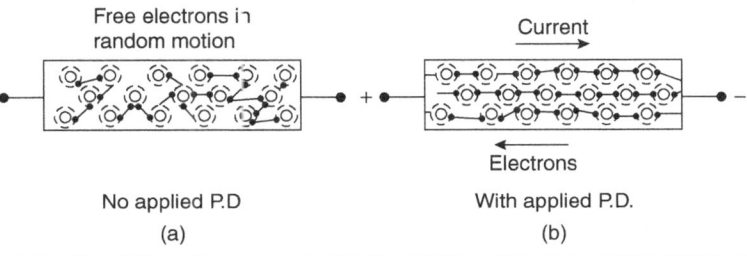

▲ **Figure 1.4** *Current vs electron flow*

Ionisation

An atom can lose or gain electrons as the result of a disturbing action or force. It then becomes electrically unbalanced having acquired charge and is called an *ion*. An atom minus an electron exhibits a +ve charge and is a +ve ion. Atoms that gain an electron exhibit a −ve charge and are −ve ions. When an electron is made to leave a parent atom by application of a force, for example, due to an electric field or the application of heat or light, it can acquire sufficient energy to detach further electrons from other atoms with which it collides. Such action causes these secondary atoms to become +ve ions and if electrons leave these atoms faster than they are regained, the state of ionisation continues. Electronic devices such as the fluorescent lamp or Cathode Ray Tube depend on ionisation to operate.

The electric circuit

A circuit is defined as a path taken by an electric *current*. A current flows through a circuit if (1) a source of electrical energy such as a battery or generator is connected, and (2) the circuit is continuous or conducting along its complete length. Figure 1.5 represents a simple circuit in which a current flows. It shows a source, from which energy is transmitted in the current, the conducting path or cable along which the current flows and the 'load'. The load is the point where energy is released or work is to be done by the flowing current.

The conditions of figure 1.5 are better represented by a circuit diagram as in figure 1.6, which illustrates the energy source as a chemical cell, the conducting path as the leads or wires and the load. A switch is shown as a vital link that when opened, interrupts a circuit's continuity and thus stops the current flowing.

Consideration of the simple circuit introduces more fundamental terms and the practical units used in electrical engineering. Flow of electricity or current is the result of pressure built up within the energy source, which manifests itself, at the circuit connecting points or terminals, as a pressure difference. One terminal, called the positive, is viewed as being at a higher pressure or potential than the other terminal, called the negative. A P.D. exists between these terminals. The current direction is from the positive (+ve) terminal through the circuit *external to the energy source*, back to the

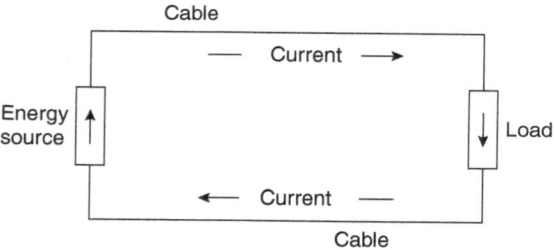

▲ **Figure 1.5** *A simple circuit with current flow*

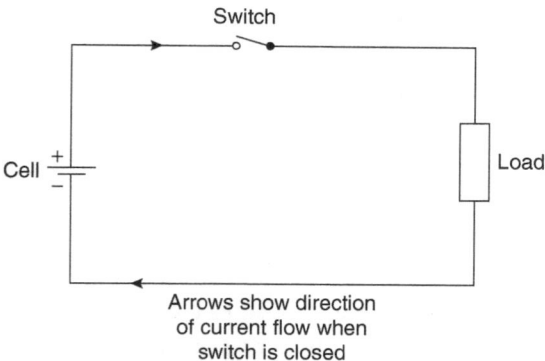

▲ **Figure 1.6** *A circuit diagram with cell and load*

negative (−ve) terminal and then through the source to the +ve terminal. Thus for the load, conventional current is from +ve to −ve terminal, but for the energy source in a cell, battery or generator, electrical current is from the −ve to the +ve terminal!

The electrical pressure generated by an energy source is termed the *electromotive force* (e.m.f.). The symbol used, E, or e.m.f., is measured as a *voltage*. The unit is the *Volt*, which is defined later, but any voltage value is represented by the letter V attached to a numerical value. Thus a voltage of two hundred and twenty volts is written as 220V. For reasons to be explained when the maths of a circuit is considered, the whole generated e.m.f. of a cell, battery or generator doesn't appear at the terminals when current flows. The P.D. across the terminals is measured in terms of the potential or voltage dropped round the external circuit. The symbol used for the terminal P.D. is V and is measured as a voltage, i.e. in volts.

Circuit Laws

1. For any circuit, current strength is found to be proportional to the voltage applied *across* its ends. Current strength is denoted by the symbol I and is measured in *Amperes*. The ampere is defined later by considering the electromagnetic effect of current flow, but any current value can be represented by the letter A added to the numerical value. Thus two hundred amperes is 200A.

Any electrical circuit is found to *oppose* the current flow. This opposition is termed the resistance of a circuit and is given the symbol R. The unit of resistance is the *Ohm*, but any value is represented by the Greek letter capital Ω (Omega) added to the numerical value. Thus one hundred ohms may be written 100Ω. The ohm is defined in terms of the volt and ampere so: a resistor has a value of 1 ohm's resistance if a current of 1 ampere passes through it when a P.D. of 1 volt is applied across its ends. An alternative definition is given in Chapter 2.

2. The current in a circuit, for a constant voltage, is found to vary inversely with resistance, i.e. the greater the resistance, the smaller the current and vice versa.

Ohm's law

The relationships stated above are summarised by the first law of an electrical circuit, which is called Ohm's law and is expressed thus: the current in a circuit is directly proportional to the voltage and inversely proportional to the resistance. This can be written as:

$$\text{Current} = \frac{\text{Voltage}}{\text{Resistance}}$$

$$\text{or } I \text{ (amperes)} = \frac{V \text{ (volts)}}{R \text{ (ohms)}} \text{ or } \frac{E}{R}$$

$$\text{Other forms are } V = IR \quad \text{or} \quad R = \frac{V}{I}$$

When using Ohm's law formulae it is essential to pay due regard to the magnitudes of the units used. Reference should be made to the appropriate table of conversions.

Example 1.1. An e.m.f. of 6V is applied across a 300Ω resistor. Find the current that will flow.

$$I = \frac{E}{R} = \frac{6}{300} = 0.02\text{A} = 20\text{mA}.$$

Example 1.2. A current of 20mA passes through a 30kΩ resistor. Find the voltage drop across the ends of the resistor.

$$V = IR = (20 \times 10^{-3}) \times (30 \times 10^{3}) = 600\text{V}.$$

Series and parallel circuits

Study of the electrical circuit shows that in its simplest form it may be built up as (1) a series circuit or (2) a parallel circuit. Resistance is considered to be concentrated in a resistor, or in more than 1 resistor, while connecting leads are assumed to have negligible resistance, unless a definite resistance value for these is stated. Similarly the cell, battery or generator is assumed to have no resistance unless otherwise stated.

Figure 1.7 shows a series circuit. Only one current path is possible and the same current passes through all the resistors. The current is thus common for such a circuit but the applied potential drops progressively as current flows along the circuit.

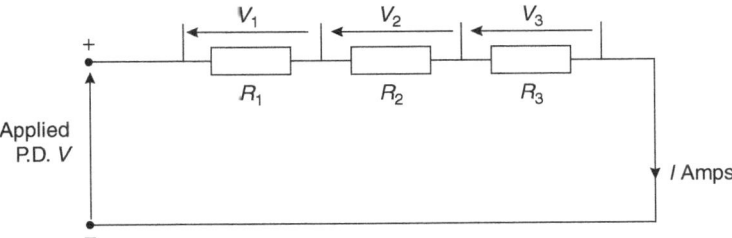

▲ **Figure 1.7** *A simple series circuit*

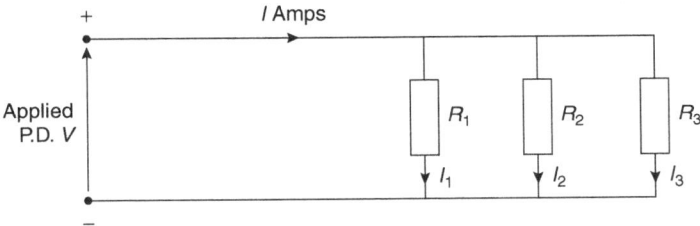

▲ **Figure 1.8** *A simple parallel circuit*

Figure 1.8 shows a parallel circuit. Here the main current is made up of 3 branch currents, but the applied P.D. is the same or common for all 3 branches. At any junction point there is no current accumulation, i.e. the total current entering a point is the same as the total current leaving the point. Simple laws based on voltage conditions for the series circuit and current conditions for the parallel circuit allow the solution of problems for such simple circuits, and also those of more complicated series–parallel arrangements or electrical networks.

Kirchhoff's laws

(1) VOLTAGE LAW. The sum of the potential or voltage drops taken round a circuit must equal to the applied P.D. Thus for figure 1.7:

$$V_1 + V_2 + V_3 = V$$

(2) CURRENT LAW. The current flowing away from a junction point in a circuit must equal the current flowing into that point. Thus for figure 1.8:

$$I_1 + I_2 + I_3 = I$$

The above laws help deduce simple formulae for series and parallel circuits in terms of the equivalent resistances of the circuits.

THE SERIES CIRCUIT. For figure 1.7, let I amperes be the common current flowing round a circuit. Then from Ohm's law, the voltage dropped across resistor R_1 is V_1 volts $= IR_1$. Similarly the voltage dropped across R_2 is $V_2 = IR_2$ etc. If R is taken as the equivalent resistance of the whole circuit then as V is the applied voltage and it will be dropped over this equivalent resistance, we can write $V = IR$.

Using Kirchhoff's voltage law then

$$V = V_1 + V_2 + V_3$$

$$\text{or } IR = IR_1 + IR_2 + IR_3 = I(R_1 + R_2 + R_3)$$

$$\therefore R = R_1 + R_2 + R_3$$

THE PARALLEL CIRCUIT. For figure 1.8, let V volts be the common voltage applied to all the parallel branches and with a total main current of I amperes. Voltage V would also cause a current of I amperes through an equivalent circuit of resistance R ohms.

Thus $I = V/R$ and using Kirchhoff's current law then

$$I = I_1 + I_2 + I_3$$

But for branch 1 $V = I_1 R_1$ or $I_1 = \dfrac{V}{R_1}$

And similarly for branch 2 $I_2 = \dfrac{V}{R_2}$ etc.

Thus $I = I_1 + I_2 + I_3$ can be written as:

$$\frac{V}{R} = \frac{V}{R_1} + \frac{V}{R_2} + \frac{V}{R_3} = V\left(\frac{1}{R_1} + \frac{1}{R_2} + \frac{1}{R_3}\right)$$

$$\therefore \frac{1}{R} = \frac{1}{R_1} + \frac{1}{R_2} + \frac{1}{R_3}$$

Note. The reciprocal of resistance is often referred to as *Conductance*, symbol $G = 1/R$.

The unit is the *Siemens*; the symbol S appended to the numerical value.

So for a parallel circuit $G = G_1 + G_2 + G_3$ etc.

Example 1.3. Three resistors of values 2, 4 and 8 ohms are connected in series across a supply of 42 volts. Find the current taken from the supply and the voltage dropped across each resistor.

$$\text{Here } R = R_1 + R_2 + R_3 = 2 + 4 + 8 = 14\Omega$$

$$\text{So supply current } I = V/R = \frac{42}{14} = 3A$$

Voltage dropped across 2Ω resistor $= 3 \times 2 = 6V$

Voltage dropped across 4Ω resistor $= 3 \times 4 = 12V$

Voltage dropped across 8Ω resistor $= 3 \times 8 = 24V$

Check. $6V + 12V + 24V = 42V$ (the applied voltage).

Example 1.4. The above resistors are connected in parallel across the same supply voltage. Find the total current and the current in each branch.

$$\text{Here } \frac{1}{R} = \frac{1}{R_1} + \frac{1}{R_2} + \frac{1}{R_3} = \frac{1}{2} + \frac{1}{4} + \frac{1}{8} = \frac{7}{8} = 0.875S$$

$$\text{or } R = \frac{8}{7} = 1.14\,\Omega \text{ and } I = \frac{42}{1.14} = 36.75A$$

$$\text{The current in Branch 1} = \frac{42}{2} = 21A$$

$$\text{The current in Branch 2} = \frac{42}{4} = 10.5A$$

$$\text{The current in Branch 3} = \frac{42}{8} = 5.25A$$

Check. $21A + 10.5A + 5.25A = 36.75A$ (the total supply current).

Internal resistance of a supply source

Until now the energy supply source was considered as having negligible resistance. In practice, a cell, battery or generator has an internal resistance which results in an internal voltage drop when current is supplied. Thus the e.m.f. generated appears at the supply terminals, shown as AB in figure 1.9, *only* when the circuit switch is open, i.e. the cell is on 'open circuit' (O.C.). When current I is supplied, an internal voltage drop of IR_i occurs, R_i being the internal cell resistance. The P.D. V at the energy source terminals is thus $E - IR_i$. V is less than the generated e.m.f. E by the P.D. required to drive the current through the cell resistance.

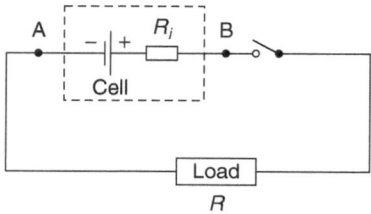

▲ **Figure 1.9** *Electrical circuit with the electrical switch open*

Electromotive force (e.m.f.) and terminal P.D. or voltage

In figure 1.9, R_i represents the internal cell resistance as shown external to the cell itself. This is diagrammatic only; sometimes this resistance is not shown, written only as a figure beside the cell e.m.f. However, internal resistance must be considered and gives rise to the difference between e.m.f. E and P.D. V. On O.C. terminal, P.D. of a source equals the e.m.f. generated; but 'on load', i.e. when current is supplied, terminal P.D. equals the e.m.f. minus the internal voltage drop. This is summarised mathematically thus:

On O.C. $V = E$

On Load $V = E - IR_i$

We also deduce that since $V = IR$ where R is the load resistance, then

$$IR = E - IR_i \text{ and } E = IR + IR_i \quad or \quad E = I(R + R_i)$$

Expressed another way:

On O.C.   Cell terminal voltage $V = E$

On Load   Cell terminal voltage $V = E - IR_i$

also   Cell terminal voltage $V = IR$

Problems are treated as a simple series circuit, if E is used as the circuit voltage and R_i included in the series resistance.

Example 1.5. A battery of e.m.f. 42V and internal resistance 7Ω supplies the series circuit of Example 1.3, i.e. 3 resistors of 2Ω, 4Ω and 8Ω in series (figure 1.10). Find the current and terminal voltage and by how much the cell voltage 'sits' or 'drops down' when supplying the load.

Note. It is appropriate here to explain that a battery is an arrangement of more than 1 cell. The methods of connecting cells are discussed in Chapter 2; here a battery is considered arrangement of cells in series. Thus a battery e.m.f. is the sum of its cell e.m.f.s and the battery internal resistance is the sum of the cell internal resistances.

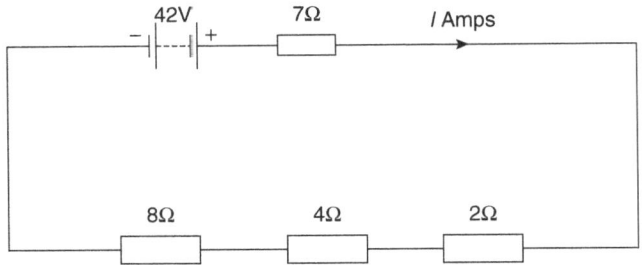

▲ **Figure 1.10** *Circuit with 3 resistors in series*

External resistance of the circuit $R = 2 + 4 + 8 = 14Ω$

Battery resistance $= 7Ω$

Total resistance of circuit $= 14 + 7 = 21Ω$

Circuit current $I = E/R = 42/21 = 2A$

Terminal voltage $V = IR = 2 \times 14 = 28V$

Voltage drop in cell $IR_i = 2 \times 7 = 14V$

Check. Terminal voltage $V = E - IR_i = 42 - 14 = 28V$

The 'Lost volts' or Cell voltage 'drops down' by 14 volts.

THE SERIES–PARALLEL CIRCUIT. In practice many circuits are built up from series and parallel groups of resistors. Solution of these associated problems, though not simple, follows a logical sequence of operations based on the methods used for simple series and parallel resistor arrangements. It is strongly urged that the idea of 'practice makes perfect' is essential and the reader should work through sufficient appropriate problems until proficiency is achieved. A solution method for a particular problem will become clear once its form is recognised and time and labour with real engineering problems will then be saved.

Figure 1.11a shows a simple series–parallel circuit, consisting of a series circuit made up of 2 sections, each comprising a group of resistors in parallel. As the main circuit or supply current may need to be found along with the current in each resistor, solution is only obtained by simplifying the problem. It is noted that the parallel groups (or banks of resistors) are called sections A and B respectively. The voltage dropped across these sections is unknown and as the voltages are essential, the procedure is set out below.

The circuit is simplified by finding the *equivalent resistance* values R_A and R_B of the parallel banks from

$$\frac{1}{R_A} = \frac{1}{R_1} + \frac{1}{R_2} + \frac{1}{R_3} \quad \text{and} \quad \frac{1}{R_B} = \frac{1}{R_4} + \frac{1}{R_5} + \frac{1}{R_6}$$

Once R_A and R_B are found, the total supply current can be obtained as is shown in figure 1.11b. The equivalent circuit is now of the simple series type.

The supply current is

$$I = \frac{V}{R_A + R_B} \quad \text{and the voltage drops across groups A and B are respectively}$$

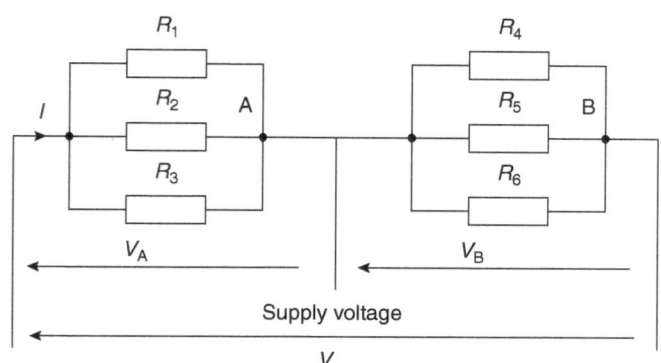

▲ **Figure 1.11a** *A simple series-parallel circuit*

$V_A = IR_A$ and $V_B = IR_B$, and substituting for current I,

$$\text{or } V_A = \left(\frac{V}{R_A - R_B}\right) R_A \quad \text{and} \quad V_B = \left(\frac{V}{R_A + R_B}\right) R_B$$

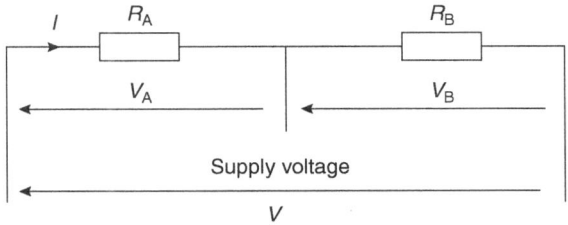

▲ **Figure 1.11b** *Finding the supply current*

Once V_A and V_B are known, indiv dual currents in each resistor are found by reverting to the original circuit.

$$\text{Thus } I_1 = V_A/R_1 \text{ and } I_2 = V_A/R_2 \text{ Also } I_4 = V_B/R_4 \text{ etc.}$$

The above method is only giver to explain the solution of Example 1.6 and illustrate a step-by-step procedure. There is no shortcut for problems of the series–parallel type. The reader should work only with the data given and not make any assumptions. Methods of solution using proportions for currents or voltages across parallel or series circuit sections are discouraged, as in practice resistance ratios are rarely simple, while adhering to and following simple, but sometimes more tedious methods will result in the correct answer.

Example 1.6. A circuit is built up from 5 resistors. Resistors of values 4Ω, 6Ω and 8Ω are connected in parallel to form a group, while resistors of 3Ω and 6Ω are connected in parallel to form another group (figure 1.12). The 2 parallel groups of resistors are connected in series across a 10V supply. Find the voltage dropped across each parallel group, the main supply current and the current in each resistor.

Let R_A be the equivalent of the first group.

$$\text{Then } \frac{1}{R_A} = \frac{1}{4} + \frac{1}{6} + \frac{1}{8} = \frac{13}{24} = 0.54S$$

$$\text{or } R_A = \frac{24}{13} = 1.85\Omega$$

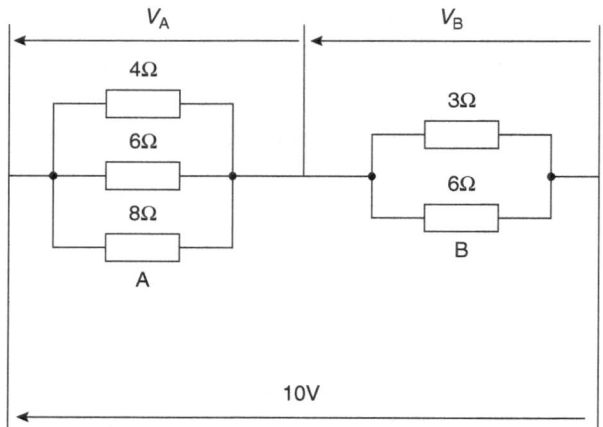

Similarly let R_B be the equivalent resistance of the second group.

$$\text{Then } \frac{1}{R_B} = \frac{1}{3} + \frac{1}{6} = \frac{3}{6} = 0.5\text{S}$$

$$\text{or } R_B = \frac{6}{3} = 2\Omega.$$

For the equivalent series circuit, total resistance is R

$$\text{or } R = R_A + R_B = 1.85 + 2 = 3.85\Omega$$

Main supply current $I = \dfrac{V}{R} = \dfrac{10}{3.85} = 2.6\text{A}$

Voltage drop across R_A or the first parallel group $= 1.85 \times 2.6$

$$= 4.8\text{V}$$

Voltage drop across R_B or the second parallel group $= 2 \times 2.6$

$$= 5.2\text{V}$$

Check. Total supply voltage is $(4.8 + 5.2) = 10\text{V}$.

Current in 4Ω resistor $= \dfrac{4.8}{4} = 1.2\text{A}$

Current in 6Ω resistor $= \dfrac{4.8}{6} = 0.8\text{A}$

Current in 8Ω resistor $= \dfrac{4.8}{8} = 0.6\text{A}$

Check. Total current is $(1.2 + 0.8 + 0.6) = 2.6\text{A}$

Similarly: Current in 3Ω resistor $= \dfrac{5.2}{3} = 1.73\text{A}$

Similarly: Current in 6Ω resistor $= \dfrac{5.2}{6} = 0.87\text{A}$

Check. Total current is $(1.73 + 0.87) = 2.6\text{A}$.

Example 1.7. A battery of e.m.f. 42V and internal resistance 7Ω feeds a circuit consisting of 3 resistors connected in parallel. The resistors have values of 2Ω, 4Ω and 8Ω. Find the battery current, the battery terminal voltage and the current in each resistor (figure 1.13).

Let R be the equivalent resistance of the parallel-connected load.

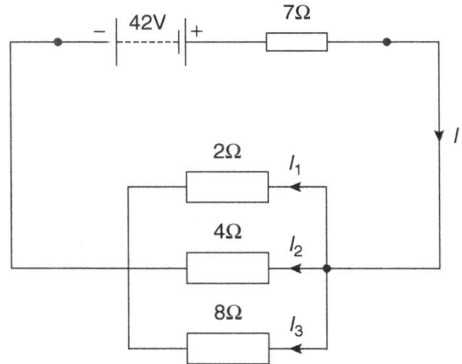

▲ **Figure 1.13** *A simple circuit composed of 3 parallel resistors*

Then $\dfrac{1}{R} = \dfrac{1}{2} + \dfrac{1}{4} + \dfrac{1}{8} = \dfrac{7}{8} = 0.875\text{S}$ and

$$R = \dfrac{8}{7} = 1.14\Omega$$

The circuit can now be considered to have a total resistance of 8.14Ω made up from 1.14Ω and 7Ω in series.

The battery current I is given by $\dfrac{42}{8.14} = 5.16\text{A}$

The terminal voltage will be $5.16 \times 1.14 = 5.88V$

or the terminal voltage will be $42 - (7 \times 5.16) = 42 - 36.12 = 5.88V$

Current I_1 in 2Ω resistor is $\dfrac{5.88}{2} = 2.94A$

Current I_2 in 4Ω resistor is $\dfrac{5.88}{4} = 1.47A$

Current I_3 in 4Ω resistor is $\dfrac{5.88}{8} = 0.74A$

Check. Total current $I = 5.16A$.

Ammeters and Voltmeters

These are the main instruments used for electrical work and figure 1.14 shows how they are connected in a circuit. Ammeters are used to measure current and voltmeters for measuring P.D. or voltage. Both instruments operate on the same principle, but ammeters must have very low resistance as they are in series with the load and must not result in appreciable voltage drop. Voltmeters on the other hand must be of high resistance, as they may be connected across points that could be at a high P.D. For most circuit purposes, an ammeter is considered to have negligible resistance and a voltmeter to have infinite resistance, i.e. it draws no current.

In figure 1.14 a generator is shown as the energy source, S may be a single-pole or double-pole switch, as is shown here, and R is the load resistance. As a practical

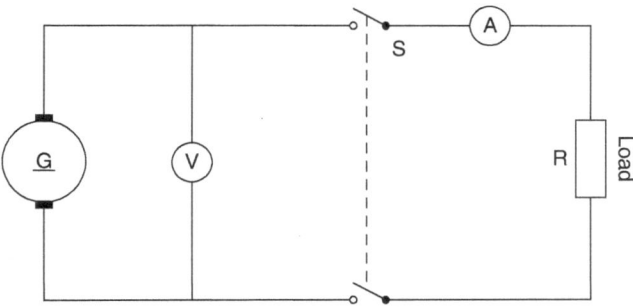

▲ **Figure 1.14** *A simple circuit with a voltmeter and ammeter*

example, the generator may have an internal resistance of 0.02Ω, the cable leads may have a total resistance of 0.03Ω, and R may have a value of 5Ω. If the generator is set to 220V on O.C., i.e. with the switch open, when the switch is closed a current of

$$\frac{220}{5 + 0.02 + 0.0.3} = \frac{220}{5.05} = 43.56A \text{ will flow round the circuit.}$$

The terminal voltage of the generator will 'sit down' to 220 − (43.56 × 0.02) volts = 220 − 0.87 = 219.13V. This wou d be shown by a voltmeter, while an ammeter would show 43.56A. If the voltmeter was disconnected and then connected directly across R it would indicate 219.13 − (43.55 × 0.03) volts = 219.13 − 1.3 = 217.83V or the voltage across R = IR = 43.56 × 5 = 217.83V. The voltage drop in the cables will be 1.3V. It is seen that the example of a simple distribution system has been worked as a simple series circuit and that the instruments perform their required functions. The ammeter shows the series circuit current, while the voltmeter indicates the potential drop across any chosen part of a circuit. It can also record the e.m.f. built up by a generator when the switch is opened, as this is the only condition when the e.m.f. appears at the terminals of the energy source.

Range of extension of ammeters and voltmeters

For practical work it isn't always possible to pass all the circuit current through the ammeter. It may be difficult to construct a suitable instrument because of size or other limitations, and to introduce standardisation, it is easier to use the ammeter with a *shunt* to measure the circuit current. There are various types of electrical measuring instruments, described by their 'movements'. Such 'movements' use different operating forces and a shunt is normally used with the 'moving-coil' type only since this is constructed to the highest accuracy and sensitivity and is ideal for working with various types of *transducer*. Transducers are devices that generate mechanical or electrical outputs for measured quantities. It is assumed in this chapter that a moving-coil ammeter or voltmeter is being considered.

A shunt is a specially constructed resistor of low ohmic value and, to make an ammeter capable of measuring a current greater than that which is passed through it, a parallel arrangement of the ammeter and shunt is used. The ammeter is designed to carry a fixed but small proportion of the main current and the rest of the current is made to 'bypass' the ammeter through the shunt, which is accurately made and set to a fixed resistance value. It is calibrated with the ammeter instrument and must always be used with it. The calibrated leads between instrument and shunt form part

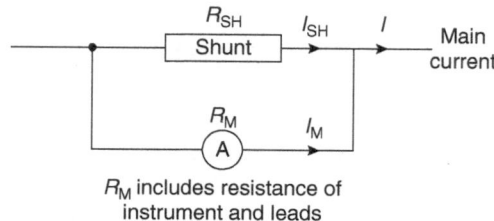

RₛH includes resistance of
instrument and leads

▲ **Figure 1.15** *A simple circuit including the resistance of instruments and leads*

of the arrangement and must not be cut or substituted for by other pieces of copper wire! Figure 1.15 shows the normal arrangement of instrument and shunt and the example shows the form of calculation necessary. It is seen that the calculation follows the pattern set for parallel-resistance circuits.

Example 1.8. Calculate the resistance of a shunt required to operate with a moving-coil milliammeter, which gives full-scale deflection (f.s.d.) for a current of 15mA and which has a resistance of 5Ω. (*Note*. 5Ω is taken to include the resistance of the connecting leads, as no specific mention of lead resistance has been made.) The combination of meter and shunt is required to read currents up to 100A.

Voltage drop across instrument when giving f.s.d. = current causing f.s.d. × resistance of instrument circuit

$$= I_M \times R_M = (15 \times 10^{-3}) \times 5 = 75 \times 10^{-3} \text{ volts}$$

$$= 0.075\text{V or } 75\text{mV}$$

Now the voltage drop across the instrument is the same as the voltage drop across the shunt.

Thus $I_{SH} \times R_{SH} = 0.075$ volts.

But the shunt current I_{SH} will be 100 – meter current

$$= 100 - 0.015 = 99.985\text{A}$$

So $R_{SH} = \dfrac{75 \times 10^{-3}}{99.985} \Omega$

$$= 0.000751\Omega$$

It is important to note the low resistance value of the shunt, which is designed to carry the current without 'heating up'. The shunt is usually mounted behind the ammeter in the main current circuit. The 'light' calibrated leads are coiled to take up any 'slack' and then brought up to the instrument. The ammeter may be marked 0–100 amperes, but in actual fact only a tiny current, some 15 mA, passes through the instrument, while the larger proportion of the current passes through the shunt. The reason for always using the instrument with its own calibrated shunt and leads is hopefully obvious!

To measure voltages higher than that for which the instrument movement is designed a *series* or *range* resistor must be used. This resistor is designed to drop excess voltage and dissipates some heat. It consists of special fine-gauge wire wound on a porcelain spool or on a mica card, and mounted inside a ventilated case. The arrangement may be mounted behind a switchboard, if not contained in the instrument's case. Thin leads for carrying the small instrument current connect the range resistor unit and the instrument to the main supply terminals, usually through fuses. The voltmeter may be scaled 0–250 volts, but in fact only 0.075V may be dropped across it, when f.s.d. occurs (see Example 1.8). By far the biggest voltage drop occurs across the range resistor, and it should be noted it always has a high ohmic value: thousands of ohms. Figure 1.16 shows the arrangement and Example 1.9 shows how the value of an appropriate range resistor is calculated.

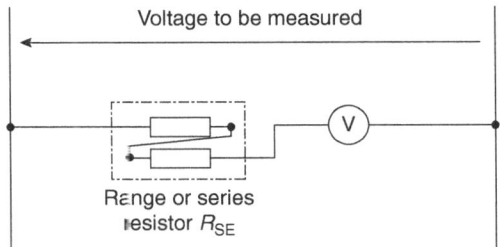

Voltage to be measured

Range or series
resistor R_{SE}

▲ **Figure 1.16** *The range or series resistor*

Example 1.9. Calculate the resistance of the range resistor required to be placed in series with the instrument of Example 1.8 to make it into a voltmeter reading 0–250V. (The instrument has a resistance of 5Ω and gives f.s.d. with a current of 15mA.)

The current through the complete voltmeter circuit must be limited to 15mA, otherwise the instrument will 'burn out'. Resistance of the voltmeter circuit will be

$$\frac{250}{15\times10^{-3}} = 16.667\times10^{3}\,\Omega = 16.667\text{k}\Omega$$

The instrument has a resistance of 5Ω, so the series or range resistor R_{SE} must have a value of (16 667 − 5) ohms = 16 662Ω.

The actual 'movement' or working unit of an ammeter or voltmeter is much the same and it is the use of a shunt or range resistor that determines whether current or voltage is measured. Multi-purpose portable test instruments are available that can make a range of measurements. A range switch, or *range multiplier*, is provided to make the appropriate connection of shunt or range resistors.

Consider an instrument movement in which 15mA at a P.D. of 75mV gives f.s.d. Its resistance $= \dfrac{75 \times 10^{-3}}{15 \times 10^{-3}} = 5\Omega$.

If a voltage range 0–15V is required, the instrument circuit resistance must be $= \dfrac{15}{15 \times 10^{-3}} = 1000\Omega$ and a range resistor of $1000 - 5 = 995\Omega$ must be switched in.

Similarly if a voltage range of 0–150V is required the range resistor must be $\dfrac{150}{15 \times 10^{-3}} - 5 = 10000 - 5 = 9995\Omega$

If a current range 0–5A is required, a shunt is used whose value can be obtained thus:

P.D. across shunt = P.D. across instrument movement for f.s.d., or
P.D. across shunt = 75mV = 75×10^{-3} volts

The current through the shunt = $5 - 0.015 = 4.985$A and the resistance of the shunt would be

$$\frac{75 \times 10^{-3}}{4.985} = 0.015\,05\Omega$$

Instrument sensitivity

This term is used to consider the suitability of a measuring instrument for a particular purpose. If, for example, a voltmeter is badly constructed so it requires a relatively large current for f.s.d., the overall circuit current will be badly affected when the instrument is connected across any particular part of a circuit. This is very important for electronic circuitry. Consider a component of resistance value 1kΩ forming part of a series circuit drawing 1mA. A voltmeter of resistance 5kΩ connected across such a component will lower the resistance of the parallel arrangement to 0.803kΩ.

$$\text{Note } \frac{1}{R} = \frac{1}{5} + \frac{1}{1} = \frac{6}{5} = 1.25 \quad \text{or} \quad R = \frac{5}{6} = 0.803\text{k}\Omega.$$

Accordingly the circuit current will rise appreciably and the overall circuit conditions would change – an unwanted effect. The higher the resistance value of a voltmeter, the less the effect and voltmeters are therefore given a 'sensitivity' figure of ohms per volt. Thus a meter rated at 20kΩ/v will require a current of $\dfrac{1}{20 \times 10^3}$ amperes or $\dfrac{1}{20}$ milliamperes or $\dfrac{1}{20} \times 10^{-3} = 50\mu A$ for f.s.d. and the range resistor required is calculated on this basis. Such a voltmeter connected across the component of the example would have little effect on the circuit current and should be the instrument used.

Practice Examples

1.1. A circuit is made up from 4 resistors of value 2Ω, 4Ω, 5Ω and 10Ω connected in parallel. If the current is 8.6A, find the voltage drop across the arrangement and the current in each resistor (1 decimal place).

1.2. One resistor group consists of 4Ω, 6Ω and 8Ω connected in parallel and a second group consists of 3Ω and 6Ω in parallel. The 2 groups are connected in series across a 24V supply. Calculate (a) the circuit current, (b) the P.D. across each group and (c) the current in each resistor (all 2 decimal places).

1.3. If the resistor arrangement of Q1.1 is connected to a 12V battery of internal resistance 0.65Ω, find the circuit current and the battery terminal voltage. Find also, the current in the 5Ω resistor (all 1 decimal place).

1.4. A moving-coil instrument has a resistance of 10Ω and requires a current of 15mA to give an f.s.d.. Calculate the resistance value of the resistor necessary to enable it to be used to measure (a) currents up to 25A (3 decimal places) and (b) voltages up to 500V (5 significant figures).

1.5. Two resistors of 60kΩ and 40kΩ value are connected in series across a 240V supply and a voltmeter having a resistance value of 40kΩ is connected across the 40kΩ resistor. What is the reading on the voltmeter (2 significant figures)?

1.6. When a 10Ω resistor is connected across a battery, the current is measured to be 0.18A. If similarly tested with a 25Ω resistor, the current is measured to be 0.08A. Find the e.m.f. of the battery (2 decimal places) and its internal resistance (1 significant figure). (Hint: consider this as 2 linear equations for E.) Neglect the resistance of the ammeter used to measure the current.

1.7. Two groups of resistors A and B are connected in series. Group A consists of 4 resistors of values 2Ω, 4Ω, 6Ω and 8Ω connected in parallel and group B consists of 2 resistors of values 10Ω and 15Ω in parallel. If the current in the 4Ω resistor is 1.5A, calculate (a) the current in each of the remaining resistors, (b) the supply voltage and (c) the voltage drop across the groups A and B (all 2 decimal places).

1.8. The voltage of a D.C. generator, when supplying a current of 75A to a load, is measured to be 108.8V at the switchboard. At the load, the voltage recorded is 105V and when the load is switched off the voltage rises to 110V. Find the internal resistance of the generator (3 decimal places), the resistance of the supply cables (1 decimal place) and estimate the fault current if a 'short-circuit' of negligible resistance occurred at the load terminals (3 significant figures).

1.9. The ammeter on a switchboard, scaled 0–300A is accidentally damaged. The associated shunt is marked 300A, 150mV. A small ammeter, scaled 0–1A with a resistance of 0.12Ω, is available, and might be used. Find if such an arrangement is possible, and if so, how it could be achieved using surplus resistors that are available?

1.10. Five resistors AB, BC, CD, DE and EA are connected to form a closed ring ABCDEA. A supply of 90V is connected across AD, A being positive. The following is known about the resistors: AB is 10Ω, BC is of unknown value R_1 ohms, CD is of unknown value R_2 ohms. DE is 6Ω and EA is 9Ω. A high-resistance voltmeter (taking negligible current) when connected across BE reads 34V with B positive and when connected across CE reads 6V with E positive. Find the values of R_1 and R_2, the current in branch ABCD and the main supply current (all 1 significant figure).

2

THE ELECTRIC CIRCUIT: UNITS

Wisdom is like electricity. There is no permanently wise man, but men capable of wisdom, who, being put into certain company, or other favourable conditions, become wise for a short time, as glasses rubbed acquire electric power for a while.

Ralph Waldo Emerson

All engineering studies stress the need for units that allow measurements to be taken and calculations to be made, essential for comparison of experimental measurements with theoretical predictions, and in derivation of formulae from theory. In Chapter 1, the ampere, volt and ohm were considered, and although these units are not yet defined, their importance in relation to basic electric circuits will be appreciated. The engineering student will recognise these units are in daily use. Electrotechnology uses the same range of units. Modern engineering technology is based on the universal adoption of SI units and several of these were encountered earlier in Chapter 1.

Before proceeding with further study of SI system units, it is helpful to introduce a historical note and consider the situation of engineering units as they developed.

Towards the end of the nineteenth century, 2 unit systems emerged in engineering; the British or foot–pound–second (fps) system and a metric or centimetre–gramme–second (cgs) system. The British or *Imperial system* had little merit since all units of the same kind, such as those of length, area, volume, etc., bore no relation to each other; there were also other units, such as the calorie and horsepower, that were arbitrarily and sometimes differently defined! The European metric system, however, was first devised as a benefit to industry and commerce but physicists soon realised its benefits and it was adopted prior to 1870 in British scientific and technical circles. In 1873 the British Association for the Advancement of Science (BAAS) selected the centimetre and gramme as basic units of length and mass for physical purposes. Measurement of other quantities called for a base time unit and adoption of the second gave the cgs system.

The metric system, in cgs form, was adopted for electrical engineering in the early days of development. The system had the benefit that all the same kind of quantities are multiples of 10 and it was readily accepted internationally. The sizes of the absolute unit of the centimetre and the gramme gave rise to difficulties for the size of the desired electrical units, which were either too large or too small for practical working. Use of these absolute units for key engineering formulae also proved difficult and so more workable units had to be defined. Such practical units include the volt, ampere and ohm. In about 1900, practical measurement in metric units began to be based on the Metre, Kilogramme and the Second (MKS) and the mentioned electrical units, constituting the unrationalised MKS system.

The next development came from a fact, repeatedly pointed out over the first half of the twentieth century, that a system of units could be devised to make the practical units of volt, ampere and ohm the absolute units of such a system. A more workable unit system known as the rationalised MKS system was recommended by the International Electrotechnical Commission of 1950. The change to the MKS necessitated revision of many reference and textbooks. For the student, the new units made learning easier and formulae more manageable.

Prior to 1970, conditions existed when both older Imperial and newer rationalised MKS systems of units were in use simultaneously. The latest extension of metric units into all branches of business and industry enabled engineering to develop the SI system, the units of which are used throughout this volume. From an electrical viewpoint, the SI system is a rationalised MKS system with units in all the other fields of measurement being fully metricated. Further refining developments of the MKS standards included adoption of atomic definitions for both distance and time, but for the practical marine engineering student this does not concern topics covered in this volume.

The SI System

All measurement is in comparison with some standard or unit. The 3 fundamental units are those of length, mass and time. In the SI system the metre is taken as the fundamental unit of length (s), the kilogramme as the unit of mass (m) and the second as the unit of time (t). From these fundamental units are derived all other units, which may further be classified as mechanical or electrical units. Thus Force is a derived mechanical unit involving a fundamental unit and a derived unit, i.e. mass and acceleration. For the SI system, a unit of force, or *Newton,* is introduced. Velocity is a derived unit involving distance and time, as is acceleration involving velocity and time. Both velocity and acceleration are mechanical units. The ampere is a derived unit involving force and length but it is used as a fundamental electrical unit. Other electrical units are the volt and the ohm, which are derived units. The *Joule* and *Watt*, although used mostly in the past in connection with electrical engineering, are both derived from mechanical relationships and will be defined later.

Once units are recognised and understood, the reader should consider them as general engineering units rather than mechanical or electrical units. This applies especially to the units of work and power. Both mechanical and electrical engineering fields are concerned with common appliances and associated problems and using appropriate units, where a correct understanding of the magnitudes of the quantities involved, is essential to the engineer.

Mechanical units

Notwithstanding my earlier comment about atomic standards, the fundamental units here require little definition as they are *accepted* working standards. The metre is the absolute standard, and is taken as the distance between 2 set marks on a certain metal bar. Similarly, the kilogramme is the mass of an accepted 'standard' of a chosen metal. The time unit is the second, defined as $\dfrac{1}{86400}$ of a mean solar day.

Most of the principal SI derived units have already been introduced to the engineering student but revision is made only to extend use into electrical engineering.

Unit of force

Force can be defined as that which tends to cause an object to move, to change its motion or to keep the object at rest. The symbol for force is F but any value of newtons can be represented by the letter N after the numerical value.

THE NEWTON. This is the force needed to accelerate a mass of 1 kilogramme at a rate of 1 metre/second^2 (i.e. F = ma N). However, if the acceleration of the mass is due to gravity alone, the rate of acceleration is a constant 9.81 ms^{-2}. This acceleration due to gravity is given the symbol 'g',

$$\therefore F = mg \text{ N.}$$

The force due to gravitation is referred to as *weight*. A mass of 1kg has a weight of 1kg × 9.81 ms^{-2} = 9.81N

$$\text{i.e. } 1\text{kg} = 9.81\text{N}$$

Unit of work and energy

THE JOULE. This is defined as the *work* done or *energy* stored when a force of 1 newton acts through a distance of 1 metre in the direction of the force. The symbol for work or energy is *W* but any value in joules can be represented by the letter J after the numerical value.

From the definition, it follows that a force of *F* newtons, acting through a distance of *s* metres, does *F* × *s* newton metres of work or *F* × *s* joules.

$$\text{Hence: } W \text{ (joules)} = F \text{ (newtons)} \times s \text{ (metres)}$$

Unit of power

THE WATT. *Power* is the rate at which work is done or energy is converted and its unit is the watt. A watt is the power resulting when a joule of energy is expended in a second. The symbol for power or rate of doing work is *P* but any value in watts can be represented by the letter *W* after the numerical value.

The definition can be more generally written as

$$P \text{ (watts)} = \frac{W \text{ (joules)}}{t \text{ (seconds)}}$$

The joule and the watt were originally used in electrical engineering and are encountered throughout electrical problems. Example 2.1 is set out here to introduce electro/mechanical relationships.

Example 2.1. An electrical pump is required to lift 1200 litres of water through 10 metres in 6 minutes. Calculate the work done in joules and the pump's power rating. Assume 1 litre of water has a mass of 1 kilogramme.

Work done = force of gravity × distance lifted

Thus $W = Fs = mgs$

$$= (1200 \times 9.81) \times 10$$

$$= 12 \times 9.81 \times 10^3 \text{ newton metres}$$

$$= 117.72 \times 10^3 = 117\,720 \text{Nm or } 117\,720 \text{J}$$

$$\text{Power} = \frac{\text{work done}}{\text{time taken}} = \frac{117\,720}{6 \times 6\text{C}} = 327\text{W}$$

Pump power rating will be 327 watts.

Note. In this problem no account has been made of machine *efficiency*. This will be introduced later, but here, the practical rating figure of the electric motor driving the pump will be larger.

Electrical units

The same fundamental units are used as for the mechanical units, namely: the metre, kilogramme and second. The primary derived unit is the ampere, which is the basic electrical unit of current and is a fourth fundamental unit. Before considering the ampere's definition, we will describe 2 related effects, which are observed when a current flows in a circuit.

(1) If the circuit's resistance is concentrated in a short length of conductive wire, a temperature rise of the wire is noted, showing a conversion of electrical energy into heat energy.

(2) If the circuit is supplied through 2 wires laid together, when the current is large and the wires flexible, a mechanical effect is observed. When a current is switched on, the wires are observed to move and this electromagnetic effect is used to define the ampere for the SI system. The factors governing the magnitude and direction of the force on the wires is described in the chapter on Electromagnetism.

Unit of current

THE AMPERE. This is the current that, when maintained in each of 2 infinitely long, straight, parallel conductors placed in a vacuum and separated by a distance of 1 metre between the wires' centres, produces on each conductor a force of 2×10^{-7} newtons per metre length of the conductor.

As stated in Chapter 1, the symbol for current is *I* and any value in amperes is represented by the letter A after the numerical value. The reader is reminded that practical circuit currents range from only a few microamperes to several thousand amperes. Attention is drawn to the Table of Prefixes of Magnitudes given at the front of this book. Full consideration *must be* given to correct use of the abbreviation that follows the numerical value.

When a current flows for a set period of time, a quantity of electricity or current is conveyed round the circuit. The quantity that passes can be shown to be related to the work done in the circuit, but before this relationship is considered further, electricity must be defined in terms of current and time.

Unit of quantity

THE COULOMB. The usual unit – sometimes called the ampere second. For practical everyday use a larger unit, in electrical engineering, the *Ampere hour*, is used in connection with the capacity of batteries for accumulator charging.

The symbol for the quantity of electricity is Q and any value in coulombs can be represented by the letter C after the numerical value. Similarly, any value in ampere hours may be represented by the letters Ah after the numerical value. As the quantity of electricity that is conveyed round a circuit varies with the strength of the flow of electricity and with time, a simple definition for the coulomb can be given:

A coulomb is the quantity of electricity conveyed by a steady current of 1 ampere flowing for a time of 1 second.

Thus Q (coulombs) = I (amperes) $\times t$ (seconds)

or Q (ampere hours) = I (amperes) $\times t$ (hours)

Thus, the following can be deduced:

1 ampere hour = 1 ampere $\times$ 1 hour

$\qquad$ = 1 ampere $\times$ 3600 seconds

$\qquad$ = 3600 ampere seconds

$\qquad$ = 3600 coulombs

Thus 1A h = 3600C.

Example 2.2. Consider Example 1.5, where a battery of e.m.f. 42V and internal resistance 7Ω is used to supply a circuit of 3 resistors 2Ω, 4Ω and 8Ω in series. If the current is switched on for 30 minutes, find the quantity of electricity that would have been transferred (1 decimal place).

Total resistance of circuit = 7 + 2 + 4 + 8 = 21Ω

Circuit current $= \dfrac{V}{R} = \dfrac{42}{21} = 2A$

Quantity of electricity = current × time in seconds

$$= 2 \times 30 \times 60 = 3600C$$

or Quantity of electricity = current × time in hours

$$= 2 \times 30/60 = 1.0A\ h.$$

Flow of an electric current results in energy being expended. This energy may appear as the work done by the rotation of an electric motor, the heating of a furnace element or that needed for electrolytic dissociation of a salt solution. However, the relation between conveying a quantity of electricity round a circuit by an applied voltage and the resulting work done helps derive the units of voltage and resistance in terms of the coulomb and the joule defined previously.

Unit of voltage

THE VOLT. This is the unit of e.m.f. and P.D. and is defined as the e.m.f. applied, or the P.D. available between 2 points in a circuit, if 1 joule of work is to be done when 1 coulomb of electricity passes between the 2 points.

As stated in Chapter 1, the voltage symbol or e.m.f. is V and any value in volts is represented by the letter V after the numerical value. The reader should again refer to the Table of Prefixes of Magnitudes, and the correct use of the abbreviations.

From the definition above it is stated that the work done by the electric circuit equals the voltage applied across that part of the circuit times the quantity of electricity conveyed.

$$\text{Thus: } W \text{ (joules)} = V \text{ (volts)} \times Q \text{ (coulombs)}$$
$$\text{or } W \text{ (joules)} = V \text{ (volts)} \times I \text{ (amperes)} \times t \text{ (seconds)}$$
$$W = VIt = I^2Rt$$

Example 2.3. Consider Example 2.2. A battery of e.m.f. 42V and internal resistance 7Ω is used to supply a circuit of 3 resistors, 2Ω, 4Ω and 8Ω in series. If the current is switched on for 30 minutes, find the energy converted (as heat) in joules by each resistor and within the battery itself.

Circuit current was found to be 2 amperes.

Using form $W = I^2Rt$ then energy converted in each resistor will be:

2 ohm resistor $= 2^2 \times 2 \times 30 \times 60 = 14\,400$ joules

4 ohm resistor $= 2^2 \times 4 \times 30 \times 60 = 28\,800$ joules

8 ohm resistor $= 2^2 \times 8 \times 30 \times 60 = 57\,600$ joules

7 ohm battery $= 2^2 \times 7 \times 30 \times 60 = 50\,400$ joules

Total energy converted by the circuit =

$14\,400 + 28\,800 + 57\,600 + 50\,400 = 151\,200$ joules

Check. The total energy converted by the entire circuit may be found from $W = VIt$ joules

$= 42 \times 2 \times 30 \times 60 = 151.2$kJ.

The definitions of Power and Energy have already been considered, but it is useful to emphasise the key points, namely that power is the *rate* at which work is done and its unit is the Watt. Thus: $P = W/t$ or $W = Pt$

From $W = VIt$ it follows that the rate of work $P = VI$ or P (watts) $= V$ (volts) $\times I$ (amperes).

The above is a most important relationship, which can also be expressed in the following forms:

$$P = I^2R \quad \text{or} \quad P = \frac{V^2}{R}$$

The student's attention is drawn to the following facts.

As Energy = Power $\times$ time   1 Joule = 1 Watt second

Now a joule is a tiny unit of energy and for practical purposes a larger electrical unit of energy is needed. This unit is the *kilowatt-hour*, abbreviated to *kW h* and is the commercial unit of electricity or 'unit'.

As 1 kilowatt-hour = 1 kilowatt $\times$ 1 hour

$= 1000$ watts $\times$ 3600 seconds

So 1 kilowatt-hour $= 3\,600\,000$ joules (3.6MJ)

Example 2.4. A 220V electric fire is rated at 2kW. Find the current taken when the fire is switched on and also how much it costs to operate the fire for 5 hours with electricity charged at 6p per unit.

$$\text{Current taken} = \frac{W}{V} \frac{2 \times 1000}{220} = 9.09\text{A}$$

Electricity used $= 2 \times 5 = 10$ kW h $= 10$ units

Cost $= 10 \times 6 = 60$p.

Unit of resistance

THE OHM. The ohm was defined in Chapter 1 as the unit of resistance in terms of the volt and ampere. A resistor has a value of 1 ohm resistance if 1 ampere passes through it when a P.D. of 1 volt is applied across its ends. Now the fundamental relationships between the ampere, volt, joule and watt are defined, it is possible to define the ohm in terms of the power or energy dissipated. Thus the ohm is defined as: that resistance which when 1 ampere passes through it dissipates energy at the rate of 1 joule per second (i.e. 1 watt).

Alternatively, the ohm is that resistance in which a current of 1 ampere flowing for 1 second generates 1 joule of energy.

For a resistor the energy produced by current flow appears as heat, and the following is of importance.

Since $P = VI$ and $V = IR$ then $P = (IR)I$

or $P = I^2R$ as seen earlier.

Power dissipated in a resistor is thus proportional to the current *squared*. If the current is doubled by raising the voltage, the power dissipation will be 4 times as large. Temperature rises in similar proportion and assuming the resistor is capable of carrying its normal current, with little capacity for working at a higher temperature, a 'burn-out' will occur. The same limitations apply to cables, electrical machines and switchgear. Electrical equipment is assigned a rating that, on full load, enables it to operate with a safe temperature rise. Increase of the normal rated current, due to overloading or an 'overvoltage', results in a temperature rise proportional to the new current squared. The total temperature will rise rapidly as the over current occurs and if this is maintained circuit damage will result. Damage to electrical insulating materials can occur because of sustained overloads, and overheating is often the main cause of electrical machine failure.

Example 2.5. A hotplate of a ship's electric galley is fitted with a control marked High, Medium and Low. The heating element consists of 2 equal sections, connected in parallel for High, and in series for Low. Only one section is used for Medium. If the plate when set at High is rated at 2kW on 220V, find the wattage rating when the control is set at Low and at Medium.

Current taken at High. Two sections in *parallel*

$$I = \frac{W}{V} = \frac{2000}{220} = 9.09A$$

Current taken by 1 section $= \dfrac{1000}{220} = 4.545A$

Resistance of 1 section $= \dfrac{V}{I} = \dfrac{220}{4.545} = 48.41\Omega$

Resistance of 2 sections in series $= 96.82\Omega$

Current taken at Low. Two sections in *series*

$$I = \frac{V}{R} = \frac{220}{96.82} = 2.27A$$

Power dissipated $= V \times I = 220 \times 2.27 = 499.4W$

or wattage rating $= 500W$ (approx.).

Current taken at Medium. One section only across 220V or $I = 4.545A$.

Wattage rating is half of the High setting $= 1kW$ or wattage rating $= 220 \times 4.545 = 1000$ Watts $= 1kW$.

Examples Relating Mechanical and Electrical Energy

A good way to understand the units discussed earlier is achieved by considering examples where mechanical work is performed by electrical means or vice versa. It is necessary first to stress that no machine is perfect and that its overall performance is measured by its *efficiency*.

Efficiency

The symbol usually used is η – the Greek letter 'eta'. In all sensors and machines, energy losses will occur due to friction, air turbulence, unwanted electrical currents, escaping heat, etc. Such losses result in the output of such sensors and machines, when measured as work, always being *less* than the input. The ratio of the output to the input is termed the efficiency.

Or since output = input – losses

Thus:

$$\text{Efficiency} = \frac{\text{output}}{\text{input}} = \frac{\text{input} - \text{losses}}{\text{input}}$$

$$\text{or Efficiency} = \frac{\text{output}}{\text{output} + \text{losses}}$$

Efficiency is expressed as a percentage.

Example 2.6. A diesel engine has a measured power of 7.5kW and a mechanical efficiency of 85%. It drives a generator that supplies a lamp load at 110V. How many 60W lamps can be supplied, if the generator's efficiency is measured to be 88%? Find the total load current to 2 significant figures.

The output of the engine = input × efficiency

$$= 7.5 \times \frac{85}{100} = 6.375\text{kW}$$

At the coupling between engine and generator, it is assumed there is no loss of energy, so the power input to the generator must be the power output of the engine. It follows that:

Output of engine = input to generator = 6.375kW

Thus generator output = input × efficiency

$$= 6.375 \times \frac{85}{100} = 5.61\text{kW}$$

Number of lamps $= \dfrac{5610}{60} = 93.5$ (i.e. 93 whole lamps at 60W output).

Load current $= W/V = \dfrac{93 \times 60}{110}$

$= 50.73A$, i.e. 51A

or alternatively

Load current $= \dfrac{5610}{110} = 51A$ (total power/voltage).

Example 2.7. A pump is required to lift 12 tonnes of water through 10m in 2 minutes. Calculate the power needed to drive the pump, and the current taken if driven by a 220V motor and the cost of pumping is 5p per kW h unit. Assume the pump efficiency is 60% and the driving motor efficiency is 85%.

Work to be done = Force opposing gravity × distance lifted. Note 1 tonne = 10^3 kg

Thus work to be done $= (12 \times 10^3 \times 9.81) \times 10$ newton metres

$$= 117.72 \times 10^4 \, Nm$$

Also 1 177 200 Nm = 1177.2kJ

This is the pump output. The input will be greater, i.e.

$1177.2 \times \dfrac{100}{60} = 1962kJ$

As pumping is achieved in 2 minutes or 120 seconds, the power input during this time $= \dfrac{1\,962\,000}{120} = 16350W$ Thus power required to drive the pump supplied by the motor is 16.35kW.

The motor's output power rating must be 16.35kW and the input power will be

$16.35 \times \dfrac{100}{85} = 19.24kW$

Current taken by motor $= P/V = \dfrac{19\,240}{220} = \dfrac{962}{11} = 87.45A$

Energy used = Power × time $= 19.24 \times \dfrac{2}{60} = \dfrac{19.24}{30} = 0.641kW \, h$

Cost = 0.641 × 5 = 3.2p.

Example 2.8. The electric motor used to drive a ship's winch has an efficiency of 86%. The winch lifts a mass of 0.5 tonnes through a distance of 22m in 22 seconds. The winch's winding gear has an efficiency of 60%. Calculate the motor's power rating and also the current taken from the 220V ship's mains to 2 decimal places.

Work done by the winch = 500 × 9.81 × 22 newton metres

$$= 107\ 910 \text{ Nm or } 107\ 910\text{J}$$

This is the winch output or the winding gear's output. The input to the winding gear will be

$$107\ 910 \times \frac{100}{60} = 179\ 850\text{J}$$

The input to the winding gear will also be the motor output = 179 850J.

As the lifting is done in 22 seconds, the motor will give out power during this time.

$$= \frac{179\ 850}{22} = 8175\text{W}$$

For a motor output of 8175W, the input power will be

$$8175 \times \frac{100}{86} = 9505.8\text{W}$$

Thus power rating of motor is 9.51kW

$$\text{Input or motor current} = \frac{9506}{220} = 43.21\text{A}$$

Example 2.9. A storage battery is provided for emergency use on board a ship. The battery is arranged to supply essential services such as lighting during the period of time taken to start up a 'standby' generator. The principal load supplied by the battery is the 'emergency' motor for an electric–hydraulic steering gear. This motor is rated at 220V, 15kW and has an efficiency of 88%. The battery must have a capacity sufficient to operate this motor and an additional lighting load of 20 60W lamps for a period of 30 minutes. Estimate the size of the battery and also its discharge current (2 decimal places).

Output of motor = 15kW

$$\text{Input to motor} = 15 \times \frac{100}{88} = \frac{187.5}{11} = 17.045\text{kW}$$

$$\text{Input current to motor} = \text{W/V} = \frac{17\ 045}{220} = \frac{852.3}{11} = 77.48\text{A}$$

Lighting load $= 20 \times 60 = 1200\text{W}$

Lighting current $= \text{W/V} = \dfrac{1200}{220} = \dfrac{60}{11} = 5.45\text{A}$

Total current $= 77.48 + 5.45 = 82.93\text{A}$

or Discharge current $= 82.93\text{A}$

Size of battery (Ah) $= 82.93 \times \dfrac{30}{60} = 41.47\text{A h.}$

Grouping of Cells

Ohm's law states that the current in a circuit can be increased by raising the P.D. applied across a circuit or by decreasing circuit resistance. If the supply source is a generator, the applied P.D. can be varied by controlling the e.m.f. generated in the machine, but if a battery is the source of energy applied voltage cannot be varied easily. As a battery consists of a group of cells and as the e.m.f. of any cell is fixed, determined by its chemical composition, then a larger e.m.f. or a greater current can only be obtained by correct arrangement of the cells connected in either series, parallel or series–parallel arrangements.

Series connection

For this arrangement the −ve cell terminal is connected to the +ve terminal of an adjacent cell as shown in figure 2.1a. The arrangement is more simply depicted in figure 2.1b showing a battery of 3 cells in series.

From Kirchhoff's voltage law, the e.m.f. of the source is equal to the sum of e.m.f.s taken round the circuit and for a battery of n cells in series, the e.m.f. $=$ e.m.f. of 1 cell $\times n$. Since this is a series circuit, the current in any 1 cell is the circuit current. The internal resistances of the cells are also in series and should be deducted as for resistances in a series circuit. This point is illustrated in figure 2.2.

Example 2.10. A battery consists of 4 cells in series, each of e.m.f. 1.5V and internal resistance 0.6Ω. Find the current flowing, if the battery is connected to 2 resistors of 2Ω and 0.6Ω connected in series. The arrangement is shown in figure 2.2.

(a)

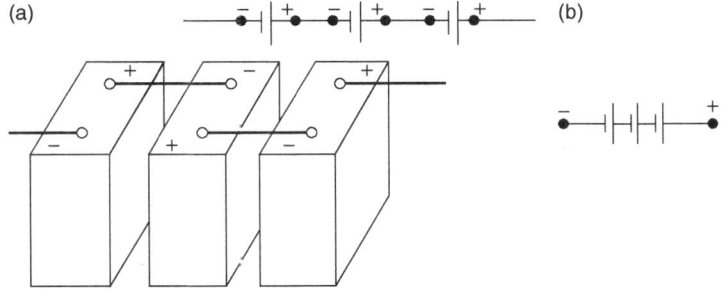

(b)

▲ **Figure 2.1** *A battery with cells in series with internal resistance*

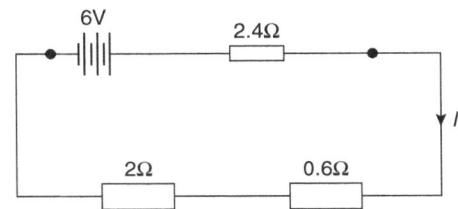

▲ **Figure 2.2** *A battery with cells and resistances in series*

Total battery e.m.f. $= 4 \times 1.5 = 6V$

Total battery internal resistance $= 4 \times 0.6 = 2.4\Omega$

Total circuit resistance $= 2.4 + 2 + 0.6 = 5\Omega$

So circuit current $= \dfrac{V}{R} = \dfrac{6}{5} = 1.2A$

Other values of interest would be

Battery terminal voltage $= iR = 1.2 \times 2.6 = 3.12V$, or

Battery terminal voltage $= V - iR = 6 - (1.2 \times 2.4)$

$$= 6 - 2.88$$

$$= 3.12V$$

Voltage drop across each resistance $= 1.2 \times 2 = 2.4V$

and $1.2 \times 0.6 = 0.72V$

Current in 1 cell $=$ circuit current $= 1.2A$.

Parallel connection

For this arrangement, the +ve terminals of all the cells are connected together as are all the −ve terminals. The arrangement is shown in figure 2.3.

From Kirchhoff's current law, the total current is the sum of the currents in each branch. The total current from the battery is equal to the sum of the currents available from each cell. For correct working, the e.m.f. of each cell should be the same and ideally have the same internal resistance. If n cells are in parallel, the total current is n times that given by 1 cell, but the battery e.m.f. is that of any one cell.

Battery internal resistance is obtained from the parallel-resistance formula, i.e. it is $\frac{1}{n}$ th of a cell's resistance. Battery resistance once determined is added to the external resistance to give the total circuit resistance as seen in Example 2.11.

Example 2.11. A battery consists of 4 cells in *parallel*, each of e.m.f. 1.5V and internal resistance 0.6Ω. Find the total current flowing if the battery is connected to a resistance of 2.6Ω in series for the arrangement shown in figure 2.4 (3 decimal places).

E.m.f. of battery = e.m.f. of 1 cell = 1.5V

Since $1/Ri_{TOTAL} = 1/Ri_1 + 1/Ri_2 + 1/Ri_3 + 1/Ri_4 = 4/0.6Ω$

Internal resistance of battery $= \dfrac{0.6}{4} = 0.15Ω$

Total resistance of circuit = 2.6 + 0.15 = 2.75Ω

Current $= \dfrac{V}{R} = \dfrac{1.5}{2.75} = 0.545A$. Other information would be

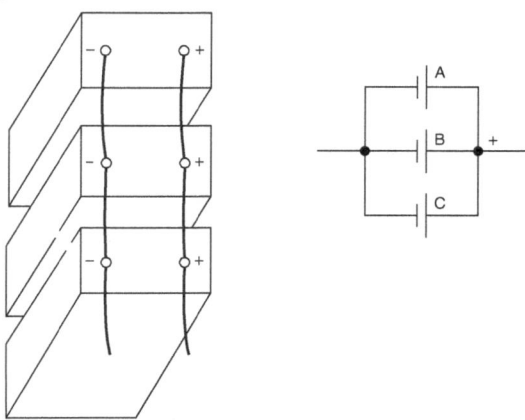

▲ **Figure 2.3** *A battery with cells in parallel with internal resistance*

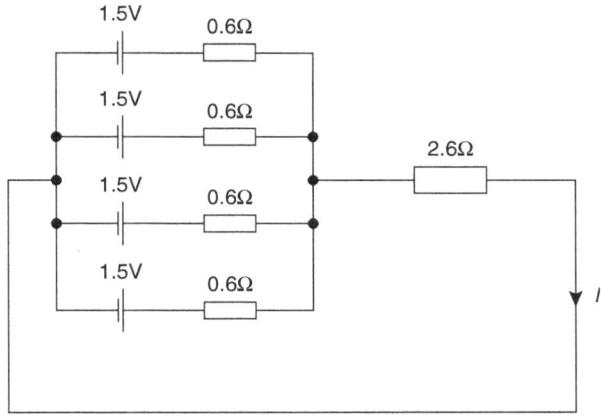

▲ **Figure 2.4** *A battery with cells in parallel and resistance in series*

Terminal voltage $= 0.545 \times 2.6 = 1.418V$

or Terminal voltage $= 1.5 - (0.545 \times 0.15) = 1.5 - 0.082$

$$= 1.418V$$

Current of 1 cell $= \dfrac{0.545}{4} = 0.136A.$

Series–parallel connection

To build this arrangement, several cells connected in series are then connected in parallel with a similar number of cells in series. The arrangement is as shown (figure 2.5) and provides respectively both increased voltage and current. Cells in series provide an increased e.m.f., while parallel banks of cells supply an increased current.

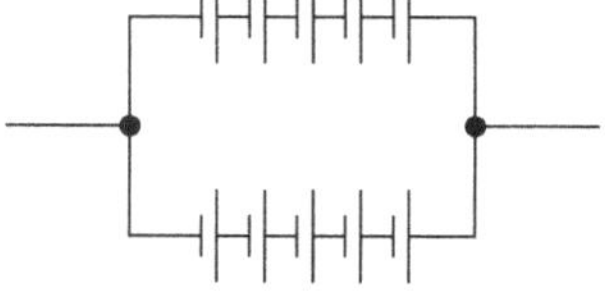

▲ **Figure 2.5** *Series parallel combination*

The procedure for solving problems follows the same sequential approach already covered for the series and parallel arrangements.

Example 2.12 (a). Ten cells each of internal resistance 3Ω and e.m.f. 2V are connected in 2 parallel banks of 5 series cells per bank. They are then connected to an external load resistance of 20Ω. Find the load current and the P.D. across the battery terminals (3 decimal places). The arrangement is shown in figure 2.6a.

E.m.f. of a bank = 5 × 2 = 10V = battery e.m.f.

Resistance of 1 bank = 5 × 3 = 15Ω

Resistance of battery $=\dfrac{15}{2}=7.5Ω$ (as 1/R = 1/15 + 1/15, 2 banks in parallel)

Total circuit resistance = 7.5 + 20 = 27.5Ω

Circuit or load current $=\dfrac{10}{27.5}=0.364A$

P.D. or terminal voltage = 0.364 × 20 = 7.28V

Current per cell = iR = current of 1 bank $=\dfrac{0.364}{2}=0.182A.$

Example 2.12 (b). If the battery is rearranged with 5 banks of 2 cells in each, find the new current and voltage (3 decimal places). The arrangement is shown in figure 2.6b.

E.m.f. of a bank = 2 × 2 = 4V = battery e.m.f.

Internal series resistance of 1 bank = 2 × 3 = 6Ω

Internal resistance of 5 banks in parallel battery $=\dfrac{6}{5}=1.2Ω$

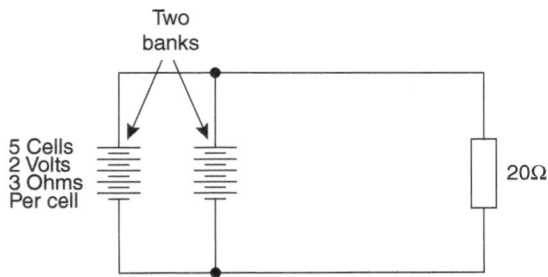

Two
banks

5 Cells
2 Volts
3 Ohms
Per cell

20Ω

▲ **Figure 2.6a** *2 parallel banks of 5 series cells*

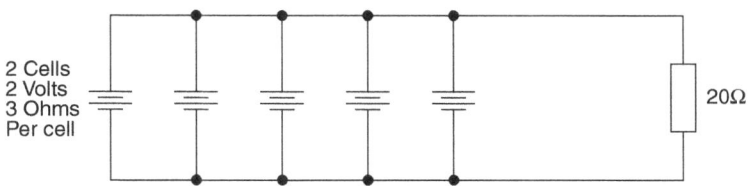

▲ **Figure 2.6b** *5 parallel banks of 2 series cells*

Total circuit resistance = external resistance + internal resistance = 20 + 1.2 = 21.2Ω

Circuit or load current $= \dfrac{V}{R} = \dfrac{4}{21.2} = 0.188A$

Terminal voltage = 0.188 × 20 = 3.77V

(or $V - iR_i$) 4 − 1.2 × 0.188 = 4 − 0.23 = 3.77V

Current/cell = current of 1 bank $= \dfrac{0.188}{5} = 0.0376A.$

Maximum power transfer condition for a loaded circuit

Consider the simple circuit of figure 1.5, Chapter 1. If I was the circuit current then the power P supplied to the external load R will be given by I^2R, where $I = \dfrac{E}{R + R_i}$, and R_i is the internal resistance of the supply source, so:

$$P = \left(\dfrac{E}{R + R_i}\right)^2 R = \left[\dfrac{E^2R}{(R + R_i)^2}\right]$$ or $P(R) = U(R)/V(R)$ where $U(R)$ and $V(R)$ represent respectively the numerator and denominator as functions of R.

Differentiating with respect to R, the maximum power condition is obtained when

$$\dfrac{d}{dR}\left[\dfrac{E^2R}{(R + R_i)^2}\right] = 0$$

Differentiation using standard methods results in a maximum for P when $R = R_i$, i.e. when the load resistance is equal to the internal resistance of the supply source.

The efficiency (as a percentage) of the supply for the condition $R = R_i$ is:

$$\dfrac{\text{Output Power}}{\text{Total Power}} \times 100 \quad \text{or} \quad \eta = \dfrac{I^2R}{I^2(R + R_i)} \times 100$$

But as $R = R_i$

$$\therefore \eta = \frac{I^2 R}{I^2 2R} \times 100 = \frac{1}{2} \times 100 = 50\% \text{ at maximum power transfer.}$$

Practice Examples

2.1. An electric hoist must lift a load of 2 tonnes to a height of 30m. The cage has a mass of 0.25 tonnes and the lifting operation must take 1.5 minutes. If the 220V motor is metered to take a current of 50A, find the installation's *percentage* efficiency (1 decimal place).

2.2. Thirty cells each having an e.m.f. of 2.2V and an internal resistance of 0.3Ω are connected to give a supply e.m.f. of 22V. If the arrangement is then connected to three 20V, 10W lamps in parallel, calculate (a) the battery terminal voltage (2 decimal places), (b) the current taken by each lamp (3 decimal places) and (c) the power wasted in each cell (3 decimal places).

2.3. A pump delivers 12 700 litres of water per hour into a boiler working at 15 bars pressure. The pump, which is 82% efficient, is driven by a 220V motor, having an efficiency of 89%. Calculate the current taken by the motor (1 decimal place). Assume 1 litre of water has a mass of 1kg and 1 bar = 10^5N/m².

2.4. A resistor of 5Ω is connected to a battery made up of 4 similar cells in series. Each cell has an e.m.f. of 2.2V and the current that flows is 1.4A. If the cells were connected in parallel, find the current that would flow through the 5Ω resistor (2 decimal places).

2.5. A 5-tonne cargo winch is required to lift a load of 5 tonnes at 36.5m/min. Calculate the power rating of the 220V driving motor if the efficiency of the winch gearing is 75% and that of the motor is 85%. Calculate also the current taken from the ship's 220V mains (1 decimal place).

2.6. A 220V diesel-driven generator is required to supply the following on full load: (a) lighting load comprising one hundred 100W and two hundred 60W lamps, (b) a heating load of 25kW, (c) miscellaneous small loads drawing a current of 30A. Calculate the required power output of the diesel engine when the generator supplies all the loads at the same time (1 decimal place). Assume a generator efficiency of 85%.

2.7. A battery is made up from 3 similar correctly connected dry cells in series. The O.C. e.m.f. is measured to be 4.3V. When the battery is connected to an unknown resistor the current is metered to be 0.4A and the battery terminal voltage as 4.23V. If one of the cells of the battery is reversed and the circuit made up as before, estimate the new current value (3 decimal places).

2.8. A 150W, 100V lamp is to be connected in series with a 40W, 110V lamp across a 230V supply. The lamps are required to operate at their rated power values. Determine the values of suitable resistors to be used with the lamps and make a sketch showing how they should be connected (1 decimal place).

2.9. A resistor of 0.525Ω is connected to the terminals of a battery consisting of 4 cells, each of e.m.f. 1.46V joined in parallel. The circuit current is found to be 0.8A. Find the internal resistance of each cell (1 decimal place).

2.10. Twelve cells, each of e.m.f. 1.5V and internal resistance 0.225Ω, are arranged 4 in series per row or bank, with 3 banks in parallel. This battery is connected to a load consisting of a series–parallel resistor arrangement, made up of a 2Ω resistor connected in parallel with a 3Ω resistor, which are in turn connected in series with a 2.5Ω resistor. Find the battery terminal voltage, the power ratings of the resistors (all these 1 decimal place) and the energy converted into heat in the complete circuit, which is switched on for 1 hour (3 significant figures).

3

CONDUCTORS, INSULATORS AND SEMICONDUCTORS

Nothing tends so much to the advancement of knowledge as the application of a new instrument.

Sir Humphry Davy

In this chapter we will look at the experimental differences between conductors, insulators and semiconductors. The reasons *why* certain materials are good conductors of electricity, while others are not, will be discussed at the electronic level. A substance that freely allows the passage of electricity is classed as a conductor, for example, metals, certain arrangements of carbon and some liquids – chiefly solution of salts, acids or alkalis. An insulator can be defined as a substance that prevents the free passage of electricity. Examples are rubber, porcelain, slate, mica, some organic materials and some liquids – notably oils. A semiconductor is defined as a material with properties between that of a conductor and an insulator due to controllable manufacturing processes to create precise atomic layered fabrication.

Resistance of a Conductor

Variation of conductor resistance with dimensions and material

The resistance or 'ohmic' value of a conductor, for example, a coil of wire, may be altered in several ways. Thus if coils of different lengths of the same wire, i.e. the same material and with identical cross-sectional area, are measured for resistance, their ohmic values vary in direct proportion to their lengths. Again, if coils of wire of the same material and length, but of *different cross-section* are measured, their resistance values will vary in inverse proportion to the areas of the wires of which they are wound.

A similar series of comparative measurements with coils of wire of the same length and cross-sectional area but different material show that resistance values vary with the conductor material.

The basic tests described above indicate that the resistance of a conductor or resistor can be changed by varying its dimensions or the material, and the relation of these factors to the actual conductor resistance will now be examined in detail.

1. DIMENSIONS. 1 (a). Resistance of a conductor is proportional to its length, for example, the conductor resistance of a 100m length of cable will be double that of a 50m length of the same cable. This can be shown as follows:

Let R_A ohms = the resistance of a 50m length. Then two 50m sections in series will have a resistance of R ohms

Hence $R = R_A + R_A = 2R_A$

But the length has been doubled

So 2 × Length = 2 × Resistance of original length. Summarising:

$$\text{Resistance is proportional to Length or } R \propto l.$$

1 (b). Resistance of a conductor is inversely proportional to its area, for example, the conductor resistance of a 1mm² cross-sectional area cable will be twice that of the same length of cable of the same conductor material, but of 2mm² or twice the cross-sectional area. This can also be shown thus:

Let the resistance of 1mm² cable be R_A and suppose an identical cable to be connected in parallel with it. The resistance of the combination will be R ohms.

$$\text{Hence } \frac{1}{P} = \frac{1}{R_A} + \frac{1}{R_A} = \frac{2}{R_A} \quad \text{or} \quad R = \frac{R_A}{2}$$

Thus the resistance of the combination is *half* the original cable resistance, but the area of the combination is twice that of the original cable

$$\text{or } 2 \times \text{Area} = \frac{1}{2} \times \text{Resistance of original length.}$$

From the above, it follows that doubling the area halves the resistance of a conductor of the same length and material. Thus:

Resistance is inversely proportional to Area or $R \propto \dfrac{1}{A}$

2. MATERIAL. The resistance of a conductor depends upon the material from which it is made, for example, the resistance of a length of iron wire is nearly 7 times greater than the resistance of a piece of copper wire of identical dimensions, i.e. same length and cross-sectional area. If resistance varies with the material, we must define a property to compare resistance values with standard dimensions of the conductor. The term *resistivity or specific resistance* (symbol ρ – the Greek letter rho) is introduced, and measured in ohm-metres or microhm-millimetres.

The Resistivity or Specific Resistance of a material is the resistance measured between the opposite faces of a cube of unit dimensions. Thus for the diagram (figure 3.1), if a cube of pure copper of sides 1 metre is taken and the resistance measured between the faces as shown by the arrows, the resistivity would be measured as $1.725 \times 10^{-8}\,\Omega$m or 17.25μΩmm.

Material temperature is recorded at the time the test is made and is often specified with the resistivity figure. Thus ρ for copper is given as 1.725×10^{-8} ohm-metre at 20°C. The reasons for specifying the temperature will be explained shortly.

As R (Resistance) $\propto$ 1 (length) and $R \propto \dfrac{l}{A}$ (Area). So $R \propto \dfrac{l}{A}$ or $R = k\dfrac{l}{A}$ or where k is a constant. If ρ is taken as this constant k, then the foregoing can be written as:

$$R = \frac{\rho l}{A}$$

Problems involving resistivity are solved by use of the above expression, but it is essential to note the units used for the resistivity. If *l* and *A* are not in these units, they must be converted as shown (Example 3.2.)

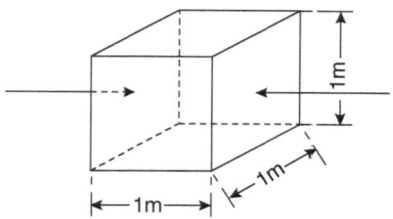

▲ **Figure 3.1** *Unit cube showing all surfaces*

Example 3.1. If 5m of manganin resistance wire, 0.1mm diameter, have a resistance of 267.5 ohms, find the resistivity of the material (2 significant figures).

$$A = \frac{\pi \times d^2}{4} = \frac{3.14 \times (1 \times 10^{-4})^2}{4}$$

$$= 0.785 \times 1 \times 10^{-8} \text{ m}^2$$

So $l = 5\text{m}$

Since $R = \dfrac{\rho l}{A}$ then $\rho = \dfrac{RA}{l} = \dfrac{267.5 \times 0.785 \times 1 \times 10^{-8}}{5}$

or $\rho = \dfrac{535 \times 0.785 \times 10^{-8}}{10}$

$$= 41.998 \times 10^{-8} \text{ ohm-metre or } 42 \times 10^{-8}\Omega\text{m}$$

$$= 420 \times 10^{-9} = 420 \times 10^{-6} \times 10^{-3}$$

$$= 420\mu\Omega\text{mm}.$$

Example 3.2. Find the length of wire required to make a 10Ω resistor, if the diameter is 1mm and the resistivity is $450\mu\Omega$ mm (2 decimal places).

Here $R = 10\Omega$. $\rho = 450 \times 10^{-6} \times 10^{-3}$

$$= 45 \times 10^{-8} \text{ ohm-metre}$$

$d = 1\text{mm} = 1 \times 10^{-3} \text{ metre}$

Since $R = \dfrac{\rho l}{A}$ hence $l = \dfrac{RA}{\rho}$

or $l = \dfrac{10 \times \pi \times (1.0 \times 10^{-3})^2}{45 \times 10^{-8} \times 4}$ metres

$$\therefore l = 17.44\text{m}.$$

Occasionally a problem involving the formula $R = \dfrac{\rho l}{A}$ can be worked by a method of proportion.

This makes for easier working than finding the resistivity values and resubstituting in the formula to obtain the answer. This is illustrated by the following example.

Example 3.3. If the resistance of 1.6km of copper wire of 0.5mm diameter is 170Ω, calculate the resistance of 1km of iron wire of 1.0mm diameter, assuming that the resistivity of iron is 7 times that of copper (3 significant figures).

Resistance of 1600m of copper wire 0.5mm diameter is 170 ohms, then resistance of 1000m of copper wire 0.5mm diameter is

$$\frac{170 \times 1000}{1600 \times 4} \text{ ohms}$$

and the resistance of 1000m of copper wire 1.0mm diameter is

$$\frac{170 \times 1000}{1600} \times \frac{1}{4} \text{ ohms}$$

Since Area $= \dfrac{\pi d^2}{4}$, it follows that wire of twice the diameter will have an area 4 times as great and the resistance therefore will be reduced by a factor of 4.

Resistance of 1000m of copper wire 1.0mm in diameter is

$$\frac{170 \times 1000}{1600 \times 4} \text{ ohms}$$

So resistance of 1000m of iron wire 1.0mm diameter

$$= \frac{170 \times 1000 \times 7}{1600 \times 4} = 186\Omega$$

Alternatively, as $R_1 = \rho_1 \, l_1/A_1$ for the copper and $R_2 = \rho_2 \, l_2/A_2$ for iron, dividing one by the other will give the ratio: $R_2/R_1 = \rho_2 \, l_2 A_1 /\rho_1 \, l_1 A$ and using the ratio $\rho_2/\rho_1 = 7$ will arrive at the same answer.

Variation of conductor resistance with temperature

Most conductors' resistance changes when their temperature changes. Usually this change or variation follows a straight-line relation, as is shown in figure 3.2. If the resistance of a resistor is measured each time its temperature is altered and the results plotted, a graph such as (1) or (2) or (3) is obtained. These graphs cover the main types of conductor and show that:

(1) For pure metals, resistance increases continuously with temperature.

(2) For certain metal alloys used for making resistors, such as Manganin or Constantan, the graph is horizontal, i.e. resistance is largely unaffected by temperature.

(3) For certain partial conductors, such as carbon, resistance *decreases* with temperature.

TEMPERATURE COEFFICIENT OF RESISTANCE. Because of the straight-line relationship between resistance and temperature, illustrated in figure 3.2, a simple law is possible and an equation can be deduced, which within normal ranges of temperature allows the resistance R of a conductor at any temperature T to be obtained in terms of the resistance R_0 at 0°C and a coefficient (symbol α – the Greek letter alpha), known as the *temperature coefficient* of the conductor material. The appropriate equation is:

$$R = R_0(1 + \alpha T) \text{ where } T \text{ is the temperature in °C.}$$

For the diagram (figure 3.3), the graph for a pure metal (copper) is illustrated and enlarged to show this formula for metals, alloys and carbon, which cuts the R axis to give a value of R_0, the resistance at 0°C. If the resistance has a value R_0 at 0°C then at 1°C it will be increased by a small amount x. The fraction $\dfrac{x}{R_0}$ is taken as the temperature coefficient α of the metal or $\dfrac{x}{R_0} = \alpha$ and $x = \alpha R_0$.

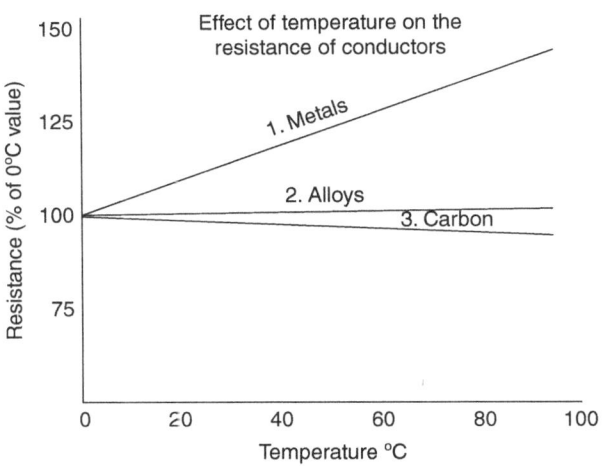

▲ **Figure 3.2** *Resistance vs temperature*

a is thus the fraction of the resistance at 0°C by which the resistance increases for 1°C rise in temperature.

If x = increase in resistance for 1° rise in temperature, then

xT = increase in resistance for T° rise in temperature

Thus $R = R_0 + xT = R_0 + R_0\alpha T$

or $R = R_0(1 + \alpha T)$

For copper, α has a value $\dfrac{1}{234.5} = 0.004\,265/°C$.

The temperature coefficient of resistance α is usually based on average conditions obtaining from 0°C to 100°C and is thus the ratio of the increase in resistance per °C rise in temperature to the resistance at 0°C. Alternatively, the temperature coefficient of resistance is defined as the increase of resistance of 1 ohm at 0°C for 1°C rise of temperature.

If the graph (figure 3.3) rises as temperature increases, the material will have a 'positive' temperature coefficient of resistance; whereas if the graph falls, the material has a

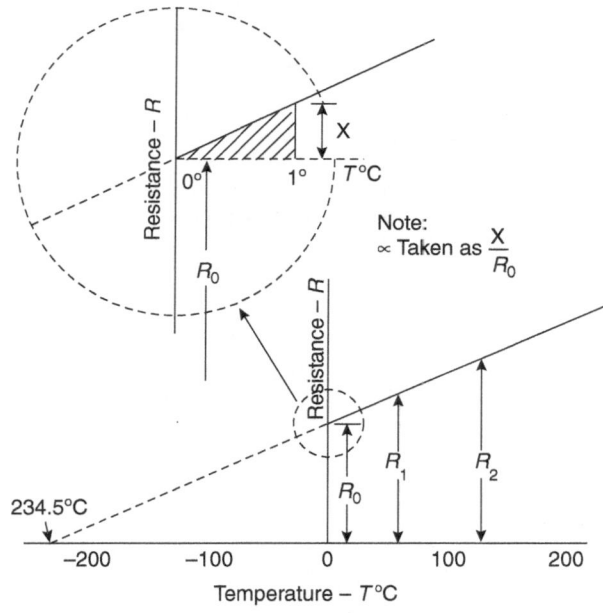

▲ **Figure 3.3** *Resistance vs temperature with gradient snowing temperature coefficient of resistance*

'negative' temperature coefficient. Usually for practical work and problems, the ohmic value as given for a resistor will be at a temperature other than 0°C and the following expression will help facilitate obtaining the resistance value at any temperature, if its resistance value is given for any other temperature condition.

Let R_1 be the resistance at a temperature T_1°C and R_2 be the resistance at a higher temperature T_2°C.

Then $R_2 = R_0 (1 + \alpha T_2)$ and $R_1 = R_0 (1 + \alpha T_1)$

Dividing, we have

$$\frac{R_2}{R_1} = \frac{R_0 (1+\alpha T_2)}{R_0 (1+\alpha T_1)} \quad \text{or} \quad R_2 = R_1 \frac{(1+\alpha T_2)}{(1+\alpha T_1)}$$

Example 3.4. The cold resistance of a coil of wire is 20Ω at 15°C. It is heated to give a resistance of 23Ω. Find its temperature rise, if the temperature coefficient of the resistance material is 0.0042 per °C (1 decimal place).

Using the expression given above.

$$\frac{23}{20} = \frac{R_0[1+(0.0042 \times T_2)]}{R_0[1+(0.0042 \times 15)]}$$

Thus $1.15 \times 1.063 = 1 + 0.0042 \times T_2$

or $1.222\ 45 - 1 = 0.0042 \times T_2$

and $0.223 = 0.0042 \times T_2$ or $T_2 = 53.1$°C

So the temperature rise is 38.1°C.

Example 3.5. The filament of a 230V lamp takes a current of 0.261A when working at its normal temperature of 2000°C. The temperature coefficient of the Tungsten filament material can be taken as 0.005/°C at 0°C. Find the current that flows at the instant of switching on the supply to the cold lamp, which can be considered to be at a room temperature of 20°C (1 decimal place).

Resistance of lamp (hot) $= V/I = \dfrac{230}{0.261} = 882$Ω at 2000°C

Here $R_2 = R_0 (1 + \alpha T_2)$.

So $R_0 = \dfrac{882}{[1 + (0.005 \times 2000)]} = \dfrac{882}{11} = 80.2\Omega$

Again

$R_{20} = 80.2 [1 + (0.005 \times 20)]$

or $R_{20} = 80.2 (1 + 0.1) = 80.2 \times 1.1 = 88.22\Omega$

and current taken at switching on when cold $= \dfrac{V}{R} = \dfrac{230}{88.22} = 2.6\text{A}.$

This example shows how resistance change with temperature affects practical working conditions and must be taken into account. The 'tripping' of a circuit-breaker or 'blowing' of a main-fuse may mean loss of supply to a large lighting or resistance load. Before any attempt is made to restore the supply, sections of the load must be isolated, so that when the main switching on takes place, only part of the load is applied to the supply and this load is gradually increased to its full value, by closing the individual circuit switches. The reason for 'shedding load' is related to the fact that the lamps will have cooled when the supply was off and when supply is restored, a current of 7–8 times the full-load value will be taken as a surge! This current will fall as the lamps heat up and may only last for a few milliseconds, but it could be sufficient to re-trip the circuit-breaker or blow the main-fuse. Thus a fault condition may be suspected, but in fact the cause of the current surge can be explained and appropriate action taken.

Resistance of an Insulator

Variation of insulation resistance with dimensions and material

An insulator is defined as a substance that prevents the free flow of electrons. In electrical devices, machines and cables, insulation helps confine electricity flow to the required circuit and to prevent any current from taking 'leakage paths'. Leakage currents are minimised by making the resistance of their paths as large as possible. Thus material with a high resistivity (ρ) is used for the insulation, the length of the leakage path is kept as large as possible and the area as small as possible. It should be realised that insulation allows some current to pass and that by measuring this current the insulation quality can be judged. Instruments such as high-resistance ohmmeters or insulation testers are availab e such as the 5kV and 15kV 'Megohmmeter' available from *Megger*.

Figure 3.4 shows the path o⁻ the leakage currents in the cables forming part of the circuit feeding a load. Note: the leakage currents flow radially from or towards the conductors of the cable. Cables can be looked upon as steam pipes radiating heat; the longer the pipe, the more the heat radiated, i.e. the greater its radiating area. Similarly, for cables, the thickness of the insulation, shown as

t, is actually the length l in the formula $R = \dfrac{\rho l}{A}$, while area A is given by the curved surface of the insulation. The larger this surface, the easier it is for leakage current to flow.

Cable insulation resistance is measured between the core and sheath, or 'earth', and is given by an approximation of the formula $R = \dfrac{\rho l}{A}$.

Here ρ would have a very high value; for vulcanised rubber it is $10^{15}\Omega$m or 10^9MΩm. The insulation thickness t and surface area a are proportional to the cable length. If the insulation resistance of 100m of cable was measured as 180MΩ, then 200m of the same cable would have a resistance value of 90MΩ. The key point is that cable-conductor resistance is *doubled* for double the length, but insulation resistance is *halved*. Doubling the length has doubled the area of the leakage paths and since $r \propto \dfrac{1}{A}$, if a is doubled r is actually halved.

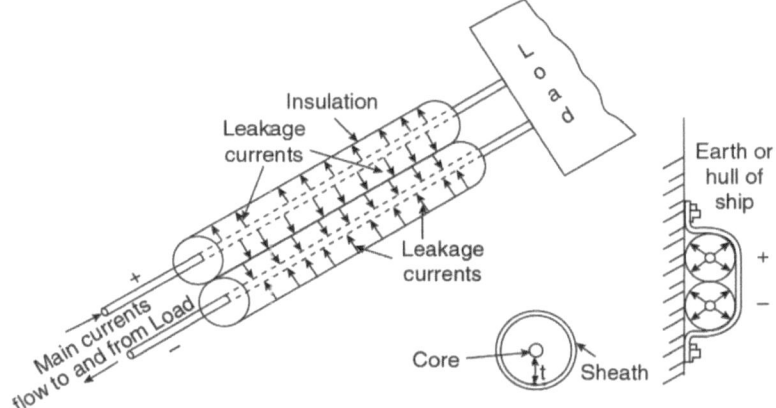

▲ **Figure 3.4** *Path of leakage current*

It must be understood why a large electrical cable installation or a machine when tested for insulation resistance may give a low figure, while the value obtained for a small installation or machine may be much larger. Insulation resistance is also affected by other factors, besides the size of the installation or machine. Site conditions such as temperature, humidity, cleanliness together with age must be taken into account and the resistance value means little unless compared with that obtained for a comparable new installation or machine. Acceptable insulation-resistance values for installations and machines are set out in the appropriate Institution of Electrical Engineers (IEE) Regulations and the points made above are stressed to show that test results must be treated with due consideration. Conductor-resistance measurements are more straightforward, although here again, special testing techniques should be employed depending on the type of resistor or apparatus being measured.

Variation of insulation resistance with temperature

For electrical devices, machines and cables, the permitted working temperature and thus the equipment's current-carrying capacity is limited largely by the restrictions imposed by the insulation. Insulation is usually made from cotton, silk, rubber or plastics, and as a general rule, if they are subjected to excessive temperatures their electrical and mechanical properties are reduced. Even if insulation such as mica or porcelain is not damaged by excess temperatures, it is seen from figure 3.5 that, like the partial conductor carbon, the insulation resistance falls with temperature rise, but here the relationship is not a straight line. The graph follows a logarithmic law and so insulation resistance falls rapidly as temperature rises. An increasing leakage current flows through the insulation as its temperature rises and the current generates more internal heat, which may eventually cause a sudden catastrophic 'breakdown' of the insulation. The allowable temperature rise for any electrical equipment, which gives a safe insulation-resistance value, has been determined by experience and the power rating of appliances is set in accordance with accepted specifications. Lloyds has its set maritime standards. There are several standards that cover marine equipment, including EN 60533, Lloyds Register specification N°1 and the most common standard IEC 945. IEC 945 is a complete test standard for the marine environment taking into account the mechanical, electrical and environmental conditions as well as the ElectroMagnetic Compatibility (EMC) phenomena. Sections 9 and 10 of IEC 945 include EMC requirements, covering a wide frequency range due to the breadth of equipment found on board ships. Section 9 covers EMC emissions, which take into account the protection of the ship's essential navigation and radio systems as well as the consideration that ships normally operate

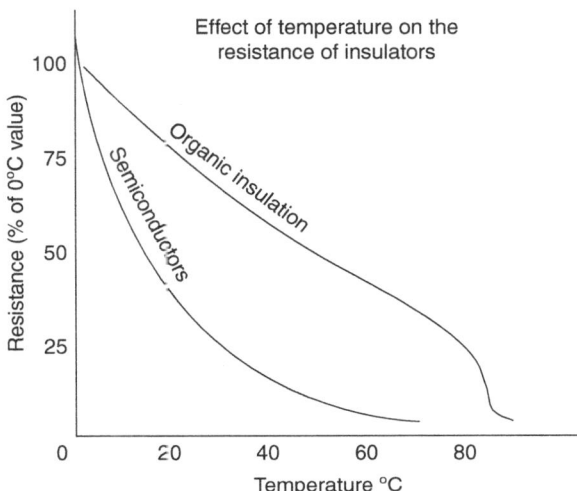

▲ **Figure 3.5** *Semiconductor and organic insulators*

at sea and must not interfere with land-based radio communications. Section 10 is an EMC immunity section, designed for the marine environment. Although IEC 945 and EN 60533 are comprehensive test specifications (in some cases the Lloyds Register Spec. N°1 applies for ships' underwriters, and may be required), this is normally part of the contractual agreement when supplying equipment integral to a particular ship.

For example, BS Specifications or Lloyd's Regulations may specify a working temperature rise of 50°C for a particular motor when performing a certain duty. This could be when it developed its rated output in an ambient or room temperature of 30°C. Thus a total temperature of 80°C will be allowed. This varies for the type of insulation with which the machine is constructed, but for the example, if the same motor is to work in an ambient of 50°C, then the allowable temperature rise will be reduced to 30°C. The motor will now only be capable of giving a reduced output and would have to be derated. Alternatively, a larger machine may be used, if the full original power output is still required. Derating of equipment is necessary to ensure a maximum *safe* working temperature for the insulation and for this condition, the insulation resistance will reach an acceptable minimum value.

Insulation-resistance values change as the temperature of the equipment alters, and it is also affected by other load factors such as: size of installation, humidity, cleanliness, age and site conditions. A true indication of the state of the installation or machine can only be gained by reference to a record or log of readings, built up over time. It should be accepted that such a log is essential for large electrical installations. Many ships are now fitted with insulation-resistance indicators, which record leakage current and the insulation resistance. Such indicators assist the keeping of a log, which shows comparative readings for the same temperature rise taken when the installation or machine was new, dry and clean. The difference between the readings helps assess the state of the equipment at the time of checking, and if an improvement in readings is deemed essential for safe working, appropriate arrangements can be made for cleaning, drying out or a more thorough inspection and overhaul.

Resistance of a Semiconductor

Electronic devices utilising semiconductor materials will be considered in some depth in Volume 7. A semiconductor can be described as a material that, for given dimensions, has a resistance value between that of a conductor and an insulator of the same dimensions. The main use of semiconductor materials is in solid-state devices such as rectifier diodes and transistors, but here we consider the resistance/temperature property in relation to the *thermistor*, which is an electrical resistor whose resistance is greatly reduced by heating, and is used for measurement and control.

Variation of semiconductor resistance with temperature

Semiconductor materials have resistance values that change when heated. Germanium and silicon are typical examples, and both have negative temperature coefficients that are not constant, but increase as the materials are heated. The relationship of resistance with temperature is an inverse variation and gives a graph similar to that shown in figure 3.5. It is observed that as a semiconductor material is heated, its resistance *falls* and if this material is used as a resistor the current passed will *increase* as the material heats up. The semiconductor, when used in this way, is known as a thermistor. It was adapted for use as a measuring or regulating device and developed for marine work as the detecting element of an electrical temperature-indicating instrument. The original thermometer head consisted of a coil of platinum wire, which, when heated, altered the resistance of an indicator circuit so the circuit could be calibrated to indicate temperature. A thermistor element, being more robust and of smaller dimensions than a platinum wire, gives a greater resistance change for a given temperature change and is thus more sensitive and accurate.

The device can be used as a regulator since it can alter the operating current to a controlling circuit when its temperature changes. If a thermistor is placed in the windings of an electric motor, any overheating adjacent to it will result in the thermistor-circuit current increasing until the connected motor protective device is operated. Such thermistor operated units are used for marine applications in conjunction with motor starters but must be correctly located and connected.

Heat and Electrical Energy

Energy exists in various forms; the mechanical, electrical, thermal and chemical forms are those most commonly used in the modern marine industry and the work done when energy is expended is put to use in various ways. Although the term 'energy expended' is commonly used, it should be remembered that energy cannot be destroyed or lost, it can only be changed from one form to another, and the convertibility between mechanical and electrical energy is seen in a machine like the electrical generator or electric motor. For the former, mechanical energy is passed in at the shaft and electrical energy obtained and utilised in a circuit connected across the machine terminals. Electrical energy may be converted into heat, light or mechanical energy. For the electric motor, electrical energy is passed in at the terminals and mechanical energy is passed out at the shaft.

Relation between mechanical and heat energy

The fact that heat is a form of energy is obvious to the practical engineer, who is aware of the dangers associated with a 'hot bearing', 'slipping belt' or 'clutch'. In these instances mechanical energy is made available by a prime-mover and converted into unwanted heat through the mechanism of friction. If this energy conversion process continues unchecked, the temperature of associated machine parts may rise to dangerous levels, when a 'burn-out' or fire may result. Examples have been given to show that an elementary deduction can be made showing that the heat energy produced is proportional to the mechanical energy being expended.

SPECIFIC HEAT CAPACITY. This is found by a simple mechanical test. The apparatus consists of a hollow brass cylinder, rotated by a belt drive. The cylinder is filled with a known amount of water and rotated against a friction surface applied with known tension. By simple calculation, the work put in at the drive pulley is related to the heat produced at the cylinder. James Joule, an English scientist, by careful experimentation showed that 4.187 joules of work is required to produce sufficient heat to raise the temperature of 1 gramme of water by 1 degree Celsius (or 1 Kelvin). In SI units, if the mass of water is taken as 1 kilogramme it follows that 4187 joules (4200J approx.) is required. The joule is also an SI unit of heat, and this constant of 4200 is taken into account by introducing the term *specific heat capacity*, defined as the quantity of heat needed to raise unit mass of a material through a temperature interval of 1 degree Celsius or 1 Kelvin. Different materials require different amounts of heat to produce the same temperature rise on the same mass. The units of specific heat capacity (symbol c) are heat units per unit mass per unit temperature. As, for SI units, the most convenient unit of mass is the kilogramme, so the kilojoule will be the appropriate size of heat unit to give specific heat capacity in kilojoules per kilogramme per Kelvin or kJ/kgK. In terms of the Celsius temperature scale, this will be in kJ/kg°C. Because the relation between energy and heat is readily determined for water it is taken as 4200 joules, and follows that the specific heat capacity value for water would be in 4.2kJ/kg°C. The values for other materials are also determined by experiment and compiled into tables of physical constants (e.g. Tables of Physical and Chemical Constants by *Kaye & Laby*). The following examples illustrate conversion from mechanical to heat units, which involves use of differing specific heat capacity values.

Example 3.6. A motor brake-test rig consists of a water-cooled, cast-iron pulley and a fixed frame made to carry 2 spring balances to which are fastened the ends of a rope, which passes round the pulley. Both spring balances hang from screwed rods, which are arranged to be adjustable to alter the tension on the rope. Tests made on a small motor running at a full-load speed of 750 rev/min give the following readings. Spring balances 16.89kg and 0.55kg. The pulley is hollow 102mm long, 380mm in diameter (these are outside dimensions). It has an average wall thickness of 6.4mm. It has a mass of 2.72kg and is designed to be half-filled with water. Estimate the output power of the motor being tested and the time for which the motor can be tested before the water starts to boil. The temperature of the pulley and water is 15°C at the start of the test and the rope diameter is 25mm. Take the specific heat capacities of water and cast iron as 4.2 and 0.42kJ/kg°C respectively.

Power output of the motor is given by the expression $\dfrac{2\pi NT}{60}$

Effective load on brake = (16.89 − 0.55) = 16.34kg

Hence restraining force F = 16.34 × 9.81 newtons

Effective radius $= \dfrac{380 + 25}{2} = 202.5$mm

Restraining torque = 16.34 × 9.81 × 202.5 × 10^{-3}Nm

Output power $= 2 \times \pi \times \dfrac{750}{60} \times (16.34 \times 9.81 \times 202.5 \times 10^{-3})$ W

$$= 2560\text{W} = 2.56\text{kW}$$

Heat energy available per minute $= 2.56 \times 60 = 153.6\text{kJ}$

Volume of water $= \dfrac{\pi d^2}{4} \times \text{width} \times \dfrac{1}{2}$

$$= \dfrac{\pi}{4}(380 - 12.8)^2 \times (102 - 12.8) \times \dfrac{1}{2}\ \text{mm}^3$$

$$= 4.72 \times 10^{-3}\ \text{m}^3$$

Since 10^3kg is the mass of 1 cub c metre of water

Then mass of water $= 4.72 \times 10^{-3} \times 10^3 = 4.72\text{kg}$

Mass of pulley $= 2.72\text{kg}$

Temp rise/min $= \dfrac{153.6}{(4.72 \times 4.2) + (2.72 \times 0.42)}$

$$= \dfrac{153.6}{20.97} = 7.3°\text{C}$$

Allowable temperature rise $= 85°\text{C}$

Heating time $= \dfrac{85}{7.3} = 11.7 = 12$ min (approx.).

Relation between electrical and heat energy

In Chapter 2 mention was made of the associated effect noted when current flowed in a circuit, being the temperature rise in any part of the circuit, where resistance was concentrated. As energy cannot be destroyed, this is another example of energy conversion from one form to another, and a simple test can examine the relation between heat and electrical energy. Such a test determines a material's specific heat capacity by an electrical method and, as water is a convenient substance, an appropriate experiment is described.

The experimental apparatus consists of a glass flow-tube surrounded by a glass water-jacket, spaced a little distance from it which is completely sealed so the space between the flow-tube and jacket is a vacuum. A heating wire runs along the tube and thermometers are placed at either end of the tube. Water, the specific heat capacity of which is to be found, flows steadily through the tube and is heated electrically by a known current passing through the wire. After a period the inlet and outlet temperatures of the water become constant and this difference in temperature is noted. The constant rate of flow of the water is measured and so the mass of liquid being heated in a given time is found. The voltage drop across the heater is also measured and the quantity of heat absorbed by the water equals the electrical energy expended by the heating element:

$$Vlt = mcT$$

where m is the mass of liquid, c the specific heat capacity and T the temperature rise.

The value of c is found to be 4.2kJ/kg°C. It is noted that the specific heat capacity of water, determined either by mechanical or electrical means, is the same. Different forms of apparatus have been developed to find the c values of various materials; the electrical method is favoured because of the accuracy with which control of the test can be made and measurements taken.

The following examples show how the specific heat capacity value is used in electrical problems.

Example 3.7. A brass calorimeter was found to have a mass of 67g. It was filled with water when the new mass was 131.7g. The temperature of water and container was 18°C. A heater coil immersed in the calorimeter is suitably lagged to minimise heat loss. Find the time taken to heat the water and calorimeter to a temperature of 33°C, if the heating is done by passing a current of 2A through the coil, the voltage drop across the coil is 7.5V. Take the specific heat capacity of brass as 0.39kJ/kg°C and that of water as 4.2kJ/kg°C.

Mass of water = 131.7 − 67 = 64.7g or 0.0647kg

Mass of calorimeter = 67g = 0.067kg

Temperature rise of water and calorimeter = 33 − 18 = 15°C

Heat required by water (mcT) = 0.0647 × 4.2 × 15kJ

Heat required by calorimeter (mcT) = 0.067 × 0.39 × 15kJ

Total amount of heat required

= 15 [(0.0647 × 4.2) + (0.067 × 0.39)]kJ

= 15 × 0.2978 × 10³J

Input power $= IV = 2 \times 7.5 = 15W$

∴ Time taken to produce the temperature rise

$\dfrac{15 \times 297.8}{15}$ seconds $= 4.96$ minutes

$= 5$ min (approx.).

Example 3.8. A 220V electric kettle has an efficiency of 90%. Calculate the resistance of the heater coil and the current reeded to raise the temperature of 1 litre of water from 15°C to boiling point in 9 minutes. Take a litre of water to have a mass of 1 kilogramme and the specific heat capacity as 4.2kJ/kg°C.

Since no information is given about the kettle, the effect of heating it is neglected.

Heat received by water $(mcT) = 1 \times (100 - 15) \times 4.2$

$= 357kJ$

Heat energy put out by the heater (kettle is only 90% efficient)

$\dfrac{357}{0.9}$ KJ

Time of heating $= 9 \times 60$ seconcs

Power rating of heater coil $= \dfrac{357 \times 10^3}{0.9 \times 9 \times 60}$

$= 735W$

Heater current $= \dfrac{W}{V} = \dfrac{735}{220} = 3.34A$

Heater resistance $= \dfrac{V}{I} = \dfrac{220}{3.34} = 65.9\,\Omega$

Example 3.9. A 120W electric soldering-iron is plugged into the 120V ship's mains for 5 minutes, the ambient temperature being 15°C. The copper mass is 0.133kg and 50% of the heat generated is assumed to be lost in radiation and heating other parts of the iron. Find whether the iron reaches the working temperature in the time specified. Take the specific heat capacity of copper as 0.39kJ/kg°C and the temperature of melting solder as 310°C.

Heat required by the iron $= mcT = 0.133 \times 0.39 \times (310 - 15)$

$= 15.34kJ$

Heat produced by the element $= \dfrac{15.34}{0.5} = 30.68 \text{kJ}$

$$= 30\,680 \text{J}$$

Power rating of the element $= 120 \text{W}$

Time for 30 680 joules to be expended

$$= \dfrac{30\,680}{120 \times 60} \text{ minutes}$$

$$= 4.27 \text{ min.}$$

Since only 4.27 min are required to achieve working temperature, then the time of 5 min as specified will be sufficient.

Example 3.10. A resistance unit has 500 turns of nickel–chrome wire, 0.5mm diameter. It is wound on a former 30 × 100mm and its resistivity is 1060μΩ mm at 15°C. At 100°C, its resistance is 2% greater than at 0°C. Determine the current taken at a temperature of 300°C, when the resistance is connected across a 250V supply.

Length of a turn $= (2 \times 30) + (2 \times 100) = 260 \text{mm}$

No of turns $= 500$

Total length of wire $= 260 \times 500 = 130\,000 \text{mm} = 130 \text{m}$

Resistance at $15°\text{C} = \rho\dfrac{l}{A}$

$$= \dfrac{1060 \times 10^{-9} \times 130}{\dfrac{\pi\,(0.5 \times 10^{-3})^2}{4}}$$

or $R = \dfrac{106 \times 13 \times 10^{-7} \times 4}{\pi \times 0.25 \times 10^{-6}}$

$$= 702\Omega$$

Also since

$R = R_0\,(1 + \alpha T)$ or $1.02 = (1 + \alpha 100)$

then $\alpha = \dfrac{1.20 - 1}{100} = 0.0002\Omega/°\text{C}$ at $0°\text{C}$

Similarly $R_1 = R_0(1 + \alpha T_1)$ and $R_2 = R_0(1 + \alpha T_2)$

or $R_2 = R_1 \dfrac{(1 + \alpha T_2)}{(1 + \alpha T_1)}$

where $R_1 = 702\Omega$ at 15°C.

Giving $R_2 = 702\dfrac{1 + (300 \times 0.0002)}{1 + (15 \times 0.0002)} = \dfrac{702(1 + 0.06)}{(1 + 0.003)}$

$$= \dfrac{702 \times 1.06}{1.003} = 742\Omega$$

Current taken $= \dfrac{250}{742} = 0.337A$

Atomic theory of conduction

In Chapter 1 reference was made to the fact that different atomic arrangements result in matter being in solid, liquid or gaseous states. Returning to atomic physics, let us consider in more detail the electron or *quantum shell*, which are considered to be concentric and 7 in number. They are identified by the letters: K, L, M, N, O, P and Q, the K shell being closest to the nucleus. The appropriate numbers of electrons in each shell are found to be: 2, 8, 18, 32, 18, 13 and 2 respectively. Figure 3.6 illustrates the arrangement of shells in a copper atom (Z = 29), the most common metal in electrical circuits.

Quantum theory postulates that shells are identified by specific potential energy levels relating to their orbital distance from the nucleus. These levels are important factors in the conduction process and an understanding is needed to distinguish between conductors and insulators.

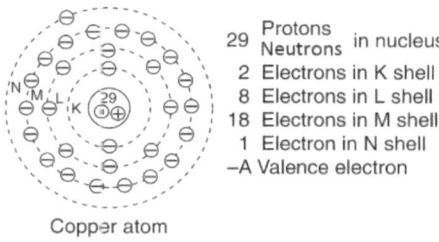

29 Protons } in nucleus
   Neutrons
2 Electrons in K shell
8 Electrons in L shell
18 Electrons in M shell
1 Electron in N shell
−A Valence electron

Copper atom

▲ **Figure 3.6** *Atomic structure of copper atom*

Energy levels

When an atom is excited by an external source such as: heat, light or some other form, energy is acquired, which can move an electron from an inner to an outer shell. Electrons only exist in the definite energy levels of a shell and cannot exist with energies *between* these levels. A sudden change in energy level can, however, result in movement from one shell to another. Electrons with low energy lie in the inner shells, i.e. nearer the nucleus. If the movement or jump is made to a lower or inner shell, the energy given out is **emitted** as radiation. When all the inner shells are filled, additional electrons can only exist in the outer shells and increase until the total −ve charge of the electrons equals the +ve charge of the nucleus, i.e. when the atom becomes stable. The unit for measuring electron energy level is the electron-volt (eV).

Energy bands

A number of fundamental laws have been determined by research into atomic behaviour. One law states that not more than 2 electrons can be in an identical energy state and these electrons must have opposing directions of *spin* (as proposed by Wolfgang Pauli 1925). Thus other electrons, while within the energy range of the shell, will occupy orbits at progressively higher energy levels. Consider a single isolated atom of copper ($Z = 29$), figure 3.6; there are 2, 8 and 18 electrons respectively in the K, L and M shells. These are filled with electrons that are very nearly, for each shell, at the same energy level. There is also one outer or *Valence* electron in the N shell with its higher energy value. The energy levels are sharp and are depicted (figure 3.7a). Consider several adjacent copper atoms whose electrons, if excited, can be given *slightly* higher energy levels. The original atom is now affected by the electric fields of the nuclei of nearby atoms as well as of its own. The shell electrons are in close proximity but are at slightly different levels. The definite shell energy lines of figure 3.7a are now broadened to a band, shown by 2 lines spaced from each other (figure 3.7b).

It is seen that the K, L and M bands, though broadened, represent *filled* shells with no free electrons and play no part in electrical conduction. The partially filled valence band indicates the presence of free electrons with an ability to move into the unfilled bands. The unfilled bands constitute a *conduction* band as these electrons are far from the nucleus and can move readily. Thus for figure 3.8a the conduction condition is shown for a conductor where the upper valency and conduction bands overlap.

For an insulator (figure 3.8b), the lower band is filled and far from the upper band, so conductivity is poor. The possibility of an electron gaining enough energy to cross the

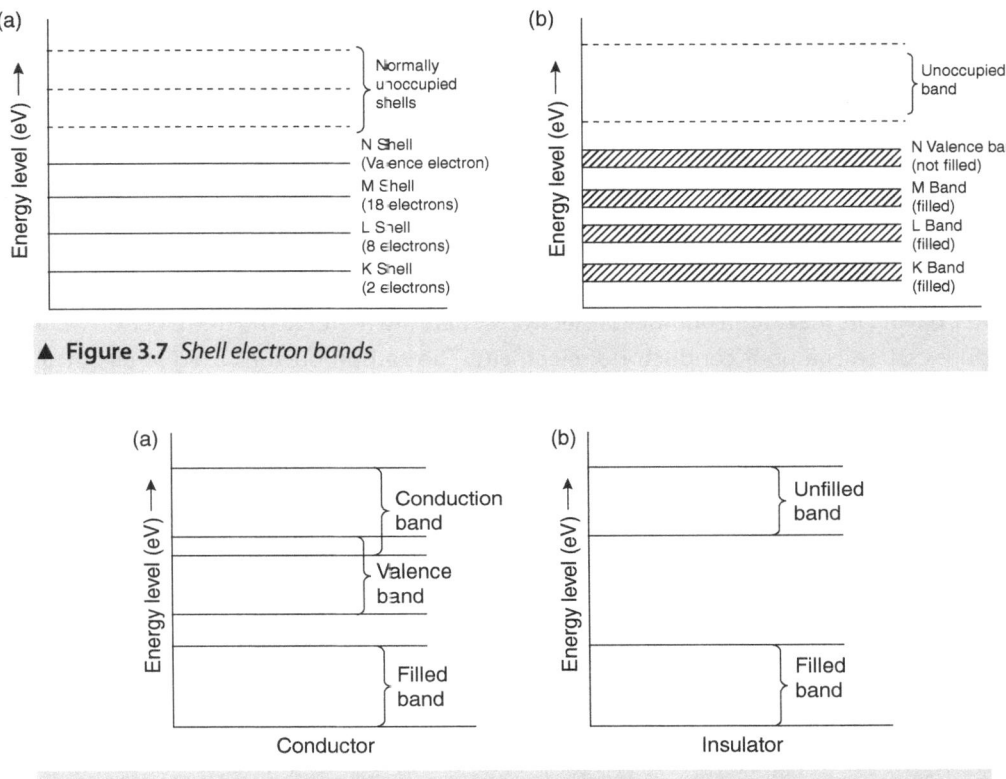

▲ **Figure 3.7** *Shell electron bands*

▲ **Figure 3.8** *Conduction and valence bands*

gap is small and will only be achieved by application of so high a P.D. that the material would be destroyed. Such movement will manifest itself in a spark or 'puncture' of the insulating material, which results in 'tracking', a burnt conducting path or area.

Crystal lattice

Some substances, including metals, in their smallest elemental form consist of many *crystals* joined together. Crystals are built up from a regular structure of atoms, which repeats continuously to form a *lattice*. The simplest crystal to examine is that of carbon but the term *covalent bonding* is introduced to allow a full understanding of the lattice concept. The idea of a covalent bond is simply shown in a molecule of hydrogen, which, though not a metal, exhibits many similar characteristics in chemical reactions.

One way in which an atom may combine with other atoms and create a change in the number of electrons in its valence shell is by covalent bonding, the *sharing* of a pair of electrons by 2 atoms, each atom contributing 1 electron to form a shared pair.

A powerful bond between these atoms results. Consider a hydrogen atom, which has only 1 valence electron. The K completed shell electrons should be 2, thus a hydrogen molecule contains 2 atoms with the nuclei linked by the valence electrons to form a pair. The arrangement is illustrated in figure 3.9.

Now consider the carbon atom, shown by the first diagram of figure 3.10a. The K shell is full but the L shell is incomplete, containing only 4 electrons. For a carbon crystal lattice, coupling occurs between the outer shells of neighbouring atoms (figure 3.10b). An atom is now produced that shares 8 electrons in the L shell. This means the shell is full, having its maximum number of electrons. There are therefore no free electrons and pure carbon is a poor conductor of electricity. The carbon brushes used in electrical machines are the result of processing to produce certain characteristics. A controlled heat treatment results in grades of natural graphite or amorphous carbon, which improves conductivity.

In crystals, atoms are arranged in an orderly geometrical pattern and all pure crystals of the same element have identical lattice structure. For carbon or diamond crystals, each

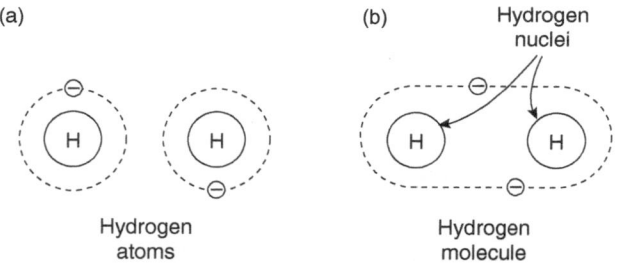

▲ **Figure 3.9** *Covalent bond*

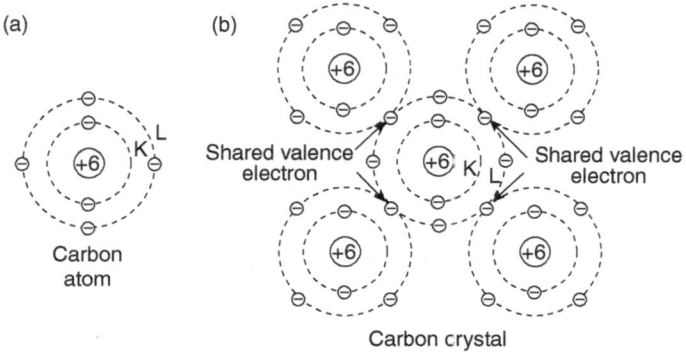

▲ **Figure 3.10** *Shared valence bonds in carbon*

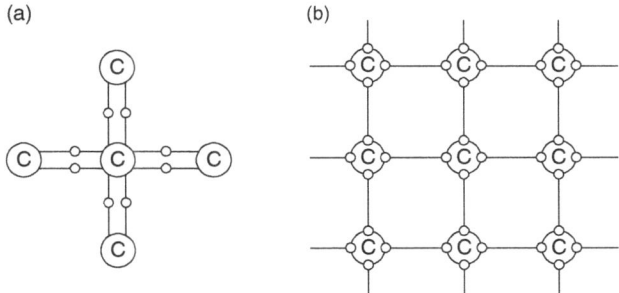

▲ **Figure 3.11** *Illustration of simple covalent 'crystal' structure*

atom is equidistant from 4 neighbours with each valence electron shared between the parent atom and 1 neighbouring atom. Two electrons, 1 from each neighbour, form a covalent bond and the arrangement is represented by the diagram of figure 3.11a, or figure 3.11b which depicts the crystal structure.

There are several basic crystal patterns and for metals it is generally believed that these consist of a lattice of +ve ions through which a cloud of negative electrons move, thus electrons are the valency electrons of the metal. The crystalline form of metals is responsible for their strength, elasticity and relative resistance to chemical reactions. Strong electrical bonds are provided by the mutual attractive forces between atoms and considerable work must be done to distort the equilibrium arrangement, irrespective of whether the metal is in tension or compression.

Conductivity

As described previously, atoms make up a crystal lattice and in these atoms the outermost electrons are influenced by adjacent atoms in the lattice, the result being that valence electrons move continuously atom to atom randomly. When a P.D. is applied to a conductor, these electrons pass through the crystal towards the positive potential. This action, representing an opposition to the passage of current, is termed conductor resistance. It is known that all atoms vibrate about a mean position, the amplitude of which is dependent on temperature. The magnitude of these vibrations is also known to increase with temperature until the melting point of the material is reached when the crystal lattice breaks up into either individual or groups of atoms. Electronic theory considers that since the vibrations increase with temperature, extra force is necessary to move free electrons through the lattice and so the resistance of a pure metal rises with temperature rise. It should be noted that since extra force is needed to move the electrons, extra energy is absorbed into the conductor and appears as heat. By similar explanation, resistance is found to decrease as temperature falls, the relationship being linear down to very low temperature values. Extrapolation of this linear law shows that as the temperature approaches absolute zero, conductor resistance reduces to a negligible value and hence conductivity rises to an immense value. This condition brings about 'superconductivity', which is at present being actively investigated by practical research in the electrical power industry for potential room temperature applications and as high power magnets for medical imaging.

Metallic conduction

Consider a short length of copper wire connected to 2 terminals or electrodes at different potentials. One electrode is given a +ve charge (deficient in electrons, and the other a −ve charge (with an excess of electrons). The +ve electrode attracts electrons from the nearest copper atoms of the wire and these atoms will be left with a +ve charge, i.e. will become +ve ions. These ions attract the mobile electrons from the next atoms down the wire, which in turn are ionised to attract electrons from the adjacent atoms further down the line, and this process continues until the atoms at the other end of the wire become −ve ions. These ions now accept electrons from the other electrode, which being negatively charged carries an excess of electrons. It is assumed that this 'jumping' action of electrons is so fast as to be almost instantaneous. It is noted that as an electron leaves an atom it may lead to formation of a *hole*. This term is often used to explain semiconductor operation. When an electron with its −ve charge leaves an atom, a gap or space is left in the atom's structure and so the atom now acquires a +ve charge. This gap or hole can be credited with this +ve charge and we now visualise a 'hole' as an entity in itself having a +ve charge. Any other mobile electron can be captured by the hole to neutralise its charge and make the original atom stable since its structure is now complete. As a result of this action, an electron has left a neighbouring atom and a hole is created there. This action continues between adjacent atoms and, as we consider metallic conduction, under the application of a P.D., electrons move towards the +ve terminal and holes move towards the −ve terminal, i.e. in the same direction as the current. The similarity with a row of occupied seats is often used. If a person (electron) is induced to leave and everyone is asked to 'move up one seat', the unoccupied space or hole can be reasoned to move down in the direction opposite to the movement of the people. The concept of a hole is useful; it is often referred to as a 'current carrier' and is considered later with electronic theory.

The conduction action described above has been treated in terms of a few atoms. It is estimated that in a cubic millimetre of copper there are as many as 10^{20} atoms so, even if we consider a wire 1/10 mm in diameter, it is obvious the number of mobile electrons and holes available for conduction is considerable!

Liquid conduction

Unlike a metal, the molecules in a liquid are less closely packed and have little or no cohesion between them. There is no crystal lattice and the electrostatic bonding between atoms is also very weak. Electrical conduction depends on the nature of the liquid and also whether it is capable of electrolytic dissociation. For the passage of current, ionic action is involved and some basic explanation of this will assist in the appreciation of the conduction process. The theory involved is detailed in Chapter 4.

Gaseous conduction

Consider a container evacuated completely and filled with a gas at normal atmospheric pressure. The gas consists of molecules of a simple structure, mainly stable, except for the occasional one that is *ionised*. Ionisation means the existence of some free electrons and molecules that have lost an electron and acquired a +ve charge. Ionisation is the result of some action by light, cosmic rays or radioactivity. Gas molecules are known to oscillate at high speed in random directions, resulting in frequent collisions. The distance between gas molecules is small and the spaces between collisions are small but if the gas pressure is reduced the distance between collisions becomes larger. When the free electrons move through the gas and meet an ionised molecule, an immediate recombination results and stability is restored. With the reduction of gas pressure, the +ve and −ve charged molecules and electrons travel a much longer distance before being 'normalised', a distance known as the 'mean free path'. As it increases, the period during which ionisation continues becomes longer. Continued reduction of gas pressure increases the free path until this approximates to the dimensions of the enclosing container. The pressure is now *so* low that the enclosed space is considered to be a vacuum and collisions with gas molecules are now practically non-existent. Since no molecules are present in a vacuum, no obstacle to electron movement exists if these are injected into the space.

In the gas-filled condition, we see that conduction is affected by pressure. If 2 electrodes, between which there is a P.D., are introduced into the gas-filled space, a directional motion of charges can be expected. Because of the low pressure, the electrons have a long free path and may attain high velocities and gain motional energy on collision. The effect of the collisions will depend upon the velocity on impact with a gas molecule. The high velocity electrons may cause the molecules to merely bounce away from each other. On the other hand, the electron velocity on impact might be large enough to cause another electron to be removed from 1 shell orbit of the neutral atom to another, thus exciting it and giving rise to the emission of light due to some of the collision energy having been absorbed. If the velocity on impact is very high, the collision energy could cause an electron to be knocked out of a neutral gas atom, thus ionising the molecule. The process results in the production of an additional free electron and an ionised molecule. The new atom is now available to join the collision action and could shock-ionise a further gas molecule. The effect is a cumulative one and, since there is a P.D. between the electrodes, a current would result, electrons making their way to the +ve plate or *anode* and ionised molecules to the −ve plate or *cathode*. Current will increase rapidly, leading to a condition that could be disastrous unless stabilised by the inclusion of a resistor in the circuit. The above action forms the basis of discharge lamp operation and will be discussed further.

When considering the conditions for ionisation it is pointed out that if electrons could be made available, ionisation would continue, especially if electrodes at a potential were inserted into the evacuated or gas-filled container. Such electrons may be obtained by either *cold-emission* – tearing them away from the electrodes – or *thermionic emission*, or the *photoelectric effect*, where incident electromagnetic waves of sufficiently high frequency lead to electron emission. Thermionic emission is an important factor in existing Cathode Ray Oscilloscope (CRO) devices, although they are now becoming replaced by modern Liquid Crystal Displays (LCDs) and Plasma Screens. However, the stability of thermionic devices against ElectroMagnetic Pulse (EMP), often associated with nuclear devices, still makes them suitable for military maritime operations.

Cold electron emission

A vacuum, although a perfect insulator, nevertheless allows current flow if a potential high enough is applied across the electrodes. Electrons contained in the atoms of the metallic electrodes can be 'extracted' by the electric force due to an applied field or potential, provided it is strong enough to overcome the internal electrical bonds of the metal atoms. Movement of electrons constitutes a current, sometimes evident as a spark. Sparks may develop into an arc if a supply source is connected to the electrodes. If the discharge is maintained, sufficient to heat the electrodes, thermionic emission can provide a large supply of electrons and a discharge will continue. If the current resulting from the initial discharge is limited by the circuit resistance then a working arrangement can result in a source of illumination. If gas at a low pressure is introduced into the container then, once the initial electron emission is achieved by the high applied potential, ionisation of gas will occur and the electrode potential can be reduced to maintain the discharge. This arrangement gives rise to the cold-cathode discharge lamp.

The cold-cathode discharge lamp

The basic principle concerning cold-cathode discharge lamp operation is use of a high voltage gradient across the gas to produce electron flow and the required collisions to cause ionisation. Electrons are initially extracted from the multiple cathodes by a large applied voltage between anode and cathode. To produce the initial ionisation, a 50% higher voltage than the running voltage is needed. When operating, a potential gradient of 1–1.3kV per metre is used, and since the total supply voltage for some lamps and signs can be in the region of 10kV, a mains transformer is used that is designed to have a high internal reactance so that the voltage settles down to a steady state value (or 'sits down' to the required running value) once the lamp is struck (figure 3.12).

The lamp, in its simple form, is used for advertising purposes and the discharge colour is determined by the gas used. For illumination purposes, mercury vapour is used together with a fluorescent coating on the inside of the tube. Long tubes (2.4m) with 3 or 4 in series are difficult to accommodate and the lamp has only limited applications for marine work. They are used for illumination and decoration on some of the larger passenger and cruise ships. A more familiar type of cold-cathode lamp is the sign lamp used for indicator purposes. This is arranged to work at mains voltage with a series resistance of about 10kΩ. Neon gas at high pressure is used and discharge is usually as a glow, which surrounds the electrodes.

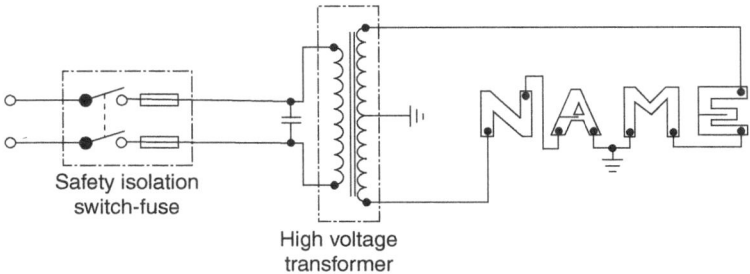

▲ **Figure 3.12** *Cold cathode neon sign*

Practice Examples

3.1. Ten thousand cubic millimetres of copper are (a) drawn into a wire 100 metres long, (b) rolled into a square sheet of 100mm side. Find the resistance of the wire and the resistance between opposite faces of the plate, if the resistance of the copper is 17μΩmm or 1.7×10^{-8} ohm-metres (2 significant figures).

3.2. A coil of copper wire has a resistance of 90Ω at 20°C and is connected to a 230V supply. By how much must the voltage be increased to keep the current constant, if the temperature of the coil rises to 60°C? Take the temperature coefficient of resistance of copper as 0.00428°C at 0°C (2 decimal places).

3.3. An electric kettle is fitted with a heater unit of 120Ω resistance. The efficiency is 84% and the voltage is 220V. How long will it take to heat 0.75 litre of water from 6°C to 100°C? Take the specific heat capacity of water as 4.2kJ/kg°C (nearest second).

3.4. Find the length of manganin wire required to make a 15.7Ω resistor, if the diameter is 0.315mm and the resistivity is 407μΩmm (nearest cm).

3.5. The cold resistance at 15°C of the field coil of a motor is 200Ω and the hot resistance is 240Ω. Determine the temperature rise, assuming the temperature coefficient of resistance to be 0.0042°C at 0°C (1 decimal place).

3.6. A 2-core cable, each core of which is 300m long and of uniform cross-sectional area of 150mm², is fed from one end at 240V. A load of 200A is taken off from the centre of the cable and a load of 100A from the far end. Calculate the voltage at each load. A single-core cable of similar material 880m in length and of uniform cross-sectional area of 50mm² has a resistance of 0.219Ω (2 decimal places).

3.7. The resistance of a 100m length of copper conductor, 1mm diameter, is 2.47Ω. Calculate the resistance of a cable 800m in length composed of 19 similar strands of conductor, but each 1.5mm diameter. Allow 5% increase in length for the 'lay' (twist) of each strand of the completed cable (3 decimal places).

3.8. A wire has a resistance of 10Ω at 0°C and 15Ω at 100°C. What is the temperature coefficient of the resistance of the material? At what temperature will its resistance be 30Ω (1 significant figure)?

3.9. A 200V, 200kW electric furnace must raise 500kg of brass from an initial temperature of 15°C to the melting point of 910°C. If overall efficiency is 0.8, calculate the operation time required. Specific heat capacity of brass is 0.39kJ/kg°C (nearest second).

3.10. A 230V electric water heater takes water at a mean temperature of 16°C and the mean temperature of the outlet water is 82°C. The cost at 2p per unit of the energy it consumes in a given period is £7.44. Determine the quantity of water used if the efficiency of the heater is 80%. Take the specific heat capacity of water as 4.2kJ/kg°C (3 significant figures).

4

ELECTROCHEMISTRY

I have been so electrically occupied of late that I feel as if hungry for a little chemistry.

Michael Faraday

The 3 main effects of current flow are those producing heat, chemical action and magnetism. In Chapter 3 the heating effect was discussed in some depth, while the magnetic effects will be covered in Chapter 5. This chapter will look at electrochemistry and studies that reveal the chemical action associated with current flow and the reversibility of this action. The electric cell was the principal source of electrical energy before the principles of electrodynamic induction were discovered and the electric generator developed – and is still important today.

The existence of static electricity, produced at a high potential and exhibiting itself as a stationary charge, had been known for centuries, and is associated with the lightning flash, the 'Van de Graaf' generator (1929), or friction effects, such as the attraction of paper by a piece of amber when the amber is rubbed. The early and sometimes accidental experiments of men like Galvani and Volta showed that electricity could be produced and controlled by various chemical means and led to the construction of the first 'voltaic piles' or batteries, consisting of several zinc and copper plates spaced with an absorbent material such as felt, soaked with acid or salt solution.

At the start of the nineteenth century the only practical way to produce electricity was by chemical means and it's therefore unsurprising that, this branch of electrical engineering science was the first to be thoroughly investigated and developed. The *laws of electrolysis*, first propounded by Faraday in 1834, summarise the basics of examining the relations between electrical and chemical action, which is little changed today.

The conversion of electrical energy into chemical energy and its reverse action is an example of the principle of the conservation of energy. We will start with the conversion of electrical to chemical energy by the conduction of a current through a liquid.

Conduction in liquids depends on the presence of ions as current carriers. Pure distilled water is deionised and is a poor conductor but a few drops of sulphuric acid added to the water produces a conducting solution when subject to a P.D. This conduction is brought about by *electrolytic dissociation* – a process we will now examine.

Electrolytic Dissociation

Consider common table salt (NaCl) in its dry crystalline form. A molecule of such a crystal consists of 1 atom of sodium (Na) and one atom of chlorine (Cl). In its normal state the sodium atom has only one electron in its valence shell instead of the possible 8. A chlorine atom has 7 electrons instead of the possible 8 in its outer shell and the combination (NaCl) is achieved by the one electron leaving the sodium atom and entering the chlorine atom. The chlorine atom thus absorbs the valence sodium electron, filling its outer shell. In this condition the chlorine atom becomes a −ve ion and the sodium atom a +ve ion. The salt molecule can be described as an *ionic* compound and is stable because of the attraction between the 2 atoms. The crystal is composed of the +ve and −ve ions arranged in regular patterns – the crystal lattice.

When the salt is dissolved in water, electrolytic dissociation occurs and some molecules break up, allowing the chlorine and sodium atoms to separate. The chlorine atom, however, still keeps the extra electron taken from the sodium. The solution has now become an electrolyte with both chlorine ions and sodium ions being able to wander. The ions in the electrolyte are continuously recombining with other oppositely charged free ions while different molecules are breaking down elsewhere. The dissociation and recombination can be shown by:

$$NaCl \rightleftarrows Na^+ + Cl^-$$

Note. The representation of the electron distribution, i.e. chlorine, has gained one electron, hence the Cl^-. Sodium on the other hand has become a +ve ion, hence Na^+. It should be noted that the electron exchange number is one, i.e. the chemical valency number.

Electrolysis

The passage of current through a solution of an acid, alkali or salt produces a chemical change, explained by the theory of *dissociation*. The solution is called an *electrolyte* and the process is termed electrolys s. The general theory is as follows. When an electrolyte is first made up, as would result from dissolving copper sulphate crystals in water, some molecules split into 2, independent of any external assistance, to form *ions*, carrying both +ve and −ve *charges*. Such ions are extremely mobile. If 2 plates, termed the *electrodes*, are immersed in the electrolyte and a P.D. applied across them, a current will pass through the solution. The +ve ions migrate, under the influence of the electric field due to the P.D., to the *cathode*, namely the electrode by which the current leaves the electrolyte. Such ions are called *cations*. The −ve ions, called *anions*, migrate to the *anode*, namely the electrode by means of which the current enters the solutions. Metal or hydrogen ions always carry +ve charges and travel with the current to appear at the cathode, whereas non-metallic ons travel in the opposite direction to the current and may appear at the anode or may engage in secondary reactions, some of which will be described shortly. If the electrodes are made from platinum, which is chemically inert, gas bubbles are produced on the electrodes. Consideration is now given to electrolysis with various liquids.

For the salt solution, ionic action is explained by the fact that Cl^- ions move to the anode and the Na^+ ions move to the cathode and current flows. The Cl^- anions revert to their normal atomic structure by giving up their surplus electrons, which are transferred round the external circuit. They are now neutral but in a very active state. Two such atoms combine to form a chlorine molecule and some of these rise to the surface as gas bubbles while others dissolve in the water. The Na^+ cations meanwhile drift to the cathode, are neutralised on reaching it, and combine with the water to form sodium hydroxide and the hydrogen liberated will appear as bubbles at the cathode. The latter chemical action is shown by the equation:

$$2Na + 2H_2O = 2Na(OH) + H_2$$

Consider next the case of dilute sulphuric acid (H_2SO_4). The dissociation and recombination can be shown by

$$H_2SO_4 \rightleftharpoons H^+ + H^+ + SO_4^{2-}$$

The hydrogen cations $2H^+$ move to the cathode and after giving up their charge are neutralised and rise to the surface as gas. The SO_4^{2-} radicals are anions that move to the anode to give up their charge, but cannot exist in this form. They combine with hydrogen atoms of the water to reform acid molecules and liberate the oxygen as gas. Thus $SO_4 + H_2O = H_2SO_4 + O$. Note that the acid is not consumed but allows easy conduction by providing a good supply of ions.

If a copper sulphate ($CuSO_4$) solution is subjected to electrolysis with copper plates used as the electrodes, action occurs as shown below. When the electrolyte is made up, dissociation and recombination can be shown by

$$CuSO_4 \rightleftarrows Cu^{++} + SO_4^{2-}$$

With the application of a potential across the electrodes, the cations (Cu^{++}) reach the cathode, give up their charges, are neutralised, and copper atoms are deposited on the copper electrode. The anions (SO_4^{2-}), being radicals, cannot exist in this form and move to the anode and combine with the metal of this electrode to form copper sulphate, which goes into solution. No gas is given off for this cell and copper is seen to be transferred from the anode to the cathode.

For the salt and acid cells, it is observed that once electrolysis proceeds, gas bubbles are released slowly and tend to blanket the electrodes, or are said to cause *polarisation*. Under this condition, if the applied potential is removed and a sensitive voltmeter connected across the electrodes, a potential of some 1.5V will be indicated, caused, it is thought, by the slow progress of ions through the electrolyte. The voltmeter will also show a P.D. value between each platinum electrode and the electrolyte.

This polarising e.m.f. rapidly becomes ineffective if used to supply current to an external circuit. The cell appears to have an internal back e.m.f., confirmed by (1) Ohm's law not being followed and (2) electrolysis only starting satisfactorily if the applied voltage is about 2V. More is said about polarisation later but it is appropriate to point out that if tests are made on several different electrolytic cells, graphs can be drawn from the results, the ordinates being voltage applied across the electrodes (V) and current passing (I). The characteristics of a composite graph are seen to follow distinct types (figure 4.1). Characteristic 1 shows that Ohm's law is followed closely and is applicable to cells where no gas is given off. Characteristic 2 is followed for cells where gas is given off. It will be noted that little current flows until the applied voltage exceeds a critical value E_b – usually 1 to 2 volts – and is linear after this. The explanation is also associated with polarisation.

A test condition can be investigated for the acid cell with different metals used for the electrodes. Consider zinc and copper with no external application of a P.D. Dissociation of the acid occurs as before, giving H^+ and SO_4^{2-} ions. Chemical action is now involved in that sulphate ions readily combine with zinc atoms but not with copper atoms. Thus zinc dissolves to zinc sulphate and leaves behind 2 electrons, i.e. $Zn + SO_4^{2-} \rightarrow Zn^{++} + 2$ electrons. The zinc electrode then acquires a −ve potential or becomes the cathode for an external circuit. The H^+ ions on reaching the copper electrode acquire 2 electrons to leave this electrode with a +ve charge and thus is able to act as the anode for an external circuit.

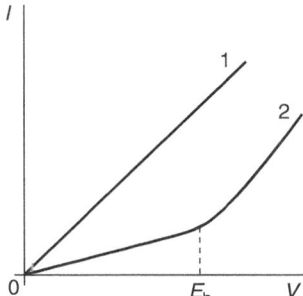

▲ **Figure 4.1** *Simple IV characteristics*

Here then we have the fundamental difference between an electrolytic and voltaic cell. For the electrolytic cell with electrodes of similar material and no applied e.m.f., chemical action is not possible as there is no apparent preference for a combination between the electrolyte and one electrode or the other. As a consequence, chemical action is only possible if an external voltage is applied.

For the latter, as a result of electrolytic dissociation and because electrodes of different metals are used, chemical action is possible if the circuit is completed. The chemical energy made available by the reaction is converted into electrical energy.

Electrolytic Cells

The whole arrangement consisting of electrodes and electrolyte, as described above, is often called an *electrolytic cell* to distinguish it from a *voltaic cell* described later. Electrolysis does not occur with solids or gases and is only possible for certain liquids. Some, like oils, are non-conductors, while others, like mercury, conduct without decomposition. The remaining liquids are electrolytes, which can be defined as liquids that decompose when current is passed through them. The electrolytic cell can be constructed to enable experiments and measurements to be made with great accuracy. In this form it is referred to as a *Voltameter*.

The water voltameter (sulphuric acid solution)

Figure 4.2 shows construction of the apparatus, which is made of glass, with platinum electrode plates placed at A and C. The lead-in wires, passed through rubber corks, are

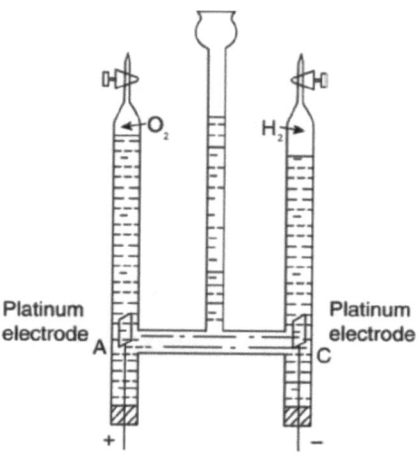

▲ **Figure 4.2** *Water voltameter*

not exposed to the solution to prevent corrosion. The voltameter is filled with acidified water and the platinum electrodes are connected to a battery of 2 volts or more. Current passes from the anode to the cathode and bubbles of gas are given off, which rise into the graduated tubes.

If care is taken before passing current to fill both tubes with acidified water by opening the taps and then closing them after all the air is expelled, certain conclusions can be made from the experiment. After a period of time, the gas collected will be found to be hydrogen at the cathode and oxygen at the anode. The ratio of the volumes of H to O will be 2:1 and the amount of gas collected would be proportional to the current strength and the time it flows or more generally the quantity of charge passed.

In the acidified water there are sulphuric acid molecules, which divide into 3 ions, 2 of hydrogen carrying +ve charges H^+, H^+ and one with −ve charge represented by (SO_4^{2-}). Note the total +ve hydrogen charge equals the −ve sulphate or sulphion charge, but the ions migrate under the influence of the electric field. So the H^+ ions give up their charges at the cathode and are liberated as hydrogen gas. Sulphions proceed to the anode, but as they cannot exist in a free state they combine with 2 hydrogen H^+ ions or atoms from the water, liberating oxygen as in the equation. Thus: $2SO_4 + 2H_2O = 2H_2SO_4 + O_2$. Oxygen rises from the anode and collects in the tube above it. The H_2SO_4 goes into solution and the electrolyte is decomposed so the water appears to be used up, but the acid content remains the same and the solution gets stronger; that is, its specific gravity rises.

The copper voltameter (copper sulphate solution)

Figure 4.3 shows the usual arrangement. A copper sulphate solution, made from crystals dissolved in pure water, is contained in a glass or glazed earthenware tank. Electrodes are made from pure copper sheet. The $CuSO_4$ molecule splits into 2 ions, Cu^{2+} and SO_4^{2-}. When a P.D. is applied across the electrodes and current is passed, copper ions migrate to the copper cathode to combine with it and give up their charges. The sulphions give up their negative charges at the anode and combine with copper from this electrode to form copper sulphate. Thus copper appears to be taken from one electrode and deposited onto the other.

The chemical equations for the electrodes are:

(1) Cathode: $CuSO_4 = Cu + SO_4$.
(2) Anode: $Cu + SO_4 = CuSO_4$.

During electrolysis a small amount of gas may be noted at the plates – due to water decomposition in the solution as described for the water voltameter. Furthermore, complex action may occur in the electrolyte due to sulphions combining with hydrogen in the water to form H_2SO_4. Oxygen from the water is released to combine with anode copper to produce copper oxide. This oxide then dissolves in the H_2SO_4 to give $CuSO_4$. Irrespective of the action, the final result is the loss in mass of the anode equals the gain in mass of the cathode.

Various forms of voltameter can be constructed for electrolysis research. A silver voltameter may consist of silver (Ag) plates and a silver nitrate ($AgNO_3$)

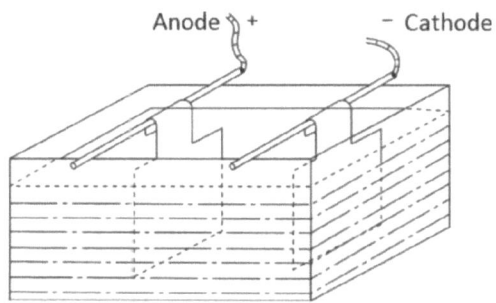

Anode + − Cathode

▲ **Figure 4.3** *Copper plate voltameter*

solution. The examples described could be connected in series and the same quantity of electricity passed through all voltameters. If the electrodes were washed and carefully weighed before electrolysis and then washed and weighed again after electrolysis, certain conclusions would be reached, which were first described by Faraday (1834) in his laws of electrolysis.

Quantitative laws of electrolysis (Faraday's laws)

(1) The mass of an element liberated from or deposited on an electrode is proportional to the quantity of electricity that has passed.

(2) The masses of elements liberated from or deposited on electrodes by a given quantity of electricity are proportional to their Chemical Equivalent Weight.

Consider the first law. It is found by experiment that the mass of any material deposited or liberated always depends on the quantity of electricity that has passed. Thus $m \propto Q$ coulombs or $m \propto It$. This proportion can be modified to:

$$m = zIt$$

where z is a constant depending on the substance deposited. z is termed the *electrochemical equivalent* of the element.

Electrochemical equivalent (E.C.E.)

The mass in grammes or kilogrammes liberated by one coulomb of electricity is called the E.C.E. of a substance. Thus, 10 amperes flowing through a copper voltameter for 1000 seconds will result in 10 000 coulombs having passed and 3.3g of copper will be deposited. Thus the E.C.E. of copper $= \dfrac{3.3}{10000} = 0.00033g/C$. Similarly, that for hydrogen would be 0.000 010 4, for oxygen 0.000 082 9 and for silver 0.001 118g/C.

With the use of SI units, it is better to think in terms of the kilogramme and the E.C.E. is defined as the mass (in kilogrammes) of a substance liberated by the passage of one coulomb. Thus the E.C.E. of copper would be $330 \times 10^{-9}kg/C$. Since the milligramme is also an accepted SI unit, the E.C.E. can be given as mg/C. Thus for copper it will be 0.33mg/C.

The first law of electrolysis leads to a method of stating the unit of current, which was considered accurate enough to allow an original definition for the International

Ampere. This was defined as that constant current which deposits silver at the rate of 1118×10^{-9}kg per coulomb when passed through a solution of silver nitrate in water.

The formula deduced above allows the solution of problems associated with electrolysis and practical electroplating. The unit in which the E.C.E. is given should be noted.

Example 4.1. Find the time taken to deposit 11.4g of copper when a current of 12A is passed through the copper sulphate solution contained in a copper voltameter (nearest minute).

The E.C.E. of copper should be taken as 330×10^{-9}kg/C.

$$\text{Since } m = zIt \text{ then } t = \frac{m}{Iz}$$

$$\text{So } t = \frac{11.4 \times 10^{-3}}{12 \times 330 \times 10^{-9}}$$

$$t = 2880 \text{ seconds} = 48 \text{ minutes.}$$

Atomic weight, valency, chemical equivalent weight

The second law of electrolysis is deduced by examining the results of tests made with a number of different voltameters in series, having been subjected to the passage of the same quantity of electricity. The results of the experiment show that the mass of the substances deposited or liberated at the electrodes is proportional to the chemical equivalent of substances. If the atomic weight of a substance is known, its E.C.E. can be found provided the valency is known and the E.C.E. of hydrogen assumed. Thus if the chemical equivalents of hydrogen, oxygen, copper and silver were 1, 8, 31.8 and 107 respectively, the masses of H, O, Cu and Ag liberated by the same quantity of electricity would be in the same proportion and therefore the E.C.E. of a substance is the E.C.E. of hydrogen multiplied by the chemical equivalent of the substance. Thus taking the E.C.E. of hydrogen as 0.0104 mg/C, that of silver will be $0.0104 \times 107 = 1.118$ mg/C.

To conclude our deductions from the second law, it is worth defining the following terms.

ATOMIC WEIGHT. Atoms are extremely small and determination of their absolute masses present, considerable difficulties. The mass of a hydrogen atom is about 1.67×10^{-24}g and it is customary, even in SI units, to practically refer to the relative *weights* of the atoms of various substances in terms of the atom of hydrogen. Thus the term atomic weight is used and is the weight of an atom of a substance in relation to the mass of an atom of hydrogen. Thus the value for oxygen is 16, i.e. it has 16 times the weight of an atom of hydrogen.

VALENCY. This is described as the combining ratio of a substance, for example, oxygen is 2, whereas that of hydrogen is 1, so water is represented by the symbol H_2O. The valency of an element can also be defined as the number of atoms of hydrogen with which one atom of the element can combine. So the valency of a sulphate is 2 since, for example, in sulphuric acid H_2SO_4, 2 atoms of hydrogen are required to combine with the sulphate.

CHEMICAL EQUIVALENT WEIGHT (C.E.Wt). It is the ratio of atomic weight to valency, or

$$\text{Chemical Equivalent Weight} = \frac{\text{Atomic weight}}{\text{Valency}}.$$

It is the weight of a substance that will combine with one part by weight of hydrogen or 8 parts by weight of oxygen. So the C.E.Wt or *combining weight* of hydrogen is 1, that of oxygen is 8, copper 31.8, etc., as found in chemical tables.

From the second law we have:

$$\frac{\text{Mass of material X liberated}}{\text{Mass of hydrogen liberated}} = \frac{\text{C.E.Wt of material X}}{\text{C.E.Wt of hydrogen}}$$

$$\text{or } \frac{m_X}{m_H} = \frac{\text{C.E.Wt of X}}{\text{C.E.Wt of H}} \text{ also from the expression } m = zIt$$

$$\frac{z_X It}{z_H It} = \frac{\text{C.E.Wt of X}}{\text{C.E.Wt of H}} \text{ or } z_X = z_H \times \frac{\text{C.E.Wt of X}}{\text{C.E.Wt of H}}$$

But the chemical equivalent of hydrogen = 1

so $z_X = z_H$ (C.E.Wt of substance X)

$$\text{or } z_X = z_H \times \frac{\text{Atomic weight of substance X}}{\text{Valency of substance X}}$$

Example 4.2. How many amperes will deposit 2g of copper in 15 minutes, if the current is kept constant? Given the E.C.E. of hydrogen as 0.0104mg/C, the atomic weight of copper as 63.56 and the valency of copper as 2 (2 decimal places).

$$\text{Then } z_{cu} = z_H \times \frac{63.56}{2} = 0.0104 \times 31.8$$

$$\text{whence } I = \frac{m}{zt} = \frac{2}{0.33 \times 10^{-3} \times 15 \times 60}$$

or $I = 6.72$A.

Example 4.3. A voltameter consists of a solution of zinc sulphate and electrodes of zinc and carbon. If current is passed in at the carbon electrode, zinc is deposited on the zinc electrode and oxygen is given off at the carbon plate. If a current 3.5 amperes is passed for 1 hour, find the mass of zinc deposited and oxygen liberated from the solution. The E.C.E. of zinc can be taken as 338×10^{-9}kg/C, the atomic weight as 65.38 and the valency as 2. Take the atomic weight of oxygen as 16 and the valency as 2.

From relationships already deduced:

$$m_{zn} = z_{zn}It \quad \text{or} \quad m_{zn} = 338 \times 10^{-9} \times 3.5 \times 3600$$

$$\text{or zinc deposited} = 425.88 \times 10^{-5}$$

$$= 4.26 \times 10^{-3}\text{kg or } 4.26\text{g}$$

$$\text{Also } z_{zn} = z_H \left(\frac{\text{At wt of Zn}}{\text{Valency of Zn}} \right)$$

$$\text{and } z_O = z_H \left(\frac{\text{At wt of O}}{\text{Valency of O}} \right)$$

$$\text{so } = \frac{z_O}{z_{zn}} = \frac{\text{At wt of O}}{\text{Valency of O}} \times \frac{\text{Valency of Zn}}{\text{At wt of Zn}}$$

$$= \frac{\text{At wt of O}}{\text{At wt of Zn}} \times \frac{2}{2}$$

$$\text{Thus } z_O = z_{zr} \frac{\text{At wt of O}}{\text{At wt of Zn}} = 338 \times 10^{-9} \times \frac{16}{65.38}$$

$$= \frac{338 \times 10^{-9}}{4.086}$$

or $z_0 = 82.5 \times 10^{-9}$kg/C or 82.5×10^{-6}g/C

and $m_0 = 82.5 \times 10^{-6} \times 3.5 \times 3600$ as $m = z/t = 1.05$g.

Back e.m.f. of electrolysis

The circuit laws established in Chapter 1 govern the conditions for most practical circuits. The loads of such circuits are mainly resistive, such as the coils of resistance wire in appliances like electric heaters and filament lamps. Such loads, termed *passive* loads, are recognised by the fact that they conform to Ohm's law. For other types of loads such as the electric motor, accumulator or storage battery when being charged, Ohm's law is not directly applicable and they represent *active* loads, of which some electrolytic cells are an example. The difference between pure ohmic resistance and that offered to the passage of current by such electrolytic cells is now considered.

Figure 4.4 shows a simple circuit for which the source of supply is a battery made up of 3 similar voltaic cells in series. Assume that the current through an electrolytic cell made up as shown is adjusted and maintained at 3 amperes by a variable resistor provided for this purpose. If the supply potential is reduced by removing one of the voltaic cells, it is assumed that, as the e.m.f. has been reduced to $\frac{2}{3}$ of the original value, the current will fall to $\frac{2}{3} \times 3 = 2$ amperes. In fact the new current strength will be well below this value, the value as expected from Ohm's law.

Experiment shows that an extra current controlling factor is present in a circuit involving an electrolytic cell and the result is explained by considering that a back *e.m.f.* is produced by the cell, so that the following equation represents the experimental conditions:

$$V = E_b + IR_i$$

Here V represents the voltage applied to the cell, E_b is the back e.m.f. of the cell, I the current causing electrolysis and R_i the internal resistance of the cell.

The magnitude of the back e.m.f., for any electrolytic cell, can be found by further experiment. Here the basic action is described, as noted for the simple water voltameter shown in the diagram (figure 4.5).

Electrodes are immersed in an electrolyte of sulphuric acid. When the switch is closed current passes from anode to cathode and both electrodes become coated with bubbles of gas, oxygen and hydrogen respectively. Once this occurs the switch is moved to a

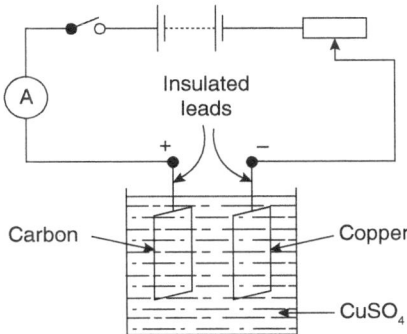

▲ **Figure 4.4** *Cell with back e.m.f.*

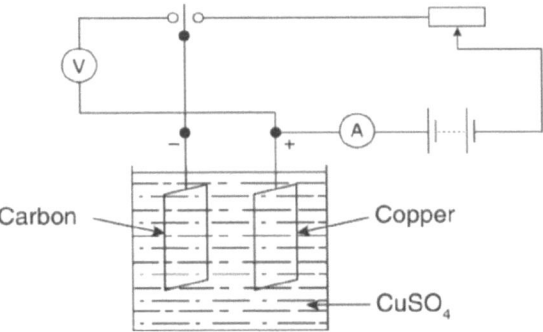

▲ **Figure 4.5** *A simple water voltameter*

second position so as to connect a sensitive voltmeter across the cell only. The main circuit current will have stopped, but the voltmeter will register a voltage across the plates, which gradually falls as the bubbles disperse, due to the flow of the small current through the voltmeter. The voltage or e.m.f. is due to the presence of the gas bubbles, and it will be noted that the cell polarity remains the same, namely that the current flows through the meter from anode to cathode in the external circuit so as to be in the opposite direction in the cell to the current flow that caused electrolysis. The value of the back e.m.f. is obviously important in that, if the applied voltage is less than this back e.m.f., electrolysis cannot take place. At start a small current flows but once polarisation begins, the back e.m.f. rises to equal the applied voltage and the current then ceases.

In the case of water the value of the back e.m.f. E_b can be calculated. It is known that 1g of water when formed by combustion of hydrogen in oxygen produces about 15.96kJ of heat. If we assume the energy required to separate H and O in 1g of water is identical, the electrical energy required will also be 15.96kJ.

Since both H and O are released by electrolysis it follows that the total mass of gas or water released by 1C will be equal to the E.C.E. of hydrogen + E.C.E. of oxygen = 0.000 010 4 + 0.000 082 9 or z_{HO} = 0.000 093 3g, i.e. 93.3 × 10^{-9}kg are released.

Thus 1C releases 0.000 093 3g of gas (or liberates this mass of water) and the electrical energy required to decompose this mass of water = 15 960 × 0.000 093 3 = 1.49J.

If this electrical energy is produced by the work done against the back e.m.f. E_b, then the applied voltage will have a value V where $V = E_b$ and the energy produced by the passage of 1C = $V \times$ 1J. Thus $V \times$ 1 = 1.49 or V = 1.49 and hence E_b = 1.49V.

During electrolysis, decomposition of the liquid produces ions, which once dissociated try to recombine and due to their slow progress through the electrolyte a back e.m.f. is produced. For water the back e.m.f. is about 1.5V and substituted for E_b in the formula: $V = E_b + IR_i$. Thus for current flow the applied voltage must be greater than the back e.m.f. by the voltage drop due to resistance of the electrolyte.

A back e.m.f. of appreciable value exists for electrolytic cells made up with electrodes of *dissimilar* materials, but if both electrodes are of the same material, as for the copper voltameter, the back e.m.f. value is *so* small that it can generally be neglected. This is explained by the fact that no difficulty is experienced in enabling the dissociated ions to recombine. They can readily combine with the electrolyte as stated in the description of the copper voltameter, and the all important result is that the passage of current does not produce an overall chemical change. All that happens is that copper is transferred materially from anode to cathode. It should be noted that for the water voltameter, although the electrodes are of the same material, namely platinum, a back e.m.f. still appears when they are coated with H and O gas bubbles. The bubbles have the effect of insulating the electrodes and retarding the passage of ions. This condition results in a back e.m.f. of about 1.5V.

General observations show that when the products of electrolysis possess chemical energy, then the equivalent electrical energy must have been supplied through electrolysis and a back e.m.f. of appreciable value must exist, for example, in the electrolysis of water. Hydrogen and oxygen will recombine in an explosive manner to form water producing heat and light. The energy latent for this recombination was derived from the electrical energy put in during electrolysis and a back e.m.f. clearly must have been present. For an electrolytic cell, such as the copper voltameter, since the product of the process possesses no chemical energy it can be assumed that a cell using electrodes of the same material has negligible e.m.f.

RESISTANCE OF ELECTROLYTES. The resistance R_i of a liquid conductor is proportional to the length and inversely proportional to the cross-sectional area (the same as for a conductive wire). It also varies with the nature of the electrolyte and the concentration; however, the temperature coefficient is a negative one. Because of the back e.m.f. effect already discussed, resistance is difficult to measure. The value obtained by dividing the voltage drop across a cell, by the current flowing will give a resistance value for the electrolyte much greater than the true figure and is erroneous because of neglecting the back e.m.f. E_b.

POWER EXPENDED DURING ELECTROLYSIS. If the voltage equation for an electrolytic cell is $V = E_b + IR_i$ then for a current flow of I amperes the power equation becomes $VI = E_b I + I^2 R_i$.

Here VI represents the power applied to a cell, $E_b I$ represents the power required to produce chemical dissociation and $I^2 R_i$ represents the heat energy produced in the cell generating a temperature rise.

Example 4.4. Find the voltage needed to pass a current of 4A through a copper voltameter with an internal resistance of 0.014Ω and a back e.m.f. of 0.25V. Find the power utilised to produce electrolysis and that wasted in heating the electrolyte. Find the overall % efficiency of the voltameter if used as a plating vat (2 significant figures).

Since $V = E_b + IR_i = 0.25 + (4 \times 0.014) = 0.306V$

Power utilised for electrolysis $= 0.25 \times 4 = 1W$

Power wasted $= I^2 R_i = 16 \times 0.014 = 0.224W$

Efficiency of the cell for electrolysis = Power used for electrolysis/Total power

$$= \frac{1}{(1 + 0.224)} = \frac{1}{1.224}$$

$$= 0.82 \text{ or } 82\%.$$

Primary and Secondary Cells

Many of the fundamentals of voltaic or *galvanic* action, as it was called originally, have been mentioned in this chapter, but we now consider in detail the theory concerned with conversion of chemical energy into electrical energy, the production of an e.m.f. by chemical action. Generation of this e.m.f. is best studied by describing the action of a simple cell.

The simple voltaic cell

If a piece of commercial zinc is dropped into a glass jar containing dilute sulphuric acid, the zinc will corrode and hydrogen gas bubbles be given off. The jar is also found to warm and it is deduced that heat is generated by the chemical action. If a piece of pure zinc is similarly experimented with, no such effects occur, nor do they happen if a dissimilar piece of pure metal such as copper replaces the pure zinc.

The chemical action noticed for commercial zinc is considered due to 'local action' and explained by the presence of impurities in the zinc, the chief of which are iron and lead. A local closed circuit is made between a particle of iron and the zinc, as both are in contact and the acid is common to both dissimilar metals. A small cell is considered to be formed and as current flow is possible, generation of an e.m.f. is thought to accompany the chemical action.

Consider 2 electrodes immersed in a solution of dilute sulphuric acid as shown in figure 4.6. Electrodes should be plates of pure dissimilar metals, such as zinc and copper, and placed to avoid touching each other. No action will take place for the arrangement described but if an ammeter and a resistance load are connected as illustrated, a current will flow when the switch is closed. The current through the external circuit will be from the +ve copper pole or anode to the −ve zinc pole or cathode, and hydrogen bubbles are given off from the copper plate while the zinc plate is slowly eaten or corroded away.

The ammeter will show the current falling slowly until it eventually ceases. The copper plate will be covered completely with hydrogen bubbles and if these are wiped off with a glass rod, the current will be found to restart and the cycle of action

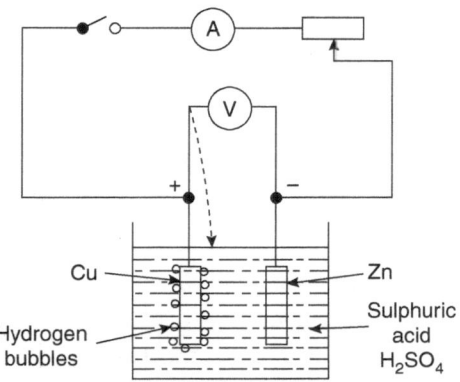

▲ **Figure 4.6** *The simple voltaic cell*

to repeat itself. Hydrogen bubbles appear to control the chemical action of the cell, which is said to be 'polarised' when the current ceases.

Further simple experiments with different combinations of electrodes and electrolytes produce different e.m.f.s, recorded by the voltmeter. When supplying current, the negative pole of each arrangement will corrode away and for every simple cell polarisation will occur. A more complete explanation of these observations can now be provided.

THE CELL E.M.F. The voltmeter in figure 4.6 shows that an e.m.f. or voltage exists across the electrodes of a cell even when on O.C. When the circuit is completed, current flows because of this e.m.f. and chemical action continues to generate an e.m.f. and maintain current. The chemical action of a voltaic cell results in a definite e.m.f., the result of the action of both electrodes or plates with the electrolyte. If the voltmeter was connected on O.C. between the zinc and the solution, the zinc will be negative to the latter by some 0.63V. When connected between the electrolyte and the copper plate, the voltmeter will record 0.47V, copper positive to the sulphuric acid. With the instrument connected across both plates, a reading of 1.1V is recorded. Production of a cell e.m.f. is explained by the electrolytic theory. When an electrolyte is made up, the molecules split into ions, which are electrically charged and mobile. An electrolyte is an ionised solution and when a metal is immersed in it, some of the metal appears to enter the solution, so there is an immediate merging of surface ions of the metal with the ions of the electrolyte. Thus the barrier between the metal and electrolyte is not the surface of the metal but along some layer of electrical potential equilibrium, which has caused ion exchange to cease. The action on immersing various metals into an electrolyte varies for the different metals. Thus for zinc in dilute sulphuric acid there is a greater tendency for +ve ions to pass to the solution than for +ve ions of the electrolyte to pass to the zinc. The zinc plate thus becomes deficient in +ve ions and becomes negative to the solution by a voltage of 0.63V. For copper in sulphuric acid, a different action takes place. This metal becomes positive to the solution because there is a greater tendency for the +ve charged hydrogen ions of the electrolyte to move to the copper than for the +ve ions of the metal plate to pass into the solution and the copper rises to a potential of 0.46V with respect to the electrolyte.

Exchange of ions results in a potential difference (P.D.) between the electrodes and solution, which gradually opposes the exchange until it eventually ceases and equilibrium is established. Thus the arrangement of zinc–copper electrodes in sulphuric acid results in a P.D. of 1.1V between the electrodes. On O.C., we see that the initial +ve Zn ions migrating into the solution combine with the sulphions to form zinc sulphate, liberating +ve hydrogen ions, which move towards and accumulate on the copper plate. Thus the chemical action is explained by the formula $Zn + H_2SO_4 = ZnSO_4 + 2H$. The potentials build up within the cell, quickly bring the ion migrations to an end and chemical action stops. If now the O.C. condition is changed to that of a closed circuit, by joining the copper electrode to the zinc electrode through an external circuit, the chemical action immediately recommences. Current is seen to flow and the formation of zinc sulphate continues with liberation of hydrogen at the copper plate.

The O.C. e.m.f. of 1.1V, which is the result of the initial ion migration, can now generate a current, the magnitude of which is determined by the circuit resistance. A flow of current is a movement of electrons, passing from the zinc cathode to the copper anode. The electrons move round the external circuit from the zinc cathode to neutralise the +ve anode charges, thus making way for further migrations of +ve hydrogen ion charges in the cell. The initial cell action, as described for the O.C. condition, can now continue and the cell functions by maintained chemical action, provided polarisation is avoided. We can conclude our study of cell action by saying that all metal electrodes produce an e.m.f. as a result of the ion exchange action with the electrolyte, and that they can be arranged in a table, in order of the value of their e.m.f.s.

THE ELECTROCHEMICAL SERIES. If any 2 elements shown in the table are used for a cell, the element *lowest* in the series is the +ve terminal, when considered with respect to the external circuit. The list comprises the usual elements, which are mainly metals, but hydrogen and carbon are found to behave like metals and are included. To illustrate use of the table, consider an ordinary dry cell used for a hand-torch, which uses carbon for the +ve electrode and zinc for the −ve electrode.

Materials well separated in table 4.1 are used for practical cells. For the torch battery the e.m.f. is about 1.5V per cell.

Table 4.1 *Electrochemical Series Common Material Potentials*

Potassium	−2.92	Lead	−0.13
Sodium	−2.71	Tin	−0.14
Magnesium	−2.37	Hydrogen	0.00
Aluminium	−1.66	Copper	+0.34
Manganese	−1.185	Zinc	−0.76
Cadmium	−0.40	Iron	−0.44
Nickel	−0.23	Mercury	+0.80
Silver	+0.80	Platinum	+1.188
Gold	+1.50	Carbon	+0.207

POLARISATION. For the electrolysis of water, it was noted that gas was liberated at the electrodes and this resulted in a decrease of current and ultimately in a change in cell action. Once polarisation (collection of gas on the electrodes) occurs, the electrodes have, in effect, changed – these now being hydrogen and oxygen. Voltaic action can now result to set up an e.m.f. operating in the *reverse* direction to the voltage applied to cause electrolysis introduced as the cell back e.m.f.

When the simple voltaic cell supplies current, polarisation will occur. The circuit current gradually falls, even though the chemical action of the cell appears to proceed. Close examination reveals that as the hydrogen bubbles make their way to the copper plate, not all are liberated here and rise to the surface. Some bubbles stick to the plate and this tendency increases until the whole plate is covered with bubbles so the cell becomes increasingly ineffective as a source of e.m.f. The layer of gas surrounding the +ve plate causes a polarising effect because (1) gas has a high resistance, so that any area of the plate covered with bubbles is almost insulated and cannot allow the passage of current. Thus the internal resistance of the cell rises as the gas layer increases and the circuit current falls as a direct result. (2) As hydrogen covers the copper plate, it begins to make its presence felt in that it effectively replaces the +ve copper electrode by a hydrogen electrode and thus reduces the e.m.f. of the cell. It will be seen from the table of the electromotive series that the spacing between zinc and hydrogen is smaller if compared with that for zinc and copper. The cell e.m.f. is thus much reduced, giving the final result as described.

Once the cause of polarisation became known it was apparent that, in order to make the simple cell an effective source of electrical energy, a method of preventing collection of hydrogen bubbles was needed.

The simplest forms of depolarisers developed operate chemically, combining with liberated hydrogen to convert it into water, preventing the gas from reaching the +ve electrode and blanketing it. The methods by which this is accomplished will be seen when examples of primary cells are studied.

Batteries

These can be dry, wet or even gel type cells.

The *dry cell* (primary cell) is made of a carbon rod surrounded by manganese dioxide with an electrolyte of ammonium chloride. The cell is sealed and activated when connected to a circuit, such as a torch light. Different chemicals can be used, but the principle is the same. The usual voltage from a cell is 1.5V.

The *wet cell* (secondary cell) is simply a container that has plates of different materials submerged together in an electrolyte. A chemical reaction takes place between the plates and positive particles will migrate through the electrolyte to the negative plate. A circuit is made between the plates and this will continue until the electrolyte has been expended. This type of battery is frequently used for motor boats.

The *gel cell* is a sealed unit very similar to the wet cell. The electrolyte is contained within the unit and it cannot be opened. The battery will only last a relatively limited lifetime, and upon discharge must be replaced. This type of battery is now fitted in many new cars.

The primary cell

Under this heading are considered practical cells, which are suitable for providing a constant e.m.f. when operating under everyday conditions. They are, however, cells that obtain their electrical energy from chemical energy, the active material being used up in the process. They differ from secondary cells in that the latter utilise materials that are *not* consumed when the cells provide electrical energy. The secondary cell can be electrically 'charged' so that its electrodes are chemically converted into materials that enable the cell to provide an active e.m.f. for supplying electrical energy. In this condition the cell discharges and the electrode materials again change chemically, reverting back to those of the uncharged state. The whole cycle of charge and discharge can then be repeated.

Primary cells suffer from the 2 main disadvantages of the simple cell: (1) polarisation and (2) local action.

Polarisation is overcome by the use of a suitable chemical depolariser, which is therefore an essential component of cell construction. Local action is minimised by using pure metal, such as zinc free from impurities like iron and lead. In its basic form the primary cell is a wet cell, which is not used to any extent today, although it was used for railway signalling in places where no electric mains was available. In the dry form the Leclanche cell is still common and attention should be paid to its construction and action.

THE LECLANCHE CELL (Dry type). One form of construction is illustrated (figure 4.7), which is a cross-sectional view of a typical practical cell.

The depolariser, manganese dioxide (MnO_2), is mixed with powdered carbon and packed round a central carbon rod. This assembly is placed in a linen bag, which serves as the porous pot of the cell. The negative electrode is a pressed zinc canister that

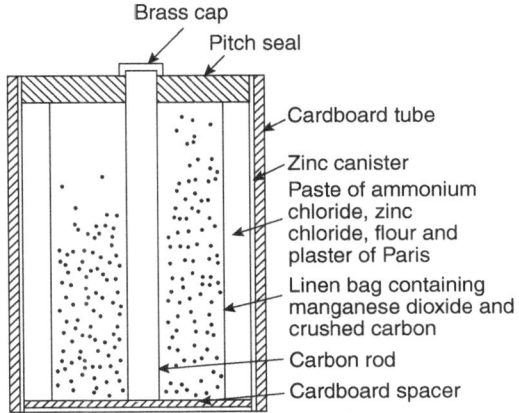

▲ **Figure 4.7** *The primary cell*

contains the linen bag assembly and the electrolyte, made up as a paste of ammonium chloride, zinc chloride, flour and plaster of Paris. One method of closing the canister is to seal it with pitch.

The following chemical formula defines the chemical action:

Action at negative electrode:

$$Zn + 2NH_4Cl = ZnCl_2 + 2NH_3 + H_2$$

Action at positive electrode:

$$H_2 + 2MnO_2 = Mn_2O_3 + H_2O$$

The form of cell as described is in most general service, but other forms have been developed for incorporating into the layered battery type, used for portable radio sets, calculators, etc. The reader should complement the information given here by referring to books specialising in the practical treatment of battery operated equipment.

The secondary cell (or accumulator)

Because of the importance of this cell as a means of storing electricity (it is sometimes called a storage cell), the reader is advised to consult other books giving more details of modern constructional methods, applications and maintenance requirements. Figure 4.8 shows only the basic construction and the description sets out only the elementary principles. The modern accumulator uses 'pasted' plates to allow maximum

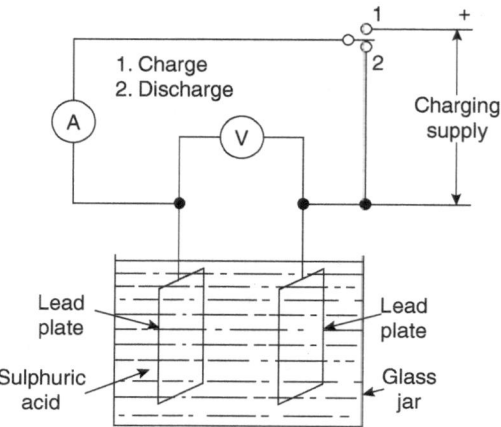

▲ **Figure 4.8** *The secondary cell*

use of available material and the process used in 'forming' the cell is too detailed for a book of basic theory. It is hoped, however, that the information given below will provide sufficient knowledge to enable the action of the lead–acid accumulator to be understood. The nickel-iron or nickel–cadmium alkaline battery also functions on similar principles, although the plate materials and electrolyte differ. This type of cell is also important and should be thoroughly investigated.

The lead–acid cell

The simple accumulator consists of 2 lead (Pb) plates immersed in dilute sulphuric acid, the whole assembly being contained in a glass or moulded ebonite container. The cell must be worked into a suitable condition before it can be used for storing electricity and the process is carried out by alternately 'charging' and then 'discharging' the cell. If a D.C. supply is connected to the plates as in the diagram (figure 4.8, switch position 1) and the cell subjected to electrolysis by passing current through it, oxygen and hydrogen gases are given off at the electrodes. As for the water voltameter, the first stage of the reaction will be decomposition of the acid (H_2SO_4). A molecule of acid dissociates to produce hydrogen ions and sulphions (SO_4). The sulphions move to the +ve plate, reacting with the water to form sulphuric acid and oxygen. The latter attacks the +ve plate only to form lead dioxide (PbO_2), which causes the original lead electrode to become a dark brown colour. The hydrogen ions discharge at the cathode and are liberated in a gaseous state. The first chemical action is thus at the +ve plate only but if the supply is switched off (switch-intermediate position), the cell is now found to have the properties of a voltaic cell and provides an e.m.f.

If the cell is short-circuited (switch position 2), it behaves like a primary cell, passing a current for a short time during which period it discharges. The solution is electrolysed in the reversed direction and the original negative plate now acts as the anode with its lead ions reacting with the sulphions of the electrolyte to form lead sulphate. Thus at −ve plate $Pb + SO_4 = PbSO_4$ (lead sulphate). The hydrogen ions from the electrolysis during discharge move to the original +ve plate, now the cathode. The hydrogen reduces the lead dioxide to lead oxide, which in turn reacts with the acid to form lead sulphate. Thus at the +ve plate the chemical action is $PbO_2 + H_2 + H_2SO_4 = PbSO_4 + 2H_2O$. Both plates are converted into lead sulphate and will be in white colour.

If the charging cycle is repeated (switch – position 1), the direction of current flow in the electrolyte is again reversed and the $PbSO_4$ on the +ve plate becomes lead dioxide (PbO_2). This is a complex result of the electrolysis of the acid. Sulphions move to the +ve plate and react with the water to form H_2SO_4 and O_2. The latter attacks the +ve plate to form PbO_2 and more H_2SO_4. The chemical action at the +ve plate is $SO_4 + H_2O = H_2SO_4 + O$ and $O + H_2O + PbSO_4 = PbO_2 + H_2SO_4$. At the negative plate, lead is produced by the hydrogen ions liberated by the acid decomposition, moving to this electrode and reducing the lead sulphate to 'spongy lead'. The chemical action at the −ve plate is:

$$PbSO_4 + H_2 = H_2SO_4 + Pb$$

After several cycles of charging and discharging, the plates become porous and the capacity of the cell is increased. When a cell is fully charged, chemical conversions are completed and hydrogen is freely given off, resulting in 'gassing' – the accepted term indicating a full charge.

The lead–acid accumulator in its practical form is provided with 'pasted' plates. Here the active material is applied to plates in the form of a paste, the backbone of the plate being a lead–antimony grid. One 'forming' charge converts the paste into lead dioxide on the +ve plate and spongy lead on the −ve plate. Irrespective of the method of production the charge and discharge action can be summarised by the following chemical equation.

Table 4.2

Charged				Discharged		
+ve Pole		−ve Pole		+ve Pole	−ve Pole	
Lead dioxide	Sulphuric acid	Lead		Lead sulphate	Water	Lead sulphate
PbO_2 +	$2H_2SO_4$ +	Pb	=	$PbSO_4$ +	$2H_2O$ +	$PbSO_4$

It will be seen that during discharge water is formed, thus diluting and reducing the specific gravity of the electrolyte. During charge, acid is formed and the tests to check a fully charged cell include:

(1) S.G. of cell charged (1.20 to 1.27); discharged (1.17 to 1.18).

(2) Voltage on O.C., charged 2.2V per cell or higher.

(3) 'Gassing' on charge.

(4) Positive plate – rich dark brown colour. Negative plate – slate grey.

The alkaline cell (nickel–iron or nickel–cadmium)

Two variations of this cell are in common use, being the results of some 110 years of developments, and patents were taken out simultaneously by Edison in America and Jungner in Sweden. Both men devised a cell with reversible action that used iron or cadmium for one plate and nickel hydrate for the other. The construction of both types is basically the same. Active materials are enclosed in steel tubes or flat interlocking pockets perforated over the whole surface area with many minute holes. The tubes (Edison form) or pockets (Jungner form) are assembled into steel retaining frames to form the +ve plate. The –ve plate, for both forms of cell, consists of a steel frame into which are assembled the flat pockets. Groups of plates of the same polarity are bolted together to steel terminal pillars. Separators of sheet ebonite are used and the plate groups assembled in a steel container. When the plates expand after the initial 'forming' process, no internal movement is possible. The cell terminals for each plate group are brought through the lid in suitable insulated glands. Batteries of cells are built up in hardwood crates, the cell containers being at a potential. Other forms of enclosure in plastic cases have been developed in recent years, leading to a lighter battery and dispensing with the need for insulated wooden crates.

Both forms of cell use the same electrolyte of dilute (21%) potassium hydroxide and have the same e.m.f. of 1.25V. The electrolyte takes no active part but functions merely as a conductor, transferring the hydroxyl (OH^-) ions from one plate to the other when charging or discharging. The cell is inert on O.C. and the electrolyte S.G. (about 1.18) doesn't alter. The battery is often described by the tradename Nife (nickel–iron, Ni–Fe). It is a robust battery that is resistant to overcharging and short-circuiting and can have very long life. It is often used in backup situations where it can be continuously charged and can last for over 20 years. Due to its low specific energy, poor charge retention and high cost of manufacture, other types of rechargeable batteries have displaced the nickel–iron battery in most applications, one of the reasons the nickel–cadmium version is favoured and has been developed. A typical cell has an outer casing, a separator between electrodes, a positive electrode and a negative electrode and is usually cylindrical to max mise energy density (figure 4.9).

The chemical reactions in an a kaline cell are complex but as a guide the following equation shows the operation. ¯his equation is given for the nickel–cadmium version. For the nickel–iron version it is similar except that Fe replaces Cd and $Fe(OH)_2$ the $Cd(OH)_2$. For a fully charged cell the nickel hydrate is at a high degree of oxidation and the −ve material is reduced to pure cadmium. On discharge the nickel triple-hydrate is reduced to a lower degree of oxidation (double hydrate) and the cadmium of the −ve plate converted to a hydroxide.

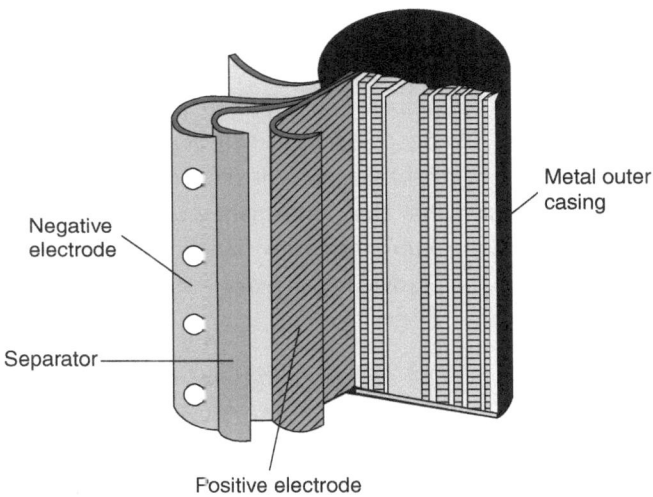

Negative electrode

Metal outer casing

Separator

Positive electrode

▲ **Figure 4.9** *Nickel–cadmium cell: outer metal casing, separator, positive electrode and negative electrode*

Table 4.3

		+ve plate				−ve plate
		Nickel hydrate		Potassium hydroxide		Cadmium
CHARGED		$2Ni(OH)_3$	+	KOH	+	Cd
	↓↑	Nickel hydroxide		Potassium hydroxide		Cadmium hydroxide
DISCHARGED						
		$2Ni(OH)_2$	+	KOH	+	$Cd(OH)_2$

When the alkaline cell is charged the e.m.f. is about 1.5V, decreasing rapidly to 1.3V then more slowly to 1.0V on discharge. The average is 1.25V so for a given voltage the number of cells is 1.7 times the number of lead–acid cells required. The cell is extremely robust and can be 'short-circuited' without damage. Chemical action is completely reversible and no fumes are given off. Provided the plates are covered with electrolyte, cells will stand almost indefinitely in any state of charge without significant deterioration. The internal resistance of an alkaline cell is 10–50% more than that of the lead–acid type but this depends upon design factors. Conductivity is adjusted by mixing finely powdered carbon with the nickel hydrate in the pockets of the +ve plate while iron powder can be mixed with the cadmium powder in the −ve plate pockets. These additions prevent the active materials solidifying and the porous requirement of the overall plate construction being lost.

CAPACITY OF A CELL. This is the *ampere hour* figure it can yield on a single discharge, until the e.m.f. falls to about 1.8V per cell (for the lead–acid cell). Generally the capacity is based on a 10-hour rate of discharge, since it decreases as the rate of discharge increases. Research has shown that the performance of a cell can be improved by working the +ve plate at a higher current density than the negative. This is achieved by keeping the plates of equal area, providing an extra −ve plate, i.e. always making the outside plates *negative*.

EFFICIENCY OF A CELL. This may be expressed in terms of (1) the Ampere hour input and output and (2) the Watt hour input and output. Thus:

$$\text{Ampere hour efficiency} = \frac{\text{Ampere hours of discharge}}{\text{Ampere hours of charge}}$$

The ampere hour efficiency neglects the varying voltages during charge and discharge. Since this is important, we have an energy efficiency compared to a quantity efficiency, and

$$\text{Watt hour efficiency} = \frac{\text{Watt hours of discharge}}{\text{Watt hours of charge}}$$

$$= \frac{\text{Average discharge volts} \times \text{Amperes} \times \text{Hours}}{\text{Average charge volts} \times \text{Amperes} \times \text{Hours}}$$

Example 4.5. A battery is charged with a constant current of 16A for 11 hours, after which time it is considered to be fully charged, its voltage per cell being recorded as

2.2V. Find its ampere hour efficiency if it is (1) discharged at a rate of 16A for 10 hours and (2) 28A for 4 hours. In either case, discharge was discontinued when the voltage per cell fell to 1.8V (2 significant figures).

(1) Ampere hour input $= 16 \times 11 = 176$

Ampere hour output $= 160$

$\therefore$ Efficiency $= \dfrac{160}{176} = 0.91$ **or** 91%

(2) Ampere hour input $= 16 \times 11 = 176$

Ampere hour output $= 28 \times 4 = 112$

$\therefore$ Efficiency $= \dfrac{112}{176} = 0.63$ or 63%.

Example 4.6. A 12V accumulator is charged by means of a constant current of 16A passed for 11 hours. The P.D. during charging varies as shown. The battery is then discharged at a constant current of 16A for 10 hours, the P.D. again varying as shown. Calculate the watt hour efficiency of the battery (1 decimal place).

Table 4.4

	Reading No.	Time (hr)	Charge (V)	Discharge (V)
Start	1	–	10.8	12.6
	2	1	11.0	12.4
	3	2	11.5	12.2
	4	3	11.8	12.0
	5	4	12.0	11.8
	6	5	12.2	11.6
	7	6	12.4	11.4
	8	7	12.6	11.2
	9	8	12.8	11.0
	10	9	13.0	10.9
	11	10	13.1	10.8
	12	11	13.2	**127.9**
			146.4	

$$\text{Average charging voltage} = \frac{146.4}{12} = 12.2\text{V}$$

$$\text{Average discharge voltage} = \frac{127.9}{11} = 11.62\text{V}$$

$$\text{Efficiency} = \frac{11.62 \times 16 \times 10}{12.20 \times 16 \times 11} = 0.867 \text{ or } 86.7\%$$

CHARGING PROCEDURE. British practice uses the 'constant current' method. This is also American practice but on the Continent the 'constant voltage' method is favoured. For this latter method the charging supply voltage is kept constant and is substantially higher than the battery e.m.f. for the discharged condition. The charging current is high initially but falls as the back e.m.f. of the battery rises. This method gives a lower charging time than the 'constant current' method, but due to the violent chemical action and heat generated in the battery there is danger of 'buckling' the plates, unless the battery is specially constructed!

For the 'constant current' method, arrangements must be provided for increasing the voltage applied to the battery as charging proceeds and the back e.m.f. rises. If a generator is used and I is the charging current, R_i the internal resistance of the battery and E_b the battery e.m.f. at start of charge, then the applied voltage must be $V = E_b + IR_i$ ... start of charge (1).

If E_{b1} is the battery e.m.f. at the end of charge, the applied voltage would have to be $V_1 = E_{b1} + IR_i$ ... end of charge (2).

Thus subtracting (1) from (2) variation of voltage would be $V_1 - V = E_{b1} - E_b$ or the applied voltage must be increased by an amount equal to the rise of the battery e.m.f.

If a constant supply voltage is used for charging, then a variable resistor is required to obtain the necessary current control, and its value will be reduced as charging proceeds.

Let V be the supply voltage, I the charging current, R_i the internal resistance of the battery, E_b the battery e.m.f. at start of charge and R the control resistor. Then:

$$V = E_b + IR_i + IR \ldots \text{start of charge (1)}$$

If E_{b1} is the battery e.m.f. at end of charge and R_1 the new value of the control resistor. Then:

$$V = E_{b1} + IR_i + IR_1 \ldots \text{end of charge (2)}$$

Subtracting (1) from (2) $0 = E_{b1} - E_b + IR_1 - IR$ or $(E_{b1} - E_b) = I(R - R_1)$. Thus the control resistance must be reduced from R to R_1 as the battery voltage rises from E_b to E_{b1}.

Example 4.7. A 24V emergency battery is to be charged from the 110V ship's mains when the e.m.f. per cell has fallen to a minimum value of 1.8V. The battery consists of 12 cells in series, has a capacity of 100A h at a 10h rate and the internal resistance is 0.03Ω/cell. If charging continues until the voltage/cell rises to 2.2V, find the value of the variable resistor needed to control the charging (1 significant figure.)

The charging current can be assumed to be equal to the maximum allowable discharge current.

$$\text{Discharge current} = \frac{100}{10} = 10A = \text{charging current}$$

At start of charge, battery voltage $= 12 \times 1.8 = 21.6V$

Battery internal resistance $= 12 \times 0.03 = 0.36Ω$

Then $110 = 21.6 + (10 \times 0.36) + (10 \times R)$

$$\text{or } R = \frac{110 - 25.2}{10} = \frac{8.48}{10} = 8.48Ω$$

At end of charge, battery voltage $= 12 \times 2.2 = 26.4V$

Then $110 = 26.4 + (10 \times 0.36) + (10 \times R_1)$

$$\text{or } R_1 = \frac{110 - 30}{10} = \frac{80}{10} = 8Ω$$

Thus the variable resistor should have a value used, 8.48Ω and be capable of being reduced to 8Ω. In practice a unit of 9Ω would be used which would be reduced by adjusting the sliding contact until the charging ammeter recorded the correct current. Further adjustments will be made periodically as charging proceeds. It is important to note that besides the ohmic value of the resistor, the wattage rating must be specified. For the unit in the example, a rating of $I^2R = 10^2 \times 9 = 900W$ is required. The control resistor must be capable of dissipating up to this power as heat during the charging, although this waste of power will decrease slightly as charging proceeds. For example, at the end of the charge the power wasted in the resistor would be $10^2 \times 8 = 800W$.

The most important point to stress is the correct connection of the battery for charging, i.e. +ve terminal of battery to +ve of mains; −ve terminal of battery to −ve of mains. It is surprising how many times this requirement is overlooked through carelessness. For incorrect connection, no control of the current will be possible with the equipment provided and damage of the ammeter, control resistor or battery could result.

The Meaning of pH

Introductory ionic theory dealt with electrolytic dissociation, which, for a solution, results in the formation of 2 separately charged ions (anions and cations). Such ionisation is assisted by a liquid that has a high 'dielectric constant' and thus can separate and support unlike charges. Pure water, being a poor conductor, serves as a good dielectric. The cations of an electrolyte are derived from the metallic part of the molecule, having +ve charge. Anions are from a non-metallic element or radical and have −ve charge. (A radical is a fundamental group of atoms, such as a sulphate (SO_4), a nitrate (NO_3) or a carbonate (CO_3) etc, that behave as individual entities and remain unchanged by most chemical reactions.) All acids produce hydrogen ions (H^+) and all alkalis produce hydroxyl ions (OH^-). Sulphuric acid (H_2SO_4) ionises to $H_2SO_4 \rightleftharpoons 2H^+ + SO_4^{2-}$. Since each hydrogen ion (H^+) can carry only one +ve charge and as each H_2SO_4 molecule is electrically neutral, the sulphate ion must have 2 negative charges (SO_4^{2-}). Similarly sodium chloride or common salt (NaCl) splits into $NaCl \rightleftharpoons Na^+ + Cl^-$. The double arrows indicate the splitting up is not complete and only a percentage of the solution is ionised, the amount depending on the physical conditions – strength of solution, temperature, type of salt, etc. The deionised portion of the solution is assumed to consist of neutral molecules. Caustic soda (NaOH) is an alkali and ionises thus: $NaOH \rightleftharpoons Na^+ + OH^-$. A hydroxyl ion ($OH^-$) is produced whereas the sulphuric acid molecule produces 2 hydrogen ions $2H^+$. Note that hydroxyl ions (OH^-) are in fact radicals.

For an electrolyte, the percentage ionisation is extremely high. It is known that the concentration of hydrogen ions present in solution determines its properties. The greater this concentration, the more acid the solution. Conversely, the smaller the concentration, the more alkaline the solution. Water is a special case since it ionises only slightly to give both hydrogen and hydroxyl ions. For many modern industrial processes, a knowledge of the acidity of the materials being used is most important. The pH value of a substance or solution is a measure of its acidity or alkalinity. A Swedish scientist, Sørensen devised a scale to indicate the hydrogen ion content. It uses the pH symbol, which stands for minus the logarithm to the base 10, for the ion concentration. This latter is expressed by [H^+].

Thus if [H^+] = 10^{-3}. pH = 3 (i.e. $-\log_{10} 10^{-3} = 3$)

Note the letter 'p' comes from the German potenz, meaning 'a power', as the word is used in mathematics. For example, $100 = 10^2$ – ten to the power two. The hydrogen ion concentration of a solution is the number of gram-equivalents of hydrogen contained in a litre of the solution. The actual extent of the ionisation of water has been determined and for pure water the ion concentration (H^+) is 0.000 000 1 gram-equivalents per litre or 10^{-7} normal. This power of 10 with the sign changed was taken by Sørensen to provide a scale up to 14 – neutrality being given as 7 (pure water). Thus by measuring its hydrogen ion concentration, it is possible to show where a solution lies on the acid/alkali scale. A strong acid will register near 1 while a strong alkali will register towards 14.

An electrical method for determining a solution's pH value involves measuring the D.C. voltage between 2 special electrode assemblies which when immersed in solution result in e.m.f.s being produced by voltaic action. The electrode assemblies are known as the *reference electrode* and the *measuring electrode*. The former is constructed so a constant potential is produced, irrespective of the pH of the solution under test. The measuring electrode assembly is constructed to allow its potential to vary with the solution under test and since the reference electrode potential is constant, the resulting potential variation between the 2 electrode assemblies is measured by a sensitive millivoltmeter. Such an instrument, suitably calibrated, provides a direct indication of the pH value of the sample or solution being tested.

Electrochemical Corrosion

The 2 main causes of electrochemical corrosion are (1) *galvanic* action and (2) *electrolytic* action. In each case the electrolyte may be water with impurities or moist earth. These corrosion causes are now considered separately.

(1) *Galvanic action.* This results in currents through the electrolyte from an anode, such as a metal structure, to some adjacent cathode. Metal is lost from the structure, which can be both costly and/or dangerous. In the case of a merchant ship or naval vessel this results in rusting by oxidation of the hull when immersed in the conductive salt water environment. Galvanic action is caused by 'local cells' set up by slight differences in the surface composition of the hull metal, pitting of the plating, welds, rivets and mill scale (figure 4.10). The corrosion occurs at local anodic areas from which currents flow through the sea to the local cathodic areas. The rate of corrosion is proportional to the currents, which in turn are affected by metal composition, electrolyte temperature and even a ship's speed.

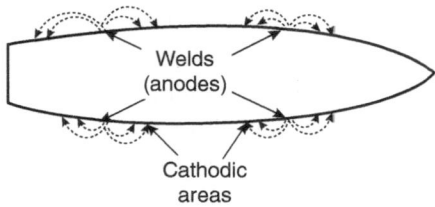

▲ **Figure 4.10** *Galvanic action on ship structure*

(2) *Electrolytic action.* This is caused by stray currents due to leakage from electrical systems, such as a D.C. distribution network – examples being an electric dockside crane rail or tramway track. Such corrosion was a severe hazard in the early days of electric traction when the 'live rail' or cable followed a curve of the track and some conductor such as a metal pipe or different cable run was situated within the bend. This condition could provide a leakage current path, so a current could leave the live conductor, enter the pipe or adjacent cable sheath and then leave the latter at an appropriate point, to rejoin the live conductor. Where the current leaves the pipe or cable sheath, an anode is formed and as a result metal is dissolved at this point with possible serious results.

CATHODIC PROTECTION METHODS. In both the above cases, corrosion is prevented by introducing an anode adjacent to the structure, pipe or cable sheath. In this way current is *forced* to enter the original anode point so that this now becomes a cathode. This current being opposed to the original local currents neutralises or reverses them. For marine work, 2 systems are in general use: (1) the 'sacrificial' anode method (2) the cathodic protection or 'impressed current' method. The latter is most favoured but both systems should be complementary to a good paint system.

(1) *The sacrificial anode method.* Counter currents are encouraged by galvanic action and the method is useful for protecting smaller structures or ships because the current adjustment range is restricted by constant potential effect. Also, as the name implies, the anodes waste away and require periodic replacement. Examples of materials that readily corrode away are magnesium or zinc. Magnesium gives a potential of some 1.8V positive with respect to iron and provided the resistance of the electrolyte path is sufficiently small, current enters the cathode (structure or hull). This is shown in figure 4.11.

(2) *The impressed current method.* For a ship, this is achieved by forcing a current through the hull, of a magnitude sufficient to nullify the effect of existing local cells. Such a current passes from the external anodes to the hull below the waterline so as to make the latter wholly cathodic. The anodes are supplied from a direct current source (generator or rectifier) and conduct the protective current into the water. The anodes are mounted on insulating plates, thus isolating them from the hull (figure 4.12).

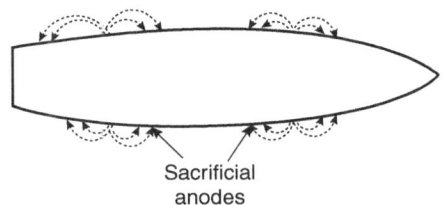

▲ **Figure 4.11** *Sacrificial anode*

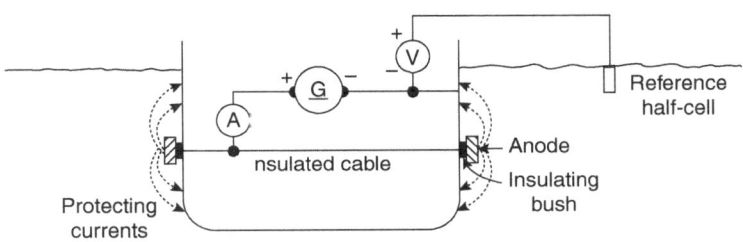

▲ **Figure 4.12** *Anodes mounted on insulation plates*

They are connected to the positive terminal of the supply through insulating feeder cables, while the hull is connected to the negative terminal. For shore work, the structure to be protected is connected to the negative terminal. The anodes are buried in the ground and suitably sited to give the desired current distribution. For a ship, the current required to give protection varies from 20 to 300 amperes, variation being dependent on paint condition, ship speed, water temperature, etc. For shore work, variation depends mainly on soil moisture content. The anode material for ship work may be a metal such as raw lead or platinum plated titanium, which is affected only a little by the discharge current. Systems operate where some anode dissolving is inherent such as with the aluminium wire system, where a wire is trailed from the ship's stern. It is suitably insulated from the hull and the wastage compensated at regular intervals by paying out a suitable amount of wire. For shore work, anodes can be made of steel scrap, graphite or ferro-silicon.

If the protecting current is too high, electrolysis will cause excessive development of alkali and hydrogen resulting in paint blistering. If the current is too low, corrosion continues and in order to monitor and control protection at the correct value the potential between the electrolyte (sea) and the hull is measured with a sensitive voltmeter and reference electrode. In actual practice, a measuring half-cell is used for checking purposes, *silver–silver chloride* or *copper–copper sulphate* cells. Unprotected iron has a potential of −0.55V against the copper–copper sulphate measuring cell but if reduced by the impressed current to −0.85V, full protection is achieved. Control is effected manually or automatically by altering the potential of the motor-generator or transformer-rectifier unit. Modern automatic systems employ a hull-mounted reference electrode and an electronic amplifier to adjust the cathodic potential of the impressed protection current.

Practice Examples

4.1. An accumulator is charged at the rate of 6A for 18h and then discharged at the rate of 3.5A for 28h. Find the ampere hour efficiency (1 decimal place).

4.2. The mass of the cathode of a copper voltameter before deposit was 14.52g, and after a steady current was passed through the circuit for 50 min, its mass was 19.34g. The reading of the ammeter was 5.1A. Find the % error of the ammeter, taking the E.C.E. of copper as 330×10^{-9}kg/C (2 decimal places).

4.3. A 90V D.C. generator is used to charge a battery of 40 cells in series, each cell having an average e.m.f. of 1.9V and an internal resistance of 0.0025Ω. If the total resistance of the connecting leads is 1Ω, calculate the value of the charging current (2 decimal places).

4.4. Nickel is to be deposited on the curved surface of a shaft 100mm in diameter and of length 150mm. The thickness of deposit is to be 0.5mm. If the process takes 8h, calculate the current that must flow. The E.C.E. of nickel is 302×10^{-9}kg/C. The density of nickel is 8600kg/m³ (1 decimal place).

4.5. A nickel–alkaline battery is discharged at a constant current of 6A for 12h at an average terminal voltage of 1.2V. A charging current of 4A for 22h, at an average terminal voltage of 1.5V, is required to recharge the battery completely. Calculate the ampere hour and watt hour efficiencies (2 significant figures).

4.6. A battery of 80 lead–acid cells in series is to be charged at a constant rate of 5A from a 230V, D.C. supply. If the voltage per cell varies from 1.8 to 2.4V during the charge, calculate the maximum and minimum values of the required control resistor. If the ampere hour capacity of the cells is 60, state the probable charging time required, assuming that the cells were in a completely discharged condition at the commencement of the charge (nearest hour).

4.7. A metal plate measuring 50mm by 150mm is to be copper plated in 30min. Calculate the current required to deposit a thickness of 0.05mm on each side (ignore the edges). The E.C.E. of copper is 330×10^{-9}kg/C and its density is 8800kg/m³ (1 decimal place).

4.8. A battery of 40 cells in series delivers a constant discharging current of 4A for 40h, the average P.D. per cell being 1.93V during the process. The battery is then completely recharged by a current of 8A flowing for 24h, the average P.D. per cell being 2.2V. Calculate the ampere hour and the watt hour % efficiencies for the battery (1 decimal place).

4.9. Thirty lead–acid accumulators are to be charged at a constant current of 10A, from a 200V D.C. supply, the e.m.f. per cell at the beginning and end of charge being 1.85V and 2.2V respectively. Calculate the values of the necessary external resistor required at the beginning and end of charge, assuming the resistance of the leads, connections, etc. to be 1Ω and that the internal resistance is 0.01Ω per cell (2 decimal places).

4.10. When a current of 3.5A was passed through a solution of copper sulphate, 4.2g of copper were deposited. If the E.C.E. of copper is 330×10^{-9}kg/C and the chemical equivalent of copper is 31.8, find the time for which the current was passed through the solution (nearest second) and also the mass of hydrogen in grammes liberated (4 decimal places).

5

MAGNETISM –

ELECTROMAGNETISM

Gilbert shall live, till Load-stones cease to draw,

Or British Fleets the boundless Ocean awe.

John Dryden

Magnets

Natural magnets

From ancient times it was known to civilisations such as the Greeks and Chinese that certain types of iron ore have magnetic properties. Pieces of iron ore can attract and repel other such pieces but can also pass on their magnetism. Another property of this ore, called magnetite or lodestone, is that if it is freely suspended, as shown (figure 5.1), it will come to rest in a roughly geographical North–South direction. The end pointing north is called a north-seeking or North (N) Pole, while the other end is a South (S) Pole. The ore is a natural magnet and if brought into contact with a quantity of iron filings, the filings will stick mainly to its ends or poles.

Further investigations made with pieces of the magnetic ore show that if 2 such magnets are each suspended as described above and their polarities determined and marked, when the *N* pole of one suspended magnet is brought near the *N* pole of

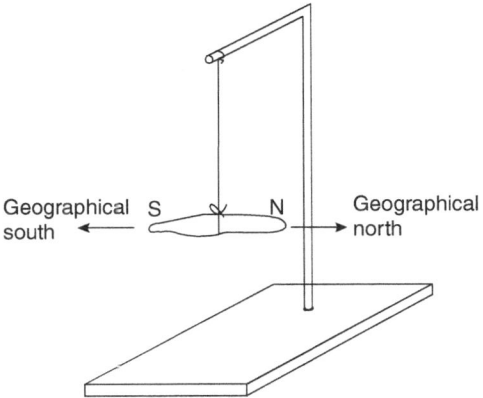

▲ Figure 5.1 *Freely suspended magnet*

the other suspended magnet, *repulsion* of poles results. Two *S* poles brought near each other will behave in a similar manner, while a *S* pole brought near the *N* pole of the other magnet produces an *attractive* effect. Thus every magnet is seen to have 2 poles of unlike polarity and that like poles repel while unlike poles attract.

Artificial magnets

A piece of iron can be converted into a magnet and exhibit properties similar to that of the iron ore. Such a piece of iron is then an artificial magnet and is said to be *magnetised*. A simple method of magnetising iron is to stroke it in one direction from end to end with one pole of an existing magnet. However, the most effective method is by use of electromagnetism.

Certain materials such as copper. aluminium, lead, brass, wood, glass, rubber, etc. *cannot* be magnetised. Thus all materials can be classified under the heading of magnetic or non-magnetic substances. A few metals such as nickel, cobalt and magnesium exhibit *slight* magnetic properties, but when alloyed with iron very strong magnetic properties result.

An artificial magnet is usually made in bar or horseshoe form. When tested, the tips are found to constitute poles of opposite polarity and, if suspended, a bar magnet will again lie on an approximate *N–S* line. The magnetic compass makes use of this principle and consists of a short, highly magnetised bar magnet, which is pivoted at its centre. A card, calibrated in degrees and/or geographic points, is mounted below and is used with the magnet to obtain a 'bearing'. It is important to realise that the *N–S* direction indicated by such a compass is not exactly *geographic N* and *S*. The angle between the lines of magnetic and geographic *N–S* is called the 'variation' and varies around the world. If the magnetic compass is used, due allowance must be made for the variation, before a map can be properly orientated and used correctly.

It is helpful to explain why a compass needle lies in the *N–S* direction. The Earth itself behaves as though it contains a magnet having its *S* pole in the region of the geographic north and its *N* pole near the geographic south. A compass needle placed on the Earth's surface will lie so that its *N* pole is attracted to the magnetic *S* (geographic *N*) pole of the Earth and its *S* pole will be attracted to the magnetic *N* (geographic *S*). Summarising the facts so far about natural and artificial magnets, every magnet has 2 poles of unlike polarity and *like* poles repel while *unlike* poles attract.

The magnetic field

This is the space around a magnet where its magnetic effects are felt. If a bar magnet is covered by a sheet of paper and iron filings are sprinkled onto the paper, after tapping the latter, the filings will align as shown (figure 5.2). The filings form a pattern that, if examined closely, shows that lines may be traced from the magnet's *N* pole to the *S* pole through the space outside and from the *S* to *N* poles inside the magnet.

The field can also be plotted using a small compass needle as shown (figure 5.3).

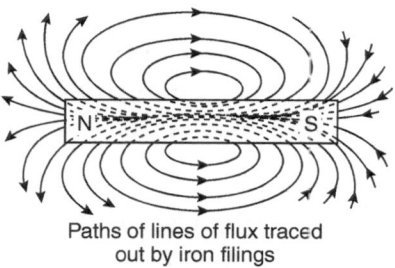

Paths of lines of flux traced
out by iron filings

▲ **Figure 5.2** *Magnet surrounded by iron filings*

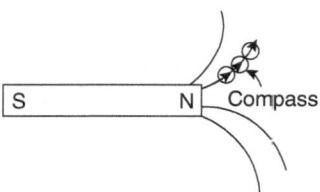

▲ **Figure 5.3** *Bar magnet with compass needle*

Field plotting with a compass needle is done as follows. Place the magnet on a sheet of paper and draw its outline. Set the compass needle against the N pole of the magnet and, with a pencil, mark a dot at the point in line with and adjacent to the N pole of the compass needle. Move the compass until the S pole of the needle coincides with the original dot.

Mark the new point in line with and adjacent to the N pole of the needle. Repeat this procedure until the S pole of the magnet is reached. Join the dots together to give a *line of force* or a *line of flux*. This can be described as the line that, when drawn through any point in a magnetic field, shows the direction of the magnetic force at that point. Using a compass needle the field can be mapped for a considerable distance around a magnet and the following deductions made:

(1) Lines of flux never cross.

(2) Lines of flux are always continuous.

If various magnetic field arrangements are plotted as shown (figure 5.4) then other conclusions can be deduced.

(3) Lines of flux are like stretched 'elastic bands' and will be as short as possible. This explains the attractive effect between 2 unlike magnetic poles, which if free to do so will move into contact, thereby reducing the length of the flux lines.

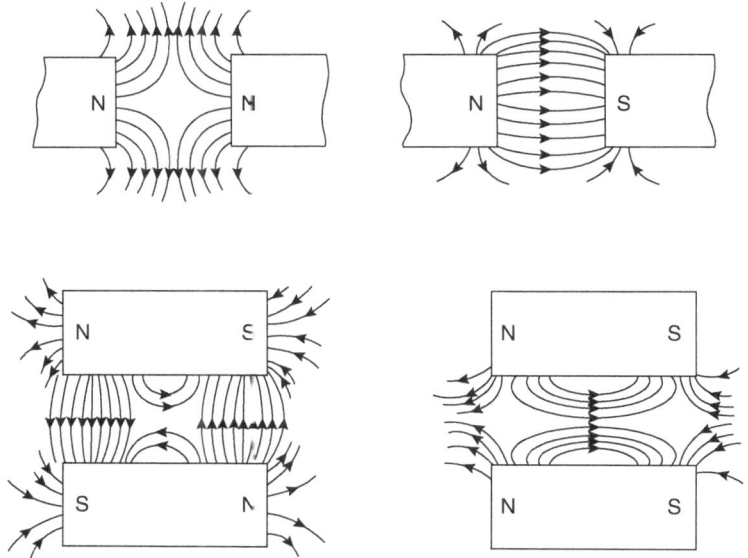

▲ **Figure 5.4** *Various magnetic field arrangements*

(4) Lines of flux that are parallel and in the same directions repel each other, for example, when 2 magnets are brought together, with like poles adjacent to each other. There is a force of repulsion between the magnets and if the field is plotted between 2 like poles, a neutral point is found where the effects of the 2 repulsive forces balance each other and the total effect is as shown by the absence of control on a compass needle placed at this neutral point.

The strength of the magnetic field around a magnet varies from point to point, but before this can be measured and methods devised for making such measurements, a system of magnetic units and terms must be introduced. Faraday conceived the idea of the line of flux, and further suggested the use of these lines to depict the strength of the magnetic field.

If a unit area at right angles to the lines of flux is considered, definitions and terms can be made.

A number of lines of flux collectively are said to constitute the magnetic *Flux* (symbol Φ – Greek letter phi) that is passing through the area studied.

Another unit of importance is *Flux Density* – and the value, at any point, is obtained from the expression:

$$\text{Flux density} = \frac{\text{Flux}}{\text{Area}}$$

Figure 5.5 illustrates the SI unit of flux or the *Weber*. For example, if 50 lines of flux are shown passing through an area of 1 square metre, for the plane considered, the magnetic flux will be 50 Webers. The symbol for flux density is *B* and the unit is the *Tesla*. Thus for any point P in the plane considered, the flux density is 50 teslas.

Note. The tesla is a name introduced for the SI system after Nikola Tesla (1856–1943), an ethnic Serb. His revolutionary developments in electromagnetism in the late nineteenth and early twentieth century formed the basis of wireless communication and radio. The original unit was the weber per square metre, i.e. Wb/m^2.

We now have Flux = Flux density × Area

or Φ (Webers) = *B* (teslas) × *A* (square metres).

This relationship will be used throughout our study of electromagnetism and magnetic circuits and should be considered a basic and important formula. It is useful to emphasise that flux lines do not exist but the properties of magnets and magnetic fields can be assessed by assuming their existence and their having definite physical

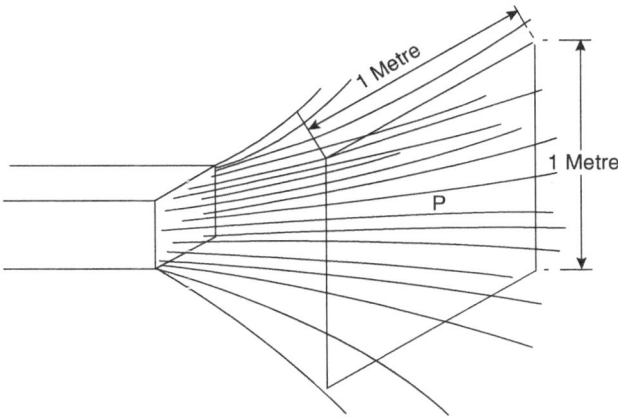

▲ **Figure 5.5** *Spreading of magnetic field lines*

properties. It must be remembered that the field of a magnet exists in all directions and is not confined to one plane.

Molecular theory of magnetism

A molecule is defined as the smallest particle of a substance that can exist separately and in any magnetic material every molecule is thought to be a complete magnet. In a piece of unmagnetised magnetic material the molecules are considered to arrange themselves in closed magnetic chains or circuits as shown (figure 5.6). Under this assumption it is considered that each molecular magnet is neutralised by adjacent molecular magnets so that no magnetism is apparent in the material. The process of magnetising a material is thought to be achieved by arranging the molecular magnets so their axes point in the direction of the magnetising force. The proof of this hypothesis is supported by the following observations. (1) There is a limit to the amount of magnetism that can be imparted to any material sample. This is explained by the supposition that, once all the molecules are 'lined up', no amount of extra magnetising force can increase the magnet's strength. (2) When a magnet is broken, the ends of the molecular magnets are exposed and the broken pieces are found to be magnets themselves. (3) If heated to about 100°C and allowed to cool, a magnet is weakened. If the magnet is heated until 'red hot', the magnetic properties are completely lost. Similarly, f a magnetic material like hard steel is cooled in a strong magnetic field then it will set as a permanent magnet. It is considered that during heating, energy is transferred to the magnet, which causes oscillations of the molecular magnets. This tends to break the 'lining up' and results in these magnets

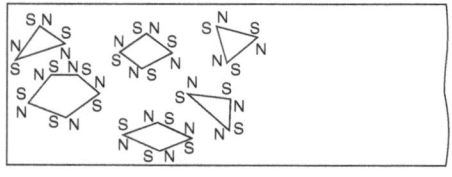

| Unmagnetised | Magnetised |

▲ **Figure 5.6** *Unmagnetised and magnetised materials*

taking up random directions. Similarly, in the cooling process, as energy is passed from the hot material, the oscillations decrease in magnitude and the molecular magnets settle in the direction of the magnetising field.

A modern theory of magnetism is based on electron theory and the concept of the atom such that an electron, the smallest known −ve charge, when rotating in an elliptical path, constitutes a circular current, which sets up a magnetic force along the axis of gyration. In a molecule the magnetic effects of the electrons of the atoms may neutralise each other, giving little resultant effect. A spinning electron also sets up a magnetic field along its spin axis. If the fields due to the effects of spin balance out, due to electrons spinning in opposite directions, the material is non-magnetic. A magnetic material is the result of the fields not balancing out, but to explain the overall apparent effect, it is thought that rather than single atoms or molecules being concerned, it is a group of molecules that act together. Such a group is called a 'domain' and is considered to function like the more molecular magnet already described.

Electromagnetism

Earlier theory has referred to an association between magnetism and electricity and this was more specifically mentioned in Chapter 2 when electrical units were defined. The discovery of a relation between an electric current and magnetism was made in 1820 by the Danish scientist Oersted (1777–1851), when he noticed that a wire arranged above and parallel to a compass needle caused deflection of the latter when a current was passed through the wire. Reversal of the current caused reversal of the deflection. Further experiments on the shape, direction and strength of magnetic fields associated with current-carrying conductors arranged in the form of loops and solenoids were the subject of much work by famous scientists such as Faraday, Maxwell and Gilbert. The result of their discoveries led to the deduction of certain fundamental relationships, which are now part of accepted basic theory. The shape of the magnetic fields due to simple arrangements of current-carrying conductors will now be considered.

Field due to long straight current-carrying conductor

The field associated with such a conductor may be determined with iron filings or a compass needle as described earlier in the magnetism section – figures 5.2 and 5.3. Assuming the current is kept constant during such a test, a field consisting of concentric lines of flux is confirmed. Figure 5.7 shows a vertical wire passing through a sheet of cardboard. The directions of the current and lines should be noted as this is a fundamental relationship.

Further tests show that if the current is reversed, the field will reverse and if the strength of the field is measured with a sensitive instrument, the results will give a graph as illustrated (figure 5.8a), which shows flux density (B) plotted to a distance (s) from the centre of the conductor.

It is seen that inside the circular conductor, the strength of field or flux density varies from zero at the centre to a maximum on the circumference. Outside the wire, flux density varies inversely as the distance from it.

Figures 5.8a and 5.8b, use the conventional method of indicating current direction. Consider an arrow, i.e. current entering the surface of the paper and receding from the viewer, with the feathered end seen as a cross. Similarly, current flow towards the viewer is shown with the tip of the arrow, i.e. a point or dot. The relation between the direction of the lines of flux and the current is summarised by Maxwell's, Right-Hand

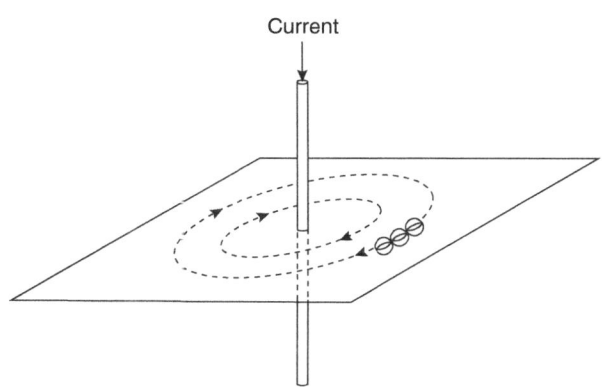

Current

▲ **Figure 5.7** *Magnetic field around long straight current-carrying conductor*

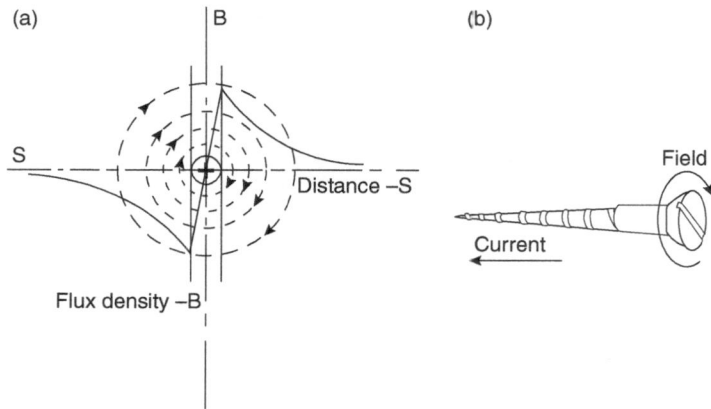

(a)   B

S   Field

Distance –S

Current

Flux density –B

▲ **Figure 5.8** *Current direction with Maxwell's, Right-Hand Screw Rule*

Screw Rule. This depicts that if current flows in the direction in which a right-handed screw moves forward when turned clockwise, then the resulting field will be in the direction of turning the screw. If the current is reversed, the screw will unscrew and the field be reversed, or the direction of turning the screw is reversed, i.e. anticlockwise.

Field due to a current-carrying conductor bent to form a single loop

Figures 5.9a and 5.9b show the loop, the current and the lines of flux encircling the conductor as deduced from condition (1) above. The resulting field can be plotted by locating the loop in a sheet of cardboard as shown. The result is considered as the field taken through section XY of the loop and the similarity with the field of a short bar magnet will be noted. The loop is considered to set up a magnetic polarity determined from first principles.

Field due to a current-carrying conductor wound as a solenoid

The next step in electromagnetic field investigations is for a coil of wire, which is a collection of several loops. A *solenoid* is a form of multi-turn coil where the axial length is much greater than its diameter. Turns of wire are wound in an open spiral or placed close together so that they touch, provided insulated wire is used. The insulation most commonly used is either a synthetic enamel or a fibrous material such as cotton or silk in the form of thread, tape or braid. The turns of a solenoid are arranged in several layers provided the current travels through the turns in the same direction. When the

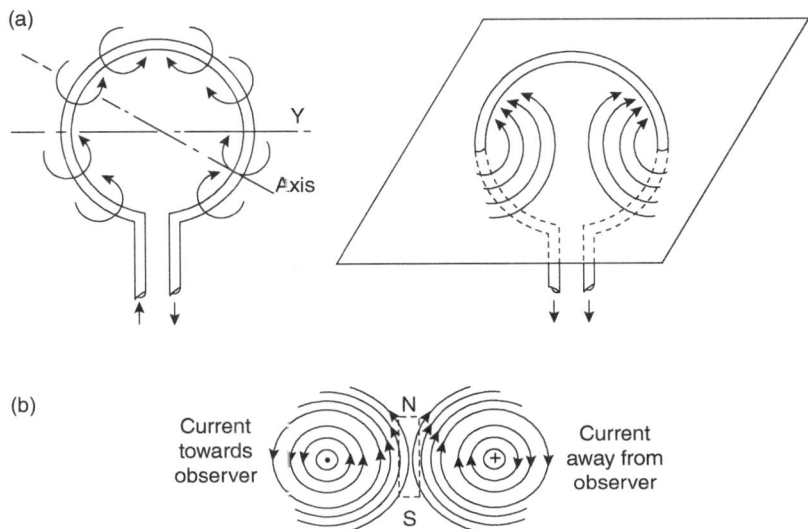

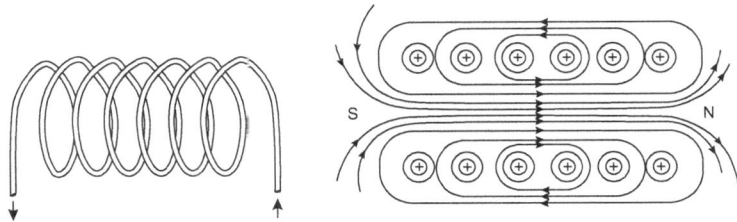

▲ **Figure 5.9** *Field surrounding a looped conductor*

▲ **Figure 5.10** *Exploring the field pattern with a compass*

field is investigated by plotting with a compass, it is found to be as in the diagram (figure 5.10). All the turns tend to produce a magnetic field in the same direction, so that this can be deduced by considering the field of a single turn or loop. The turns unite to send a straight field up the centre, which comes out at the ends, opens and spreads out to return at the other end, giving the same distribution of lines of flux as obtained from a bar magnet.

Again, a definite polarity is attributed to the solenoid when carrying current. Polarity is determined by finding the direction of the lines for any one turn by applying the right-hand screw rule but additional aids are useful, the easiest of which is the *right-hand rule*. This is explained as follows, and is shown in the diagram (figure 5.11). Place the right hand on the coil with the fingers pointing in the direction in which current flows. Then the thumb will point in the direction of the N pole.

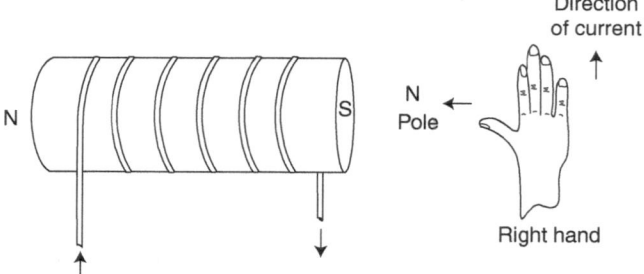

▲ **Figure 5.11** *Right-hand rule*

Introduction of an Iron Core

The iron core of a solenoid strengthens the field by concentrating flux and better defining the poles. A magnetic core allows the passage of flux more readily than air. Experiment shows that the best flux path is where the whole of the magnetic circuit is formed from magnetic material. Where this is not practical the air gaps or air paths are kept as short as possible and good examples are found in the electromagnetic paths for the flux in the electric bell and the electric motor or generator (figures 5.12a and 5.12b).

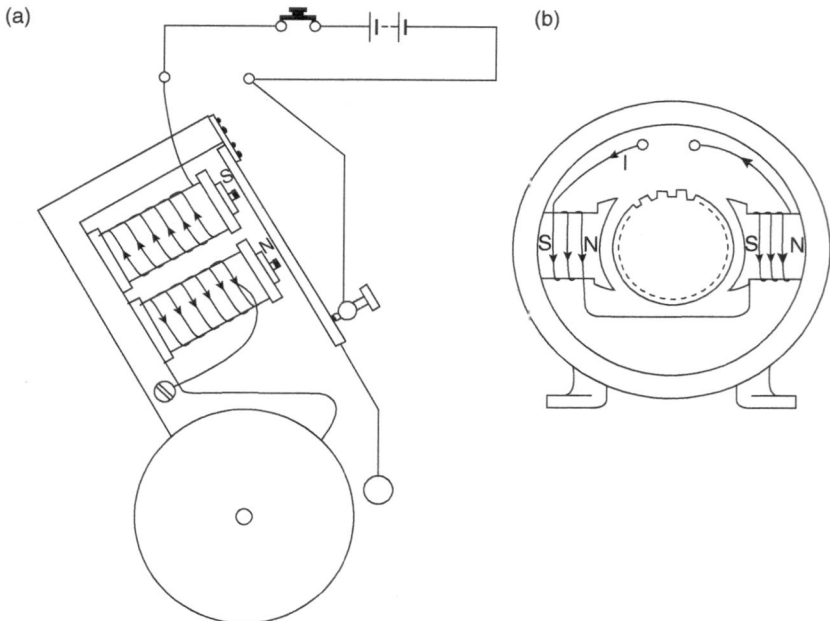

▲ **Figure 5.12** *a) Electric bell, and b) electric motor or generator*

Electromagnets are preferred to permanent magnets in industry for 2 main reasons. (1) They are made more powerful than permanent magnets by providing the desired magnetising force, i.e. solenoid coils with sufficient turns and energising current. (2) The magnetism is controlled, i.e. it can be switched on and off or varied gradually by controlling the current. In summary, permanent magnets are made of *hard steel* because this material retains its magnetism and the material is said to have a high 'retentivity'. Electromagnets have a core of *soft iron*, which is more readily magnetised but loses its magnetic properties more quickly. The material is said to have a high 'susceptibility' and soft iron is more susceptible than steel.

Force on a current-carrying conductor in a magnetic field

Oersted's experiment with a compass needle and current-carrying wire show that a force is produced when a current is switched on, bringing about a deflection of the needle. Similarly, if a needle was fixed and the wire sufficiently flexible, wire movement will be noted when a current is switched on. Further investigations led to an accepted rule – that a force acts upon a conductor when it is carrying current and situated in a magnetic field provided it is at right angles to the lines of flux. Let us now consider the electromagnetic effects that allow the ampere to be defined as a fundamental unit of the SI system.

In Chapter 2, the phenomena leading to definition of the ampere were mentioned and the points made previously are revisited here in the light of electromagnetic theory. If a circuit is supplied through 2 wires laid together side by side, then if the current is large and the wire flexible, a mechanical effect is noted, especially when the current is switched on and off, as the wires will be seen to move. This action is explained with our knowledge of the field associated with a long straight conductor.

Consider the diagram (figure 5.13), which shows 2 conductors carrying current as shown. When the current in both conductors is in the same direction, the resultant magnetic field is such as to enclose both conductors. If the current in each conductor is of the same magnitude then, by Maxwell's right-hand screw rule, the fields between

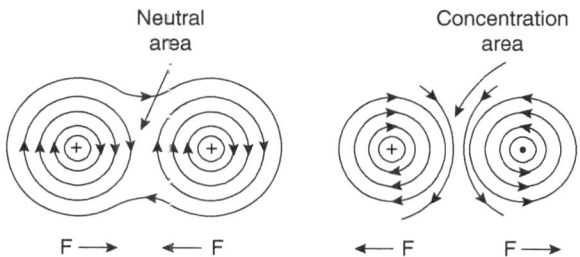

▲ **Figure 5.13** *Field surrounding two adjacent current-carrying conductors*

the wires will cancel and the outside lines of flux unite to make a field that encircles both conductors. If flux lines are likened to elastic threads then the lines of flux are stretched and forces act to move the conductors together.

Lines of force through air will keep to the shortest possible paths. Figure 5.13 shows currents in opposite directions in the 2 parallel conductors. Here the resultant flux is concentrated between the conductors, forcing them apart.

The ampere

This is defined in accordance with the electromagnetic effects described and is the accepted definition for the unit of current. Thus the ampere is that value of current that, when flowing in each of 2 infinitely long parallel conductors, situated in a vacuum and spaced 1 metre between centres, causes each conductor to have acting on it a force of 2×10^{-7} newton per metre length of conductor.

Magnitude of force (on a current-carrying conductor in a magnetic field)

Figure 5.14 shows a conductor situated in and at right angles to a magnetic field. Assume that the conductor carries current in the direction shown and that the arrangement is illustrated (a) and (b). There is a magnetic field due to the current that interacts with the main field, which distorts it so that a strong field exists on one side of the conductor and a weak field on the other. Lines of flux first appear to stretch and then return to their shortest length, while a force is exerted on the conductor, pushing it out of the way. This action forms the basis of operation of the electric motor and the

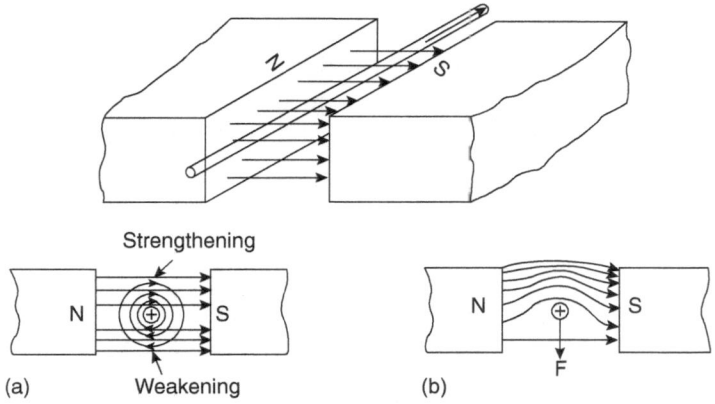

▲ **Figure 5.14** *Force acting upon a current-carrying conductor in a magnetic field*

reader should pay attention to the points made here. It can be shown that the force acting on the conductor varies directly with (1) the strength of the magnetic field, (2) the strength of the current in the conductor and (3) the length of the conductor in the magnetic field.

Summarising: Force ∝ Strength of field × Current in conductor × length of conductor in the magnetic field or $F \propto Bl\ell$ where F is the force on the conductor in newtons, B is the flux density in teslas, I is the current in amperes, ℓ is the length of the conductor in the field in metres.

The above relationship is converted to the expression $F = Bl\ell$ if the correct unit of flux density is chosen for this equality. This unit is the tesla and is defined below, giving the important formula:

$$F(\text{newtons}) = B(\text{teslas}) \times I(\text{amperes}) \times \ell(\text{metres}).$$

It is noted from now on that the expression *flux density* will be used in preference to the strength of magnetic field. This is because if lines of flux depict a magnetic field, the magnetic field strength is represented by the density of the lines. Lines well spaced apart create a weak field, while a strong field is represented by lines closely packed together. Field strength is measured by the *density* of these lines or by the flux density defined in SI units.

Unit of flux density

This unit is defined in accordance with the relationship $F = Bl\ell$ because the units for F, I and ℓ are known. Thus B is defined in terms of the other 3 factors and the unit of flux density or the tesla is the density of magnetic field such that a conductor carrying 1 ampere at right angles to the field experiences a force of 1 newton/metre length acting on it.

Unit of flux

The terms flux and flux density were introduced earlier with flux density determined by dividing the total flux by the area through which it passed. So:

$$\text{Flux density} = \frac{\text{Flux}}{\text{Area}}$$

Hence Flux = Flux density × Area. Using our definition for flux density, it follows that the weber is the unit of flux and it is the flux within an area of 1 square metre where the flux density has a value of 1 tesla. A more complex definition of the weber will follow in Chapter 6, emphasising the importance of this unit in the study of electromagnetism.

For the correct use of Φ (webers) = B (teslas) × A (square metres), see the example below.

Example 5.1. If the flux density inside a solenoid coil is measured to be 140mT and the inside diameter of a solenoid is 40mm, find the value of the total flux produced (4 significant figures).

Note. The main purpose of this example is to stress the importance of correct substitution in the formula, with attention given to the correct numerical magnitude. Thus 140mT is 140 milli teslas, or 140×10^{-3}T. Similarly 40mm must be converted to metres before substitution.

$$\text{Thus } A = \frac{\pi d^2}{4} = \frac{3.14 \times 40^2 \times 10^{-6}}{4}$$

$$= 12.56 \times 10^{-4} \, \text{m}^2$$

$$\Phi = 140 \times 10^{-3} \times 12.56 \times 10^{-4}$$

$$= 175.8 \times 10^{-6} \, \text{Wb}$$

$$\text{or } \Phi = 175.8 \mu\text{Wb}.$$

Example 5.2. Find the force exerted on a conductor 160mm long when carrying 125A and placed at right angles to the lines of flux of a magnetic field of flux density 4×10^{-3} teslas (2 decimal places).

$$\text{Substituting in } F = Bl\ell \text{ we have } F \text{ (newtons)} = 4 \times 10^{-3} \times 125 \times 160 \times 10^{-3}$$

$$= 0.08\text{N}$$

The force will be 0.08 newtons or 0.08N.

The Magnetic Circuit

Magnetising force, magnetic field strength or magnetic field intensity

As a magnetic field is produced by a coil of wire carrying a current, we must deduce a relationship that correlates the flux density at any point with the electromagnetic effort required to produce it. To allow this derivation, the electromagnetic effort is defined as the *magnetising force, magnetic field strength* or *magnetic field intensity* (symbol H) and measured in terms of the factors producing it: the current and number of turns of the coil. The magnetising force H at any point in the field is measured in

Ampere-turns (symbol *IN*) of the coil acting over 1 metre length of the flux path and is considered to cause a flux density of *B* teslas. *IN/ℓ* ampere-turns/metre length of the magnetic circuit is a measure of *H*, but in SI units it is considered that the same value of *H* is created by a current of *IN* amperes passing through 1 turn so the numerical value of ampere-turns may be replaced with a current A as an alternative to At. Thus *H* is measured in At/m or in A/m. In this book the original and common method is used, i.e. 200 ampere-turns, for example, appear as 200At rather than 200A. In line with this duality, a magnetising force value should be read as ampere-turns per metre although it may be given as amperes per metre. Summarising:

$$H \text{ (magnetising force)} = \frac{IN}{\ell} \text{ (ampere-turns/metre or amperes/metre).}$$

A magnetic circuit is taken as the complete length of the path through which the flux produced by the coil passes. We are concerned, in practical engineering or physics, with the flux path of machines and electromagnetic devices, so let us take a simple path for the simple 2-pole generator or motor shown in figure 5.12b.

We treat the field coils as the energising ampere-turns, spread over the poles of the machine and wound to produce a continuous solenoid effect. This practical arrangement gives a more symmetrical layout. Each coil has 2000 turns of thin wire and a coil current of 1.5 amperes. The *total* magnetising force producing the flux for this machine will be (2000 × 2) turns × 1.5 amperes or 6000 ampere-turns. By symmetry the flux through the poles and armature splits (figure 5.12b) and returns through both halves of the yoke of the machine.

So far it is noted that we consider a flux density to exist at a point by virtue of the magnetising force producing it. An analogy can be made with the electrical circuit and allows a clearer understanding of the magnetic circuit and associated problems.

Magnetomotive force or m.m.f.

A complete path is followed by a group of lines of magnetic flux and this path is the magnetic circuit. In an electric circuit current is due to an e.m.f. and in a magnetic circuit, flux is thought to be due to a *magnetomotive force (m.m.f.)* (symbol *F*) caused by current flowing through a coil of wire. Thus the *m.m.f.* is the total magnetising force produced by a solenoid coil and measured in ampere-turns (*IN*). From now on, the terms: magnetising force, magnetic field strength or magnetic field intensity are used for the force or m.m.f. acting over 1 metre length of a circuit and the total force for the circuit is called the m.m.f. Thus magnetising force *H* is the m.m.f./metre length.

$$\text{or } H = \frac{F}{\ell} = \frac{IN}{\ell}$$

The passage of flux through a magnetic circuit is restricted by the circuit's *reluctance*. Reluctance (symbol S) is comparable with the resistance in an electrical circuit and is proportional to the length of the magnetic circuit, and inversely proportional to the area and the absolute *permeability* (symbol μ).

Table 5.1

Electric Circuit		Magnetic Circuit	
Quantity	Unit	Quantity	Unit
e.m.f. (E)	Volt	m.m.f. (F)	Ampere-turn
Current (I)	Ampere	Flux (Φ)	Weber
Resistance (R)	Ohm	Reluctance (S)	Amp-turn/ Weber or A/Wb
Also		Also	
$I = \dfrac{E}{R}$		$\Phi = \dfrac{F}{S}$	
Other comparisons are:			
$R = \dfrac{\rho l}{A}$		$S = \dfrac{l}{\mu A}$	
Electric force (E)	Volts/metre	Magnetising force (H)	Amp-turns/metre
$= \dfrac{V}{d}$		$= \dfrac{F}{l}$	
Current density (J)	Ampere/metre^2	Flux density (B)	Tesla
$= \dfrac{I}{A}$		$= \dfrac{\Phi}{A}$	

The above concept of a magnetic circuit allows formulae to be found for the magnetising force in the fields of various current-carrying conductor arrangements, such as the long straight conductor, single loop and multi-turn coils like solenoids and toroids.

Permeability

We now can say that a magnetis ng force (*H*) produces a flux density (*B*); the magnitude of this flux density depends upon the type of material in the magnetic circuit (e.g. air, steel, soft iron, etc.). For any material the ratio of flux density to magnetising force is called the absolute permeability (μ) and is measured in henries or henrys per metre (H/m), named after Joseph Henry (1797–1878).

$$\text{Thus: } \mu = \frac{B}{H}$$

Permeability of free space (μ_0)

In vacuum and most non-magnetic materials the ratio between B and H is a constant. This can be shown by considering a long straight current-carrying conductor in a vacuum. Consider the diagram (figure 5.15) with a conductor of infinite length carrying a current of 1A.

The conductor forming the return path to the supply source is considered to be an infinite distance away so its current will not affect the magnetic field near the conductor.

The conductor arrangement constitutes a single turn and the m.m.f. F is then 1 turn × I amperes or *F* = I ampere-turns. Consider any point on a line of flux distant r metres from the centre of the conductor. The magnetising force H at this point will be the m.m.f./metre length of flux or $H = \dfrac{F}{\ell} = \dfrac{F}{2\pi r}$ So

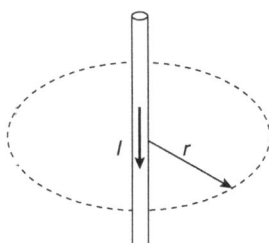

▲ **Figure 5.15** *Conductor of infinite length*

$H = \dfrac{I}{2\pi r}$ ampere-turns/metre or amperes/metre, where ℓ is the circumference for a radius r.

This result helps us to find the flux density for a certain magnetising force and permeability (μ) of the medium in which the field is established.

Consider figure 5.16, the plan view of our previous diagram.

The conductor, in vacuum, is represented by A carrying a current of 1 ampere flowing *away* from the observer. The magnetising force, at any point P 1 metre from A, is given by

$H = \dfrac{1}{2\pi}$ ampere-turns/metre as both I and r are unity in the formula derived earlier.

Next assume the flux density at point P is B tesla. Then (1) the force on a metre length of conductor placed at P, parallel to A and carrying a current of 1 ampere will be 2×10^{-7} newtons. This is known from the definition of the ampere. Also (2) the force on a metre length of this conductor is given by $BI\ell$ newtons, or is

$$B \text{ (teslas)} \times 1 \text{ (ampere)} \times 1 \text{ (metre)} = B \text{ (newtons)}.$$

Thus equating expressions (1) and (2) for the force on the conductor we see that the value of B for the condition considered will be 2×10^{-7} newtons.

Hence:

$$\mu = \frac{\text{Flux density at point P}}{\text{Magnetising force at point P}} = \frac{B}{H} = \frac{2 \times 10^{-7}}{1/2\pi} \quad \therefore \quad \mu = 4\pi \times 10^{-7}$$

In this case we consider a vacuum as the medium in which the field is established, so:

Permeability of free space $= 4\pi \times 10^{-7}$ henry/metre or $\mu_0 = 4\pi \times 10^{-7}$ H/m.

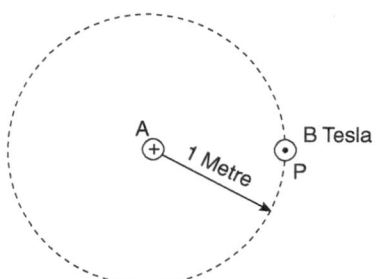

▲ **Figure 5.16** *Field around a conductor of infinite length*

Example 5.3. We must produce a flux of 0.018Wb across an air gap 2.54mm long with effective area $24 \times 10^{-3}m^2$. Find the ampere-turns required (3 significant figures).

$$\text{Area of gap} = 24 \times 10^{-3} \text{ square metres}$$

$$\text{Required flux density } B = \frac{0.018}{24 \times 10^{-3}} = 0.75T$$

$$\text{Also } H = \frac{B}{\mu_0} = \frac{0.75}{4 \times \pi \times 10^{-7}}$$

$$= 59.7 \times 10^4 At/m$$

The length of the air gap = 2.54mm = $0.254 \times 10^{-2}m$.

So total ampere-turns needed = $59.7 \times 10^4 \times 0.254 \times 10^{-2} = 1515At$.

Example 5.4. A wooden ring with a mean diameter of 200mm and a cross-sectional area of 400mm² is wound uniformly with a coil of 300 turns. If the current passed through the coil is 5A, calculate the value of flux produced in the coil (2 significant figures). The m.m.f. of the coil = $5 \times 300 = 1500At$.

The mean circumference = $\pi D = \pi \times 200 = 628mm = 0.628m$

The magnetising force H = At/m = $\dfrac{1500}{0.628} = 2380$ At/m

The flux density $B = \mu_0 H$

$$= 4 \times \pi \times 10^{-7} \times 2380$$

$$= 0.003T$$

Total flux $\Phi = BA$

$$= 0.003 \times 400 \times 10^{-6} \text{ Wb}$$

$$= 1.2\mu Wb.$$

Example 5.5. The magnet system of a moving-coil instrument provides a flux density in the air gap of 0.25T. The moving coil, of 120 turns, is carried on a former of (active side) length 25mm and width 18mm (between air-gap centres). If the coil carries a current of 2mA, calculate the turning moment on it (2 significant figures).

$$F = BI\ell \text{ newtons}$$

$$= 0.25 \times 2 \times 10^{-3} \times 120 \times 2 \times 25 \times 10^{-3}$$

$$= 3 \times 10^{-3}\,\text{N}$$

Torque $= F \times$ radius of coil

$$= 3 \times 10^{-3} \times 9 \times 10^{-3}$$

$$= 27\mu\text{N m.}$$

Practice Examples

5.1. A conductor carrying a current of 100A is situated in and lying at right angles to a magnetic field having a flux density of 0.25T. Calculate the force in newtons/metre length exerted on the conductor (2 significant figures).

5.2. A coil of 250 turns is wound uniformly over a wooden ring of mean circumference 500mm and uniform cross-sectional area of 400mm². If the current passed through the coil is 4A, find (a) the magnetising force (1 significant figure) and (b) the total flux (3 decimal places).

5.3. A current of 1A is passed through a solenoid coil, wound with 3200 turns of wire. If the dimensions of the core air gap are length 800mm, diameter 20mm, find the value of the flux produced inside the coil (exactly in Webers).

5.4. Two long parallel bus bars each carry 2000A and are spaced 0.8m apart between centres. Calculate the force per metre acting on the conductors (1 significant figure).

5.5. A moving-coil permanent-magnet instrument has a resistance of 10Ω and the flux density in the gap is 0.1T. The coil has 100 turns of wire, is of mean width 30mm and the axial length of the magnetic field is 25mm. If a P.D. of 50mV is required for f.s.d., calculate the controlling torque exerted by the spring (3 significant figures).

5.6. An air gap of length 3mm is cut in the iron magnetic circuit of a measuring device. If a flux of 0.05Wb is required in the air gap, which has an area of 650mm², find the ampere-turns required for the air gap to produce the necessary flux (4 significant figures).

5.7. A straight horizontal wire carries a steady current of 150A and is situated in a uniform magnetic field of 0.6T acting vertically downwards. Determine the magnitude of the force per metre length and the direction in which it acts (1 significant figure).

5.8. An armature conductor has an effective length of 400mm and carries a current of 25A. Assuming that the average flux density in the air gap under the poles is 0.5T, calculate the force in newtons exerted on the conductor (1 significant figure).

5.9. In an electric motor the armature has 800 conductors, each carrying a current of 8A. The average flux density of the magnetic field is 0.6T. The armature core has an effective length of 250mm and all conductors may be taken as lying on an effective diameter of 200mm. Determine the torque and mechanical power developed when the armature is revolving at 1000 rev/min (4 significant figures).

5.10. Two long straight parallel bus bars have their centres 25mm apart. If each carries current of 250A, calculate the mutual force/metre run (1 decimal place).

6

ELECTROMAGNETIC CIRCUITS

Something is as little explained ... as the attraction between iron and magnet is explained by means of the name magnetism.

Jacob Schlelden

Magnetising Force

In Chapter 5 the fundamental concepts, terminology and relationships of an electromagnetic circuit were introduced and developed. Before proceeding to consider further the effects of ferromagnetic materials, it is useful to revise some of these basic relationships.

The m.m.f. F is the force that causes magnetic flux Φ to be created in a magnetic circuit with reluctance S:

$$\text{i.e. } \Phi = \frac{F}{S} \text{ Wb.}$$

The m.m.f. is usually created by passing a current through a number of coil turns:

$$\text{i.e. } F = IN \text{ ampere-turns.}$$

Reluctance S depends upon the dimensions of the magnetic circuit and its permeability:

$$\text{i.e. } S = \frac{1}{\mu A} \text{ At/Wb.}$$

Flux density B is a measure of the magnetic flux Φ in a given area A:

$$B = \frac{\Phi}{A} \text{ Tesla.}$$

The magnetising force (magnetic field strength) is a measure of the m.m.f. per metre length of magnetic circuit required to maintain flux in that circuit:

$$H = \frac{IN}{\ell} \text{ At/m.}$$

Permeability is the ratio of flux density to the magnetising force producing it:

$$\mu = \frac{B}{H} \text{ H/m.}$$

For air, vacuum and most magnetic materials we use the permeability of free space μ_0, with a constant value:

$$\mu_0 = 4\pi \times 10^{-7} \text{ H/m,}$$

which for air $\mu_0 = \frac{B}{H}$ such that $B \propto H$.

If values of B against H are plotted for air, a straight-line graph is obtained (figure 6.1). If measurements of flux density B are made at a point outside, but near to, a long straight current-carrying conductor, for various values of magnetising force H, by changing the current (noting that $H = \frac{1}{2\pi r}$ where r is the radius from the point to the centre of the conductor), then the straight-line B/H relationship will be confirmed.

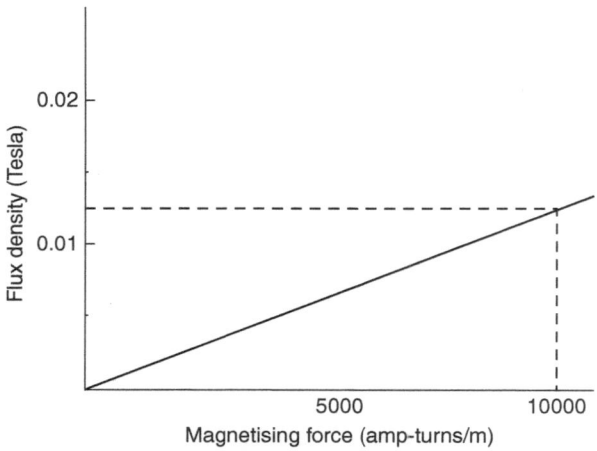

▲ **Figure 6.1** *Flux density vs magnetising force*

Magnetising force due to a long, straight current-carrying conductor

The magnetising force outside, and near to, a current-carrying conductor is given by

$H = \dfrac{1}{2\pi r}$. This expression was deduced in Chapter 5 and it should be remembered

that H is the m.m.f./metre length. M.m.f., F is measured in ampere-turns and the total m.m.f. for any magnetic circuit outside the conductor is found from $F = H\ell$.

Magnetising force inside a solenoid

If a parallel field of flux lines is assumed inside a solenoid as illustrated (figure 6.2), its length can be taken as ℓ metres, the number of turns on the coil as N and the current passed as ℓ amperes. Lines of flux will spread out at the ends and for their return path they also spread out into space. This external return path has negligible magnetic reluctance and the whole m.m.f. of the coil 'sets up' the field inside the solenoid. Thus the m.m.f. per unit length is, by definition, H – the magnetising force.

$$\text{Thus } H = \frac{F}{\ell} = \frac{IN}{\ell} \text{ ampere-turns per metre.}$$

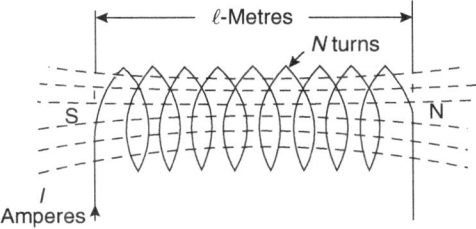

▲ **Figure 6.2** *Magnetising field inside a solenoid*

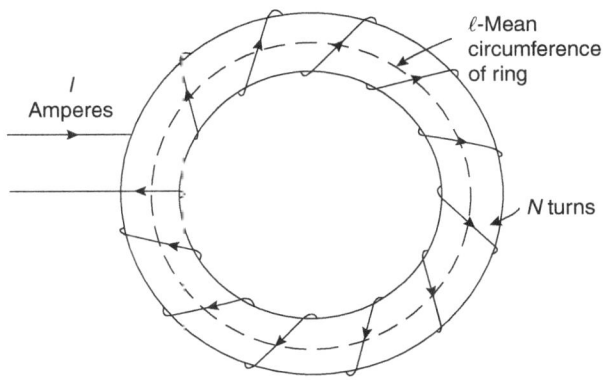

▲ **Figure 6.3** *Toroidal magnetic ring coil*

Magnetising force inside a toroid

Figure 6.3 shows a simple electromagnetic arrangement. It consists of a solenoid bent back upon itself so that the lines of flux are confined inside the coil. We consider a non-magnetic ring (or toroid) wound uniformly with a coil of N turns, carrying a current of ℓ amperes. The mean circumference is ℓ metres and as the flux is confined *inside* and the path is uniform, the magnetising force or m.m.f. per unit length is given by:

$$H = \frac{IN}{\ell} \text{ ampere-turns per metre.}$$

Example 6.1. A wooden ring with a mean circumference of 300mm and a uniform cross-sectional area of 400mm² is wound uniformly with 300 turns of insulated wire. If the current is 3A, calculate (a) the magnetising force (1 significant figure), (b) the flux density inside the toroid (4 significant figures) and (c) the total flux produced (2 significant figures).

(a) The total m.m.f. produced $F = 3 \times 300 = 900$At

The mean circumference is 300mm = 0.3 metres

$$\therefore \text{ The magnetising force } H = \frac{F}{\ell} = \frac{900}{0.3}$$

$$= 3000\text{At/m}$$

(b) The flux density is given by $B = \mu_0 H$

$$= 4\pi \times 10^{-7} \times 3000 = 3.768 \times 10^{-3}\text{T}$$

$$= 3.768\text{mT}$$

(c) The total flux produced $\Phi = B \times A$

$$= 3.768 \times 10^{-3} \times 400 \times 10^{-6}\text{ Wb}$$

$$= 1.5 \times 10^{-6}\text{ Wb}$$

$$\text{or } \Phi = 1.5\mu\text{Wb}$$

Ferromagnetism

When iron is used as the core of an electromagnet, the field is *intensified* so that a greater flux than expected results from the magnetising ampere-turns of the energising coil.

As the only change in the relation $\Phi = \dfrac{F}{S}$ is due to the reluctance S, if the dimensions of the core l and A are kept the same as for the air path, it follows that the permeability of iron must be much greater than that of air. Thus we can refer to the permeability of a magnetic material, termed the *relative permeability*.

Relative permeability (μ_r)

This is the ratio of the flux density produced in a magnetic material to the flux density produced in air by the same m.m.f.

$$\therefore \text{ Relative permeability } = \frac{\text{Absolute permeability}}{\text{Permeability of free space}}$$

$$\mu_r = \frac{\mu}{\mu_0}$$

$$\mu = \mu_0 \mu_r = \frac{B}{H}$$

$$\therefore\ B = \mu_0 \mu_r H.$$

For materials such as iron, nickel, cobalt, etc., μ_r can be between 1000 and 2000 or even higher for some special electrical steels.

The *B–H* or Magnetisation Curve

If a specimen of magnetic material is made into a ring and wound with an energising coil, measurements of flux density for various values of magnetising force can be made by winding on a secondary coil and using the *principle of transformer action*. This is an accepted industrial method for determining the magnetic properties of various materials. It is observed that if flux density B is plotted against magnetising force H for air, a straight line is obtained, but for magnetic materials, typical curves result (figure 6.4).

At first all the graphs are approximately straight lines, with B proportional to H. Then the curves begin to flatten out forming a 'knee' and finally become horizontal, exhibiting little increase in B for a large increase in H. In this state, the material is said to 'saturate'.

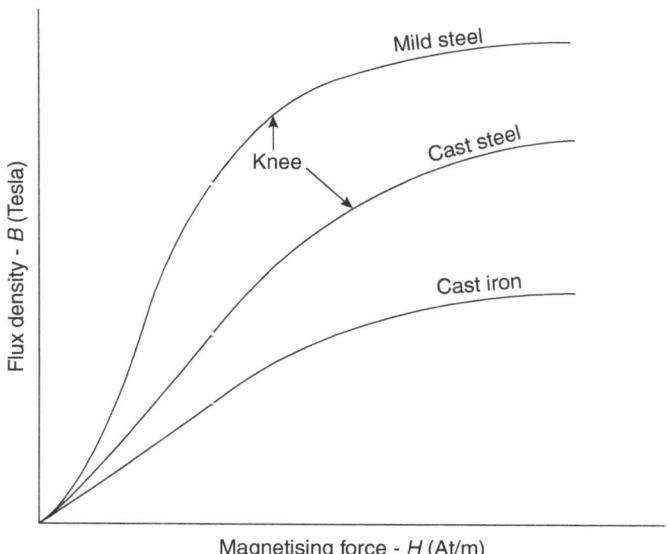

▲ **Figure 6.4** *Flux density vs magnetising force*

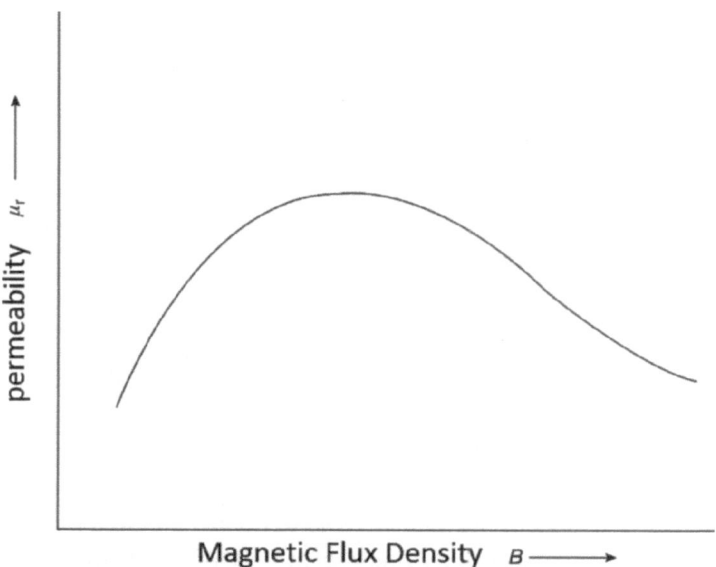

▲ Figure 6.5 *Permeability vs flux density*

If permeability (μ_r) is plotted against B, a curve (figure 6.5) will result. The permeability curve has a peak corresponding close to the 'knee' on the B–H curve where the tangent goes through the origin. Beyond this peak, permeability drops off fairly rapidly.

An examination of the B–H and μ_r–B curves shows how the properties of various magnetic materials differ. For machine design, lower working B values necessitate larger section and therefore greater mass to obtain a required flux value as $\Phi = B \times A$. The effect of high permeability materials is also apparent and the shape of the B–H curve with the saturation effect shows the limits of machine field systems. Magnetic properties depend on the actual composition, for example, manganese-steel is practically non-magnetic, but small quantities of carbon or silicon when added to steel vary the shape of the B–H curve. Sheets of commercial steel marketed under trade names like Stalloy or Lohys (a transformer iron) are available to suit different design requirements.

Reluctance (symbol S)

This term is likened to the resistance of an electrical circuit. As flux is proportional to the m.m.f. and is restricted by the reluctance, further investigation shows that reluctance is proportional to the length ℓ of the magnetic circuit and inversely proportional to its area A. Furthermore it is inversely proportional to permeability, as the greater a material's permeability, the greater its flux and hence the smaller its reluctance. We

thus write, $S = \dfrac{\ell}{\mu A}$ and with absolute permeability $(\mu = \mu_0 \mu_r)$ $S = \dfrac{\ell}{\mu_0 \mu_r A}$.

Calculations on magnetic circuits with magnetic materials are now possible, but unlike electrical circuit calculations, which use $I = \dfrac{V}{R}$, it is not always necessary to use the comparable relationship of $\Phi = \dfrac{F}{S}$. The solution of most problems associated with a magnetic circuit can be made without determining the reluctance, and experience will show the best solution method. The following typical examples indicate the alternative way of treating simple problems.

Example 6.2. A solenoid is made up from a coil of 2000 turns, carries a current of 0.25A and is 1m long. An iron rod of diameter 20mm forms the core for the solenoid and is also 1m long. Calculate the total flux produced if the iron has a relative permeability of 1000 (4 significant figures).

$$\text{Coil m.m.f. is given by } F = H\ell = \frac{IN}{\ell} \times \ell = IN.$$

$$= 0.25 \times 2000 = 500 \text{At}$$

$$\text{Area of iron} = \frac{\pi d^2}{4} = \frac{3.14 \times 400 \times 10^{-6}}{4}$$

$$= 3.14 \times 10^{-4} \text{ m}^2$$

$$\text{Reluctance of iron, } S = \frac{\ell}{\mu A} = \frac{\ell}{\mu_0 \mu_r A}$$

$$\text{or } S = \frac{1}{4 \times \pi \times 10^{-7} \times 1000 \times 3.14 \times 10^{-4}}$$

$$= 2.533 \times 10^6 \text{ At/Wb}$$

$$\text{Flux } \Phi = \frac{F}{S} = \frac{500}{2.533 \times 10^6}$$

$$= 197.3 \times 10^{-6} = 197.3 \mu\text{Wb}.$$

Alternative solution

m.m.f. of coil, $F = H\ell$ so magnetising force $H = \dfrac{F}{\ell}$

Thus $H = \dfrac{IN}{\ell} = \dfrac{0.25 \times 2000}{1} = 500\text{At/m}$

Also $B = \mu H = \mu_o \mu_r H = 4 \times \pi \times 10^{-7} \times 1000 \times 500$

$$= 0.628\text{T}$$

Total $\Phi = BA$

$$= 0.628 \times 3.14 \times 10^{-4} \text{ webers}$$

$$= 197.3 \times 10^{-6} = 197.3\mu\text{Wb.}$$

Example 6.3. A cast-steel ring has a cross-section of 400mm² and a mean diameter of 240mm. It is wound with a coil having 200 turns. What current is required to produce a flux of 400μWb, if the relative permeability of the steel is 1000?

Area of steel = 400 × 10⁻⁶ m²

$$\therefore B = \dfrac{\Phi}{A} = \dfrac{400 \times 10^{-6}}{400 \times 10^{-6}} = 1\text{ telsa}$$

Also $B = \mu_o \mu_r H = 4 \times \pi \times 10^{-7} \times 1000 \times H$

$$B = 4 \times \pi \times 10^{-4} \times H$$

So $H = B / (\mu_o \mu_r) = H = \dfrac{B}{4 \times \pi \times 10^{-4}} = \dfrac{1 \times 10^4}{4 \times \pi}\text{At/m}$

m.m.f. of ring $F = H\ell = \dfrac{10^4}{4 \times \pi} \times \pi \times 240 \times 10^{-3}$

or $F = 600\text{At}$

$$F = IN \quad \therefore I = \dfrac{F}{N} = \dfrac{600}{200} = 3\text{A.}$$

Alternative solution

$$\text{Reluctance of ring } S = \frac{\ell}{\mu_0 \mu_r A}$$

$$\text{or } S = \frac{\pi \times 240 \times 10^{-3}}{4 \times \pi \times 10^{-7} \times 1000 \times 400 \times 10^{-6}}$$

$$= 1.5 \times 10^6 \text{ At/Wb}$$

$$\text{then required m.m.f.} = \Phi S = 400 \times 10^{-6} \times 1.5 \times 10^6$$

$$= 600\text{At}$$

$$\text{Required current } I = \frac{600}{200} = 3\text{A}.$$

The Composite Magnetic Ring

The series arrangement

Consider a magnetic circuit built up as shown (figure 6.6). It is obvious that the toroid sections are in series with each other and that the same flux passes through them all.

The total m.m.f. = m.m.f. across section 1 + m.m.f. across section 2.

If the total flux is Φ then $\Phi S = \Phi S_1 + \Phi S_2$ where S is the reluctance of the composite circuit. Thus:

$$S = S_1 + S_2.$$

Summarising: Total reluctance = the sum of the individual reluctances of the sections for a series arrangement.

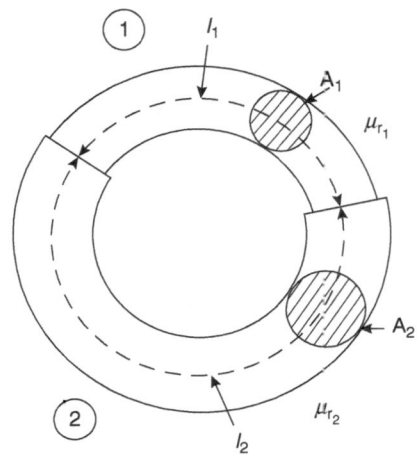

▲ **Figure 6.6** *Composite magnetic ring*

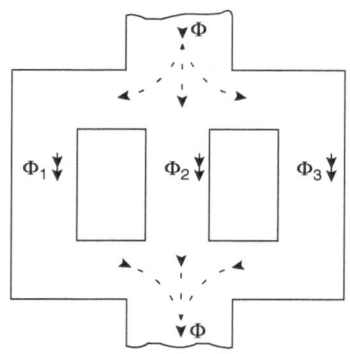

▲ **Figure 6.7** *Series flux arrangement*

The parallel arrangement

Such a magnetic circuit is not often encountered but is considered here, as it is complementary to the series circuit. The arrangement is shown (figure 6.7).

If the different paths of the magnetic circuit are in parallel, then the m.m.f. is that which will produce the required flux in each part of the circuit separately. Let F = the m.m.f. required to produce fluxes Φ_1, Φ_2, Φ_3, etc. F also produces the total flux Φ.

$$So\ \Phi = \frac{F}{\text{Total reluctance of circuit}} = \frac{F}{S}$$

$$\text{But } \Phi_1 = \frac{F_1}{S_1} \qquad \Phi_2 = \frac{F_2}{S_2}$$

$$\text{and since } \Phi = \Phi_1 + \Phi_2 + \Phi_3$$

$$\therefore \Phi = \frac{F_1}{S_1} + \frac{F_2}{S_2} + \frac{F_3}{S_3}$$

But F_1, F_2, F_3 are the m.m.f.s across the same points of the magnetic circuit and are equal to F.

$$\therefore \frac{F}{S} = F\left(\frac{1}{S_1} + \frac{1}{S_2} + \frac{1}{S_3}\right) \quad or \quad \frac{1}{S} = \frac{1}{S_1} + \frac{1}{S_2} + \frac{1}{S_3}$$

$\dfrac{1}{S}$ is referred to as the *permeance* of a magnetic circuit and the above is summarised by stating that the reluctance of a divided magnetic circuit (sections in parallel) is found by knowing that its permeance is equal to the sum of the permeances of the individual circuits.

Example 6.4. An iron ring has a mean diameter of 200mm and a cross-section of 300mm². An air gap of 0.4mm is made by a radial saw-cut across the ring. Assuming a relative permeability of 3000 for the iron, find the current needed to produce a flux of 250μWb, if the energising coil is wound with 600 turns (3 decimal places).

$$\text{Reluctance of iron } S_1 = \frac{(\pi \times 200 \times 10^{-3}) - (0.4 \times 10^{-3})}{4 \times \pi \times 10^{-7} \times 3000 \times 3 \times 10^{-4}}$$

$$= \frac{10^{-3}(628 - 0.4)}{4 \times \pi \times 9 \times 10^{-8}}$$

$$= \frac{62.76}{36\pi \times 10^{-6}}$$

$$= 555.2 \times 10^{3} \text{ ampere-turns/weber}$$

$$\text{Reluctance of air gap } S_A = \frac{0.4 \times 10^{-3}}{4 \times \pi \times 10^{-7} \times 3 \times 10^{-4}}$$

$$= 1061.5 \times 10^3 \text{ ampere-turns/weber}$$

$$\text{Total reluctance } S = S_1 + S_A = (555.2 + 1061.5) \times 10^3$$

$$= 1616.7 \times 10^3 \text{ At/Wb}$$

$$\text{Total m.m.f. } F = \Phi S = 2.5 \times 10^{-4} \times 1616.7 \times 10^3$$

$$= 404.15 \text{At}$$

$$\text{Current} = \frac{404.19}{600} = 0.674 \text{A}.$$

Alternative solution

$$\text{Since } \Phi = 250 \times 10^{-6} \text{ weber} \quad \text{then } B = \frac{2.5 \times 10^{-4}}{3 \times 10^{-4}}$$

$$= 0.833 \text{T}$$

Now H for air is given by:

$$H_A = \frac{0.833}{\mu_0} = \frac{0.833}{4 \times \pi \times 10^{-7}}$$

$$= 663.2 \times 10^3 \text{ At/m}$$

$$\text{Length of air gap} = 0.4 \times 10^{-3} \text{ metre}$$

$$\text{Ampere-turns for air} = 663.2 \times 10^3 \times 0.4 \times 10^{-3}$$

$$= 265.28 \text{At}$$

H for iron is given by:

$$H_1 = \frac{0.833}{\mu_0\mu_r} = \frac{0.833}{4 \times \pi \times 10^{-7} \times 3 \times 10^3} \text{ At/m}$$

$$= 221.066 \text{At/m}$$

Now length of iron path $= (628 - 0.4)10^{-3}$

$$= 627.6 \times 10^{-3} \text{metre}$$

Ampere-turns for iron $= 221.066 \times 0.6276 = 138.74\text{At}$

Total ampere-turns $= 265.28 + 138.74 = 404.02\text{At}$

$$\text{Current} = \frac{404.02}{600} = 0.673\text{A}.$$

For the previous examples, **alternative solutions** were given in which the reluctances for the various sections of the magnetic circuit considered were not found. This **alternative solution** method is used when the relevant *B* and *H* data for a magnetic material is given in tabular or graphical form. The relative permeability is not given as a specific value and must be found before the reluctance is calculated. Obviously such solutions are tedious and the following example is recommended to the reader on how to solve this type of problem!

Example 6.5. An iron ring of square cross-section has an external diameter of 140mm, and an internal diameter of 100mm. A radial saw-cut through the cross-section of the ring forms an air gap of 1mm. If the ring is uniformly wound with 500 turns of wire, calculate the current required to produce a flux of 0.35mWb in the gap (1 decimal place). Magnetic data of the material of the ring is given (figure 6.8). Take μ_0 as $4\pi \times 10^{-7}$ H/m.

Solution uses the graph (figure 6.9) obtained from the above data.

Table 6.1

Flux density (T)	0.65	0.89	1.06	1.18
Magnetising force (At/m)	200	300	400	500

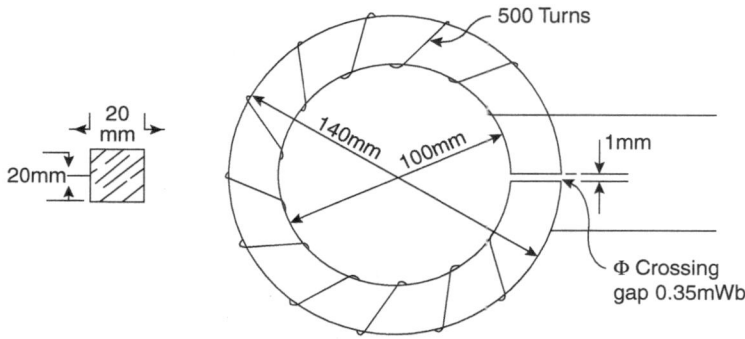

▲ **Figure 6.8** *Dimensions of the ring*

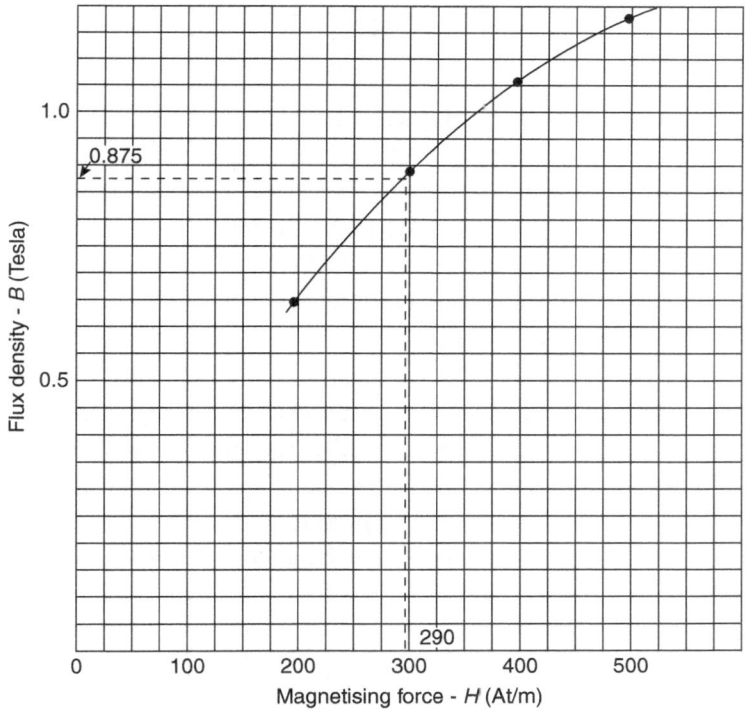

▲ **Figure 6.9** *Flux density vs magnetising force*

Area of iron and air gap $= 20 \times 20 \times 10^{-6} = 4 \times 10^{-4}$ m²

Length of iron $= \pi \times (\text{mean diameter}) - \text{air gap}$

$$= (\pi \times 120 \times 10^{-3}) - (1 \times 10^{-3}) \text{ metre}$$

$$= 375.8 \times 10^{-3} \text{ metre}$$

Length of air gap $= 1 \times 10^{-3}$ metre

Flux density for iron and air $\dfrac{\Phi}{A} = \dfrac{0.35 \times 10^{-3}}{4 \times 10^{-4}}$ $\therefore B = 0.875T$

From graph for the iron $H = 290At/m$ when $B = 0.875T$

But $H = \dfrac{IN}{\ell} = \dfrac{F}{\ell}$ $\therefore F = H\ell$

So for iron, m.m.f. $F_{IRON} = 290 \times 375.8 \times 10^{-3} = 108.88At$

For air, since $H = \dfrac{B}{\mu_0} = \dfrac{0.875}{4 \times \pi \times 10^{-7}}$ At/m

and for air, m.m.f. $F_{AIR} = \dfrac{0.875}{4\pi \times 10^{-7}} \times 1 \times 10^{-3} = 696.7At$

Total m.m.f. $F_{IRON} + F_{AIR} = 108.9 + 696.7 = 805.6At$

Current is deducted from $F/N = \dfrac{805.6}{500} = 1.6112A$

$\therefore$ Energising current $= 1.6A$ (1 cecimal place).

MAGNETIC FRINGING. Figure 6.10 illustrates how magnetic flux bridges an air gap, especially if the gap is comparatively large.

Due to spreading, flux in air occupies a larger area than that of the iron, and the flux density is thus reduced. An allowance can be made for this effect in problems when required, but unless told otherwise the area of the air gap is taken as the area of the iron.

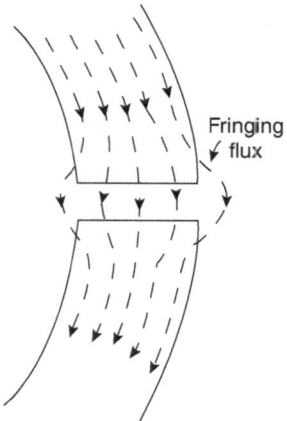

Fringing flux

▲ **Figure 6.10** *Fringing flux*

MAGNETIC LEAKAGE. For some magnetic circuits, due to the shape of the iron core and placing of the energising coil, a small amount of flux leakage may occur (figure 6.11). Some flux lines are not confined to the iron and complete their paths through air. For practical purposes, a factor called the *leakage coefficient* is added, which increases the required working flux value sufficiently to allow for this leakage.

Thus: the required total flux = the useful or working flux × leakage coefficient. The leakage coefficient is typically between 1.1 and 1.3. This leakage flux will result in reduced efficiency and increased energy losses.

Example 6.6. (a) A magnetic circuit has an iron path of length 500mm and an air gap of length 0.5mm, having uniform square cross-section, 1000mm² in area. Calculate the number of ampere-turns needed to produce a total flux of 1mWb in the air gap. Ignore fringing, and assume a leakage coefficient of 1.3. The B–H curve for the iron is given by the following (table 6.2):

(a) A conductor is passed through the air gap at a speed of 100m/s. If the length of the conductor is greater than the length of the side of the gap, calculate the e.m.f. induced.

The solution uses the graph, obtained from the above data, as shown in figure 6.12.

Area of iron and air gap $= 1000 \times 10^{-6} = 10^{-3}$ m²

Length of iron $= 500 \times 10^{-3} = 0.5$m

Flux density (B) for air $= \dfrac{1 \times 10^{-3}}{10^{-3}} = 1\text{T}$

Flux in po e = Useful flux × Leakage coefficient

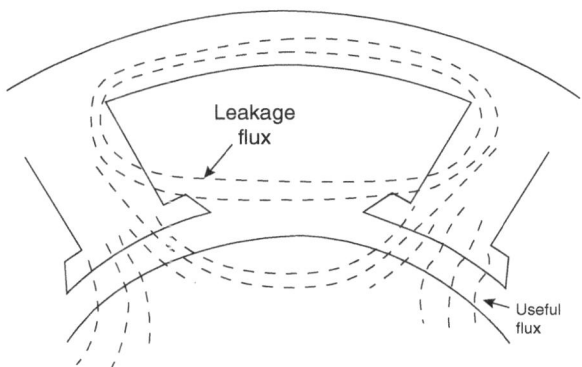

▲ **Figure 6.11** *Leakage flux*

Table 6.2

H (At/m)	100	200	300	400	600	800	1000	1200
B (T)	0.42	0.8	0.98	1.08	1.22	1.3	1.36	1.4

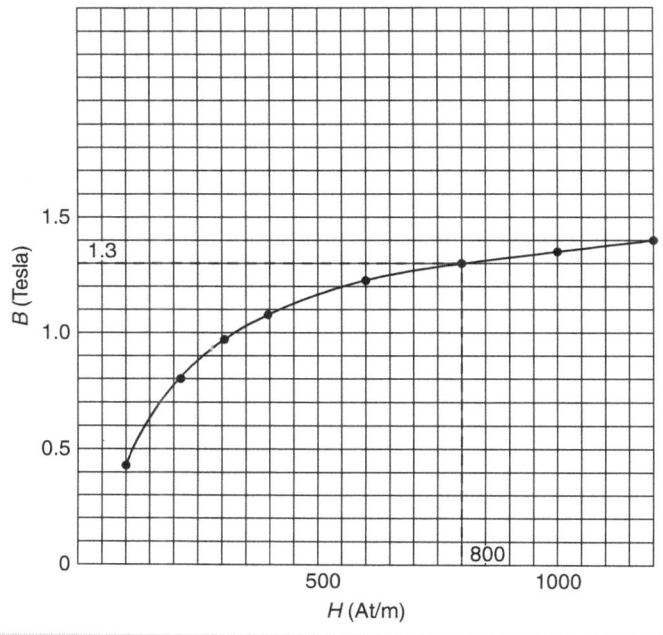

▲ **Figure 6.12** *Typical B-H curve*

Flux density (B) in iron $= \dfrac{1 \times 10^{-3} \times 1.3}{10^{-3}} = 1.3T$

For air, since $H = \dfrac{B}{\mu_0} = \dfrac{1}{4 \times \pi \times 10^{-7}}$ At / m

and m.m.f. $F_a = \dfrac{1}{4\pi \times 10^{-7}} \times 0.5 \times 10^{-3} = 398.8At$

From graph for iron, $H = 800At/m$ for a B value of 1.3T

and m.m.f., $F_1 = 800 \times 0.5 = 400At$

Total ampere-turns required for iron $= 398.8 + 400 = 798.8At$

(b) Induced e.m.f. given by $E = B\ell v$ volts

Since the area of the air gap is that of a square, the side of the

square is $\sqrt{1000 \times 10^{-6}} = \sqrt{10 \times 10^{-4}}$

$$= 3.162 \times 10^{-2} \text{ m}^2$$

Thus $E = 1 \times 3.162 \times 10^{-2} \times 10^2$

$$= 3.162 \text{ volts}$$

Note. In the above $B = 1$ tesla. $\ell = 3.162 \times 10^{-2}$ m and $v = 100m/s$.

Iron Losses

Electrical machine and transformer efficiencies are reduced by any losses in them. Apart from *Mechanical Losses* such as *Friction* and *Windage* and the *Copper Losses* (due to the resistance of the conductors), an additional loss occurs when a magnetic material is taken through cyclic magnetisation. This loss is the *Iron Loss* and is made up of 2 component losses: (1) the *Hysteresis Loss* and (2) the *Eddy-Current Loss*.

The hysteresis loop

If the magnetising force applied to an iron sample is increased from zero to its maximum value, in exactly the same way as when making the test for a *B–H* curve, and is then reduced to zero again, it is found that the new *B–H* curve, for decreasing values of *H*, lies *above* the original ascending curve, and that when *H* is zero *B* is left at some new value. The effect of the descending curve being above the ascending one is called 'hysteresis', as the *B* values lag behind those for the corresponding *H* strength when increasing. The word hysteresis comes from the Greek meaning 'to lag'. Figure 6.13 shows these effects.

The value of *B* when *H* is zero is the *remanence* and is a measure of the residual magnetism (distance OD). In order to demagnetise the iron and remove the residual magnetism it is necessary to apply a *negative* magnetising force known as the *coercive force*. If *H* is increased in the negative direction to its previous maximum value, the curve will reach a value equal to the previous maximum *B* and if *H* is next gradually reduced up to zero, then increased to its original maximum, a closed loop is traced. This is a hysteresis loop and is a measure of part of the iron loss.

To take the iron through the various stages represented by the loop, an alternating magnetising force is applied. One method of achieving this is to connect the energising

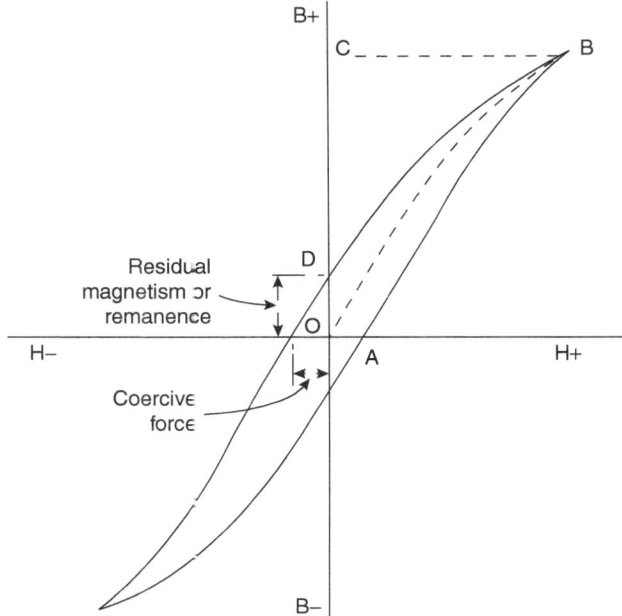

▲ **Figure 6.13** *Typical B-H remanence curve*

coil to an A.C. supply, when the iron continues to go through the same series of changes or magnetic cycles. To confirm energy is being expended, it is found that the iron core registers a temperature rise. The area of the loop is a measure of the power loss due to hysteresis. The energy absorbed per cubic metre per cycle, due to hysteresis, is given in joules by the area of the loop, provided the scales used for the graph are in appropriate SI units. The energy stored in the magnetic field in figure 6.14 can be represented by the area OABCDO in figure 6.13. When the field collapses, energy is returned to the supply, which is represented by the area DBCB. The area of the loop OABDO represents the energy lost as heat through hysteresis and is the difference between the energy put into the magnetic circuit when setting up the field and that recovered when the field decays.

If the iron sample was non-magnetic, i.e. air, then the *B–H* curve will be a straight line, as shown (figure 6.14), and the energy stored in the field when it is set up, represented by the area of the triangle OBC, will be recovered when the field collapses.

For air, the area of the right-angle-triangle $\text{OBC} = \dfrac{1}{2}$ height $\times$ base $= \dfrac{1}{2}$ OC $\times$ CB $= \dfrac{1}{2} B_m \times H$

where B_m is the maximum flux density value attained for the H value, which was impressed.

As $B = \mu_0 H$. Therefore the area of the triangle $= \dfrac{1}{2} \times B_m \times \dfrac{B_m}{\mu_0} = \dfrac{B_m^2}{2\mu_0}$ and as the area of the triangle represents the energy stored in air per cubic metre (in joules), it follows that:

For air: Energy stored per cubic metre $= \dfrac{B_\mu^2}{2\mu_0}$ joules.

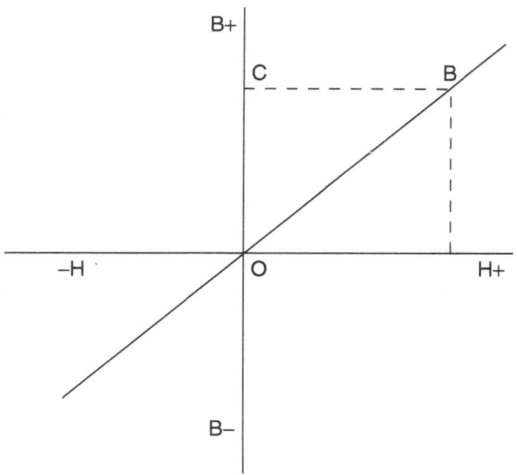

▲ **Figure 6.14** *B-H curve for energy considerations*

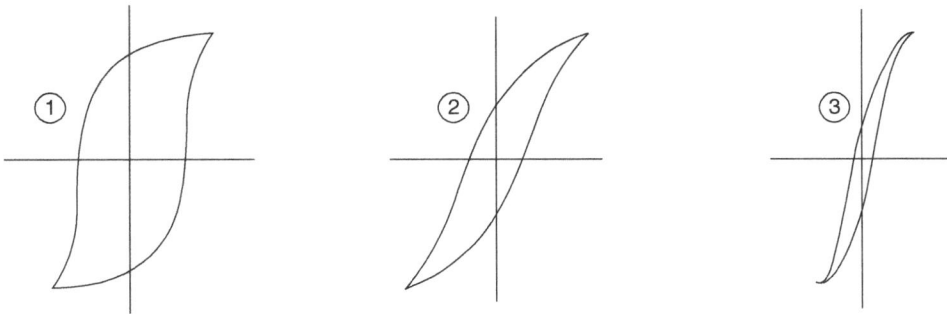

▲ **Figure 6.15** *B-H hysteresis curves*

The types of hysteresis loop, as obtained from various magnetic materials, can be grouped into 3 classes as shown (figure 6.15).

Loop 1 is for hard steel. The large value of the coercive force indicates that the material is suitable for permanent magnets. The area, however, is large, showing that hard steel is not suitable for rapid reversals of magnetism.

Loop 2 rises sharply, showing a high μ and a good retentivity (large intercept on B axis). The loop is typical of cast steel and wrought iron, which are suitable materials for electromagnet cores and electrical machine yokes.

Loop 3 has a small area and a relatively high μ. The material (mainly alloyed sheet-steels) is suitable for rapid reversals of magnetism and is used for armatures, transformer-cores, etc.

1. HYSTERESIS LOSS. Since this is a function of loop area, the effect of varying B on the area must be considered. When the value of H is increased, for example, doubled, B is *not* doubled and thus the ratio of the loop area also is not quadrupled. It is found to increase about 3.1 times. The Area of Loop is actually proportional to $B_m{}^x$ with x somewhere between 1 and 2 – where B_m is the maximum value to which flux density has been taken.

2. EDDY-CURRENT LOSS. When an armature rotates in a magnetic field, an e.m.f. is induced in the conductors. Since the conductors are let into slots, the armature teeth are considered as conductors with e.m.f.s induced across them. Moreover, as the electrical circuit is complete for these e.m.f. currents will flow from one end of a tooth through the armature end-plate, along the shaft and back to the other end of the tooth through the opposite armature end-plate. Such 'eddy currents' produce a power loss, due to the resistance of the iron circuit, which is $\propto I^2R$ or $\dfrac{E^2}{R}$.

Eddy-current loss depends on several factors, with every reasonable attempt taken to minimise such losses. The principal methods by which this is achieved are (1) laminating the iron circuit and insulating the laminations from each other by varnish, cellulose or paper, (2) using iron with a high specific resistance and (3) keeping the frequency of the magnetic alternations or cycles to a minimum.

Since a generated voltage is proportional to flux and speed, then $E \propto \Phi N$ or $E \propto B_m f$ where B_m is the maximum flux density and f is the frequency of alternation.

Again, since power loss $\propto \dfrac{E^2}{R}$ we can write:

Power loss $\propto B_m^2 f^2$ or $P_E = K_E B_m^2 f^2$ watts per cubic metre, where K_E is an *eddy-current coefficient*, which is dependent upon the type of material used, its thickness and other dimensions.

Pull of an Electromagnet

The energy stored in a magnetic field in air is given by $\dfrac{B^2}{2\mu_0}$ joules per cubic metre, where B is in teslas. Consider 2 poles arranged as shown (figure 6.16).

Each has an area A square metres and let F be the force of attraction (in newtons) between the poles.

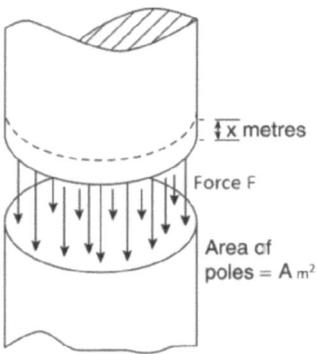

▲ **Figure 6.16** *Force between magnetic poles*

Let one pole move a small distance x (metres) against the force F, so the work done is Fx newton metres or joules. The volume of the magnetic field will increase by Ax cubic metres and the energy stored in the field is increased by $\dfrac{B^2}{2\mu_0} \times Ax$ joules. This is equal to the work done in separating the poles so that $Fx = \dfrac{B^2}{2\mu_0} Ax$ or $F = \dfrac{B^2 A}{2\mu_0}$ newtons,

where A is in square metres and B in teslas.

Example 6.7. An electromagnet is wound with 500 turns. The air gap has a length of 2mm and a cross-sectional area of 1000mm². Assuming the reluctance of the iron to be negligible compared with that of the air gap, and neglecting magnetic leakage and fringing, calculate the magnetic pull when the current is 3A.

M.m.f. of coil $F = N \times I = 500 \times 3 = 1500At$

This m.m.f. is used to pass the flux through the air gap, since the reluctance of the iron is negligible.

The magnetising force for the air is given by 'the ampere-turns per metre' or

$$H = \frac{F}{L} = \frac{1500}{2 \times 10^{-3}}$$

Also the flux density in air is B where:

$$B = \mu_0 H$$

$$\therefore B = \frac{4 \times \pi \times 10^{-7} \times 1500}{2 \times 10^{-3}} \text{ tesla}$$

$$= 0.942T$$

Now the pull $F = \dfrac{B^2 A}{2\mu_0} = \dfrac{0.942^2 \times 1000 \times 10^{-6}}{2 \times 4 \times \pi \times 10^{-7}} \text{ newtons}$

Thus $F = 353.3N$.

Example 6.8. A 4-pole D.C. generator has a cast-steel yoke and poles and has a laminated steel armature. The dimensions of the component parts of the magnetic circuit are as follows in table 6.3:

Table 6.3

Yoke. Total mean circumference = 3.04m	CSA = 0.04m²
Pole. Total mean length = 0.24m	CSA = 0.065m²
Air gap. Total mean length = 2mm	CSA = 0.065m²
Armature. Total mean path between poles = 0.4m	CSA = 0.025m²

The magnetisation curves are given in table 6.4.

Table 6.4

	H (At/m)	400	800	1200	1600	2000	2400
Cast steel	B (T)	0.45	1	1.2	1.3	1.37	1.43
Laminated steel	B (T)	1	1.34	1.48	1.55	1.6	1.63

Calculate the ampere-turns per pole, for a flux per pole of 0.08Wb in the air gap. Figure 6.17 illustrates the problem and the appropriate magnetic characteristics are shown by the graphs for the diagram (figure 6.18).

AIR GAP

Length 2×10^{-3} m

Area $0.065 = 6.5 \times 10^{-2}$ m²

$$\Phi = 0.08 \text{Wb} \quad B_A = \frac{0.08}{6.5 \times 10^{-2}} = \frac{8}{6.5} = 1.23 \text{T}$$

$$H_A = \frac{1.23}{\mu_0} = \frac{1.23}{4 \times \pi \times 10^{-7}} \text{ ampere-turns / metre}$$

Total ampere-turns or m.m.f. for air gap is given by:

$$F_A = \frac{1.23}{4 \times 3.14 \times 10^{-7}} \times 2 \times 10^{-3} \text{ ampere-turns}$$

$$F_A = 1955 \text{At}$$

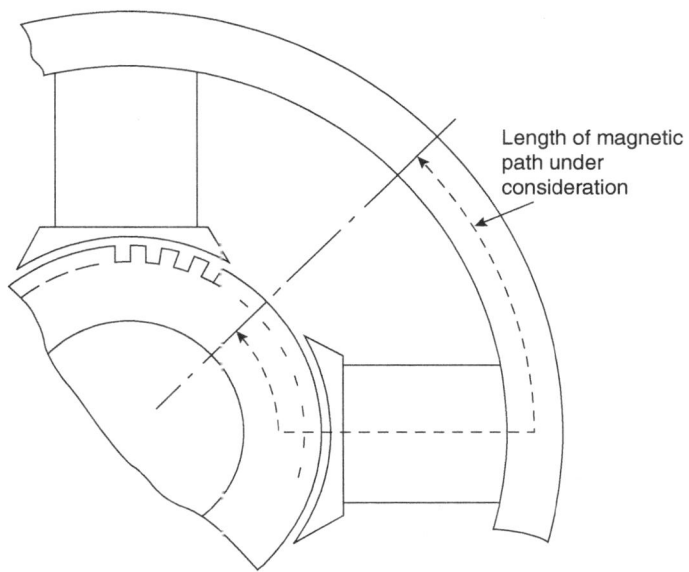

▲ **Figure 6.17** *Dimensions of magnetic pole arrangement*

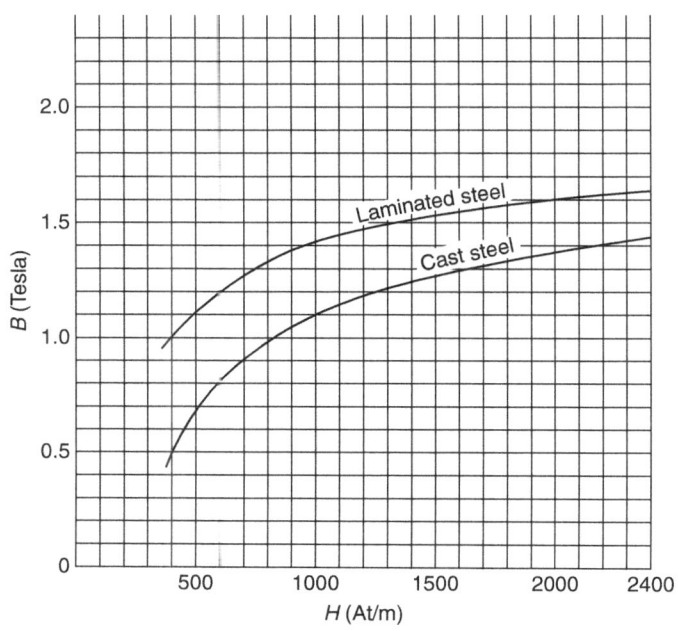

▲ **Figure 6.18** *B-H magnetic characteristics*

POLE (Cast steel)

Length 24×10^{-2} m

Area $0.065 = 6.5 \times 10^{-2}$ m²

$\Phi = 0.08\text{Wb} = B_p = \dfrac{\Phi}{A} = \dfrac{0.08}{6.5 \times 10^{-2}} = 123\text{T}$

From graph, $H_p = 1370\text{At/m}$ or

Total $F_p = 1370 \times 24 \times 10^{-2} = 330\text{At}$

YOKE (Cast steel)

Length $\dfrac{3.04}{4} = 0.76 = 76 = \times 10^{-2}$ m (between poles) or 0.38 magnetic length

Area $= 2 \times 0.04 = 0.08$ m² but Area $= 8 \times 10^{-2}$ m²

(Note the doubling of area since full pole area is taken for the flux)

$\Phi = 0.08\text{Wb} \quad B_Y = \dfrac{0.08}{8 \times 10^{-2}} = 1\text{T}$

From graph $H_Y = 800\text{At/m}$

Total F_Y for yoke $800 \times 38 \times 10^{-2} = 304\text{At}$

ARMATURE (Laminates)

Length $\dfrac{0.4}{2} = 20 \times 10^{-2}$ metre (magnetic length)

Area $= 2 \times 0.025 = 0.05 = 5 \times 10^{-2}$ m²

$\Phi = 0.08\text{Wb} \quad B_L = \dfrac{0.08}{5 \times 10^{-2}} = 1.6\text{T}$

From graph $H_L = 2000\text{At/m}$

Total F_L for armature $= 2000 \times 20 \times 10^{-2}$

$$= 400\text{At}$$

Total m.m.f. per pole $= 1955 + 330 + 304 + 400$

$$= 2989\text{At.}$$

Practice Examples

6.1. A brass rod of cross-section 1000mm² is formed into a closed ring of mean diameter 300mm. It is wound uniformly with a coil of 500 turns. If a magnetising current of 5A flows in the coil, calculate (a) the magnetising force (4 significant figures), (b) the flux density (2 significant figures) and (c) the total flux (2 significant figures).

6.2. An electromagnetic contactor has a magnetic circuit of length 250mm and a uniform cross-sectional area of 400mm². Calculate the number of ampere-turns required to produce a flux of 500μWb, given that the relative permeability of the material under these conditions is 2500. Also $\mu_0 = 4\pi \times 10^{-7}$ H/m (nearest whole ampere-turns).

6.3. In a certain magnetic circuit having a length of 1m and a uniform cross-section of 500mm², a magnetising force of 500 ampere-turns produces a magnetic flux of 400μWb. Calculate (a) the relative permeability of the material (4 significant figures) and (b) the reluctance of the magnetic circuit, $\mu_0 = 4\pi \times 10^{-7}$ H/m (3 significant figures).

6.4. An iron ring having a mean circumference of 1.25m and a cross-sectional area of 1500mm² is wound with 400 turns of wire. An exciting current of 2.5A produces a flux of 0.75mWb in the iron ring. Calculate (a) the permeability (relative) of the iron (1 decimal place), (b) the reluctance of the iron (3 significant figures) and (c) the m.m.f. of the exciting winding (1 significant figure).

6.5. A U-shaped electromagnet has an armature separated from each pole by an air gap of 2mm. The cross-sectional area of both the electromagnet and the armature is 1200mm² and the total length of the iron path is 0.6m. Determine the ampere-turns necessary to produce a total flux in each air gap of 1.13mWb, neglecting magnetic leakage and fringing (4 significant figures).

The magnetisation curve for the iron is given by:

B (T)	0.5	0.6	0.7	0.8	0.9	1.0	1.1
H (At/m)	520	585	660	740	820	910	1030

6.6. A circular ring of iron of mean diameter 0.2m and cross-sectional area 600mm² has a radial air gap of 2mm. It is magnetised by a coil having 500 turns of wire. Neglecting magnetic leakage and fringing, estimate the flux density in the air gap, when a current of 3A flows through the coil. Use the magnetic characteristics as given by the graph of Q6.5 (4 significant figures).

6.7. A built-up magnetic circuit without an air gap consists of 2 cores and 2 yokes. Each core is cylindrical, 50mm diameter and 160mm long. Each yoke is of square cross-section 47 × 47mm and is 180mm long. The distance between the centres of the cores is 130mm. Calculate the ampere-turns necessary to obtain a flux density of 1.2T in the cores (3 decimal places). Neglect magnetic leakage. The magnetic characteristics of the material are given:

B (T)	0.9	1.0	1.05	1.1	1.15	1.2
H (At/m)	200	260	310	380	470	650

6.8. An iron rod 15mm diameter is bent into a semicircle of 50mm inside radius and is wound uniformly with 480 turns of wire so as to form a horseshoe electromagnet. The poles are faced so as to make good magnetic contact with an iron armature 15 × 15mm cross-section and 130mm long. (a) Calculate the current required to produce a pull of 196.2N between the armature and pole-faces. Neglect magnetic leakage (3 decimal places). (b) Calculate the ampere-turns necessary to obtain a flux density of 1.15T in the air gap, if the armature is fixed so as to leave uniform air gaps 0.5mm wide at each pole-face. Neglect leakage and fringing. Use the magnetic characteristics as given by the graph of Q6.7 (2 decimal places).

6.9. Two coaxial magnetic poles each 100mm diameter are separated by an air gap of 2.5mm and the flux crossing the air gap is 0.004Wb. Neglecting fringing calculate (a) the energy in joules stored in the air gap (1 significant figure) and (b) the pull in newtons between the poles (3 significant figures).

6.10. Calculate the ampere-turns per field coil required for the air gap, the armature teeth and the pole of a D.C. machine working with a useful flux of 0.05Wb/pole, having given: Effective area of air gap 60 000mm². Mean length of air gap 5mm. Effective area of pole 40 000mm². Mean length of pole 250mm. Effective area of teeth 25 000mm². Mean length of teeth 45mm (4 significant figures).

Magnetic leakage coefficient = 1.2. Magnetic characteristics of the materials are:

B (J)	1.3	1.4	1.5	1.6	1.8	2.0
H (At/m)	1200	1500	2000	3000	8500	24 000

7

ELECTROMAGNETIC INDUCTION

Sir Humphry Davy's greatest discovery was Michael Faraday.

Sir Henry Paul Harvey

Electrochemistry was the first branch of science to play a full part in electrical investigations of the early nineteenth century. At that time electricity research was of interest to scientists only and could not be put to use for real engineering processes. Chemical cells as they were then known were unable to produce sufficient energy or e.m.f. for practical purposes, nor had any electromagnetic devices been invented for engineering applications. As mentioned in Chapter 5, it was only after the relation between current and magnetism was discovered that this would change. These discoveries revealed the related phenomena of electromagnetic induction and led to the development of machines that enabled engineers to produce electrical and mechanical energy.

The first electromagnetic *induction* experiments were attributed to Michael Faraday (1791–1867) who in 1821 showed that when magnetic flux linked with an electrical circuit is changing, an e.m.f. is induced in the circuit. This e.m.f. lasts only while the flux change takes place and the faster the change, the greater the e.m.f. This effect can be noted today by sliding a power neodymium magnet down a sloped block of copper and noting the induced current flow and force acting against the 'falling' moving magnet. Michael Faraday proposed the first laws concerning induced e.m.f., winding the wires of the world's first electric generator himself and building the first ever transformer induction ring to change the voltage of an electric current (1831). These original devices are today on permanent display at the Royal Institute in London.

Variations of both these revolutionary electric machines are now used in almost every power station: water, wind, gas or nuclear powered around the globe.

The magnetic flux linked with a circuit, usually a coil of insulated wire, is found to change in several different ways. Thus:

(1) A magnet could move near a coil of wire. This principle is used for the A.C. generator or alternator.

(2) A coil of wire could move near a magnet. This principle is used for the D.C. dynamo or generator.

(3) The flux could change by varying the current in the energising coil of wire. The ampere-turns are varied and the flux produced varies accordingly. This principle is essential to the operation of the transformer and familiar spark-coil of a petrol-engine ignition system. For these 3 methods of e.m.f. generation, it is seen that cases (1) and (2) involve relative physical movement between magnet and coil. Case (3), however, involves <u>no</u> such movement and the generated e.m.f. is achieved in a stationary coil with which only the linked flux changes. Thus there are 2 distinct forms of e.m.f. generation or induction referred to under 2 basic headings: (1) Dynamic Induction and (2) Static Induction.

Before these 2 methods are considered, let us look at what is meant by *flux-linkages*.

FLUX-LINKAGES. Earlier studies on magnetism showed that a magnet's field can be represented by lines of flux emanating from its poles. The strength of the flux is represented by the number of lines and is measured in webers, while flux density is measured in teslas. The flux lines make complete loops and the associated conductor or coil of wire in which the e.m.f. is induced is considered to consist of many turns. As the number of lines of flux associated with the turns are referred to as flux-linkages, a magnet with poles of flux strength 3.4μWb linked with a coil of 500 turns will result in $3.4 \times 10^{-6} \times 500 = 0.0017$ weber-turns. Figure 7.1 shows the basic concept.

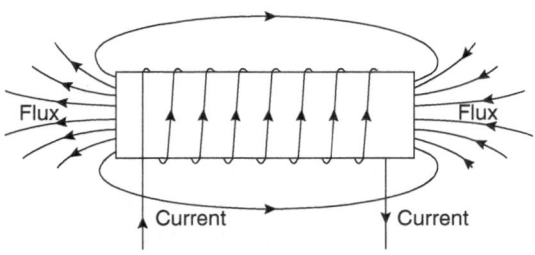

▲ **Figure 7.1** *Flux linkages*

Laws of Electromagnetic Induction

Faraday's law

This summarises the known relationship deduced for generation of e.m.f. by electromagnetic induction and is stated: the magnitude of the e.m.f. produced, whenever there is a change of flux linked with a circuit, is proportional to the *rate of change* of flux-linkages.

Lenz's law

This identifies a phenomenon noted for e.m.f. produced by induction. The law is stated as the direction of the current due to the induced e.m.f. always set up an effect tending to *oppose* the change causing it.

Thus if flux-linkages increase, the field produced by the induced current resulting from the induced e.m.f. tends to oppose this effect, i.e. it opposes flux-linkage build-up. Similarly, if the flux-linkages reduce, as when the current in a coil is switched off, then the induced e.m.f. will induce a current that, if allowed to flow, will keep up the flux-linkages to their original value. The action of the induced current will not be able to prevent the change, but it will try to do so during the period the change occurs. Faraday's law can be expressed n mathematical form and formulae deduced for static and dynamic electromagnetic induction, which will be considered separately.

Static Induction

Consider a coil connected to a D.C. power supply. At the instant of switch-on the current produces flux, which grows from the centre of the coil outwards, but this flux 'cuts' the coil turns and induces an e.m.f. that opposes the current growth (Lenz's law). Such a circuit in which a current change causes a change of flux and therefore produces an induced e.m.f. is said to be inductive or to possess *self-inductance*. There is a 'resistance' or 'inertia' of the circuit to change taking place. A circuit has an inductance of 1 henry if an e.m.f. of 1 volt is induced in a circuit when current changes at a rate of 1 amp/second.

Self-inductance

From this definition, if a circuit has an inductance of L henries and current changes from I_1 to I_2 amperes in t seconds, as shown in figure 7.2, then the average induced e.m.f.:

$$Eav = \frac{-L\,(I_2 - I_1)}{t} \text{ volts}$$

$$\text{or } Eav = -L \times \text{rate of increase of } I = -L\,\frac{dI}{dt}$$

Note: The negative sign indicates that the direction of induced e.m.f. *opposes* that of the current increase.

Example 7.1. The current through a coil having an inductance of 0.5H is reduced from 5A to 2A in 0.05s. Calculate the average e.m.f. induced in the coil (2 significant figures).

$$Eav = \frac{-L(I_2 - I_1)}{t} \text{ volts}$$

$$= \frac{-0.5(2 - 5)}{0.05} \text{ volts}$$

$$\therefore Eav = +30 \text{ volts.}$$

Note. The positive sign indicates that the induced e.m.f. tries to maintain the current flow.

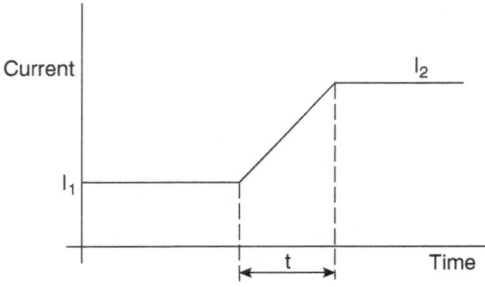

▲ **Figure 7.2** *Current vs time*

As a change of current produces a change of flux-linkages, the e.m.f. induced can be expressed in terms of the number of coil turns and the rate of change of flux.

$$Eav = \frac{-N(\Phi_2 - \Phi_1)}{t} \text{ volts} = -N\frac{d\Phi}{dt}$$

Example 7.2. When a D.C. current passes through an iron-cored coil of 2000 turns, a magnetic flux of 30mWb is produced. The supply switch is *opened* and the current falls to zero amperes in 0.12s, leaving a residual flux of 2mWb. Find the average value of induced e.m.f. (1 decimal place).

$$Eav = \frac{-N(\Phi_2 - \Phi_1)}{t}$$

$$= \frac{-2000(2 - 30) \times 10^{-3}}{0.12}$$

$$Eav = +466.6 \text{ volts}$$

Note. Again, the +ve sign indicates that the induced e.m.f. tries to maintain the current flow.

From these equations:

$$Eav = \frac{-L(I_2 - I_1)}{t}$$

$$\text{and } Eav = \frac{-N(\Phi_2 - \Phi_1)}{t}$$

and setting these 2 equations equal:

$$\frac{-L(I_2 - I_1)}{t} = \frac{-N(\Phi_2 - \Phi_1)}{t}$$

$$\therefore \text{ Inductance } L = \frac{N(\Phi_2 - \Phi_1)}{(I_2 - I_1)} = \frac{Nd\Phi}{dI}$$

$$\text{or Inductance } L = \frac{N\Phi}{I}$$

Hence Inductance L = Flux-linkages/current.

In Chapter 6 flux was expressed in terms of m.m.f. (F) and reluctance

$$F = \Phi S \quad \text{or} \quad IN = \Phi S \quad \text{and} \quad \Phi = \frac{IN}{S}$$

$$\text{Thus } L = \frac{N}{I} \times \frac{IN}{S}$$

$$\therefore \text{ Inductance } L = \frac{N^2}{S} \text{ henries}$$

E.m.f. due to static induction

Consider figure 7.3, which shows 2 coils A and B of insulated copper wire. Coil A can be connected to a battery through a switch, while B is wound over or adjacent to coil A and connected to a sensitive centre-zero voltmeter. This instrument is used because, as the pointer is positioned at the centre of the scale, deflection to the left or right depends on the polarity of the supply.

At the instant of switching on the current in coil A, flux is imagined to grow out and cut the turns of coil B. The initial growth is shown by the dotted flux lines becoming fuller until the final condition (full lines) is reached. The cutting of coil B by the flux of A results in an induced e.m.f., its magnitude and direction governed by Faraday's and Lenz's laws. The flux-linkages, i.e. flux linking with the turns (N_B) of coil B, increase and if the linking flux grows to a value of Φ webers from its original zero value, then the rate of change of flux-linkages will be the flux-linkages divided by the time (t_1) taken for them to grow, i.e. the time taken for the current to reach its final value.

$$\text{Thus e.m.f. induced in coil B} = \frac{\text{flux-linkages}}{\text{time}} = \frac{-N_B\Phi}{t_1} \text{ volts.}$$

For this equation N_B = turns of coil B, Φ is the flux in webers linking with it and t_1 is the time taken for the energising current to reach its final value I. It might be assumed that value I is reached immediately the switch is closed, because the electricity flow is considered to be instantaneous, but the current takes an appreciable time to reach its full value – due to the circuit 'inertia' or inductance. It is seen that when the switch for the *primary* coil A is closed, the voltmeter pointer gives a 'kick', for example, to the

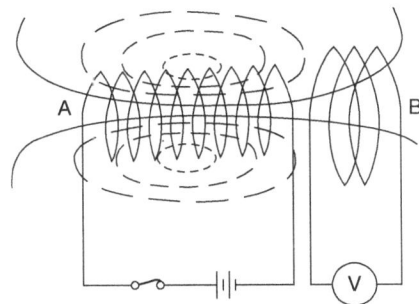

▲ **Figure 7.3** *E.m.f. due to static induction*

left, showing an e.m.f. is induced in coil B – the *secondary* circuit. The value of e.m.f.

$E_B = \dfrac{-N_B \Phi}{t_1}$ and the voltmeter will show the polarity of coil B to be such that the

current, flowing through the instrument, is in a direction through the coil as to set up a secondary flux, opposite to the original flux Φ, and will try to stop it growing. Although a kick of the voltmeter pointer is seen, it returns to the zero position even though current in coil A is allowed to flow indefinitely. Thus an e.m.f. is induced only during the time when the flux-linkages change. Further experiments with coil B show that if the number of turns of wire were doubled, then the induced e.m.f. will be twice as large, even though the flux Φ of coil A is the same. The flux-linkages have increased and the induced e.m.f. rises proportionately.

Consider next the instant of switching off the current in coil A. The voltmeter kicks to the right this time, showing an induced e.m.f. of reversed polarity. The flow of current in coil B is such as to try to maintain the flux to its original value Φ and again $E_B = \dfrac{-N_B \Phi}{t_2}$

where t_2 is the time taken for 'switching off'. It is noted here that t_2 need not equal t_1. If the switch is opened quickly, the current of A will be interrupted quickly and E_B can be larger at switching off than at switching on, i.e. the rate of growth of the flux is controlled by the inductance and resistance R of the circuit.

Up to now we have only considered the induced e.m.f. in coil B and this is said to be due to *Mutual Induction*, i.e. the mutual action of coil A on B. We now turn our attention to *Self-Induction*, i.e. the conditions within coil A itself. At the instant of switching on, the flux grows outwards and in so doing cuts the turns of coil A – the primary circuit. An e.m.f. is induced given by $E_A = \dfrac{-N_A \Phi}{t_2}$. Here N_A is the turns of coil A, Φ is the linked flux and t_1 the time taken for the current to reach its full value. As before, the direction of the self-induced e.m.f. E_A will be such as to cause a current to flow in the opposite direction through the battery to produce a flux opposing the build-up of flux Φ. We can now see the reason for the opposition to the growth of current in coil A at the instant of switching on and why the current *I* takes a little time to reach its full value. As before when the switch is opened, flux collapses and in doing so, again cuts the turns of coil A, inducing a voltage of reversed polarity, which tries to keep the current flowing. This self-induced e.m.f. at 'switching off' can be extremely large in some instances where a large number of turns of an energising winding are associated with a strong magnetic flux. For example, the opening of the field circuit of a large alternator or D.C. generator. Special arrangements help to limit the e.m.f. to a safe value and prevent 'breakdown' of insulation by large induced voltages.

Mutual inductance

Two coils have a mutual inductance of 1 henry when a change of current at the rate of 1 ampere/second in 1 coil produces an e.m.f. of 1 volt in the other.

Consider once again figure 7.3. If coils A and B have a mutual inductance of M henries, and the current in coil A increases from I_1 to I_2 amperes in t seconds, then

$$\text{Average e.m.f. induced in B} = \frac{-M(I_2 - I_1)}{t}$$

$$= -M \times \text{Rate of increase of current in coil A}$$

The e.m.f. induced into coil B can be expressed in terms of flux-linkages in the same way as was applied to self-inductance.

Let the flux change from Φ_1 to Φ_2 Webers in t seconds due to a change of current from I_1 to I_2 amperes in the primary, and let coil B have N_B turns.

Thus average e.m.f. induced is:

$$E_{\text{e.m.f.}} B = \frac{-N_B(\Phi_2 - \Phi_1)}{t}$$

$$\text{Hence} = \frac{-M(I_2 - I_1)}{t} = \frac{-N_B(\Phi_2 - \Phi_1)}{t}$$

$$\therefore M = \frac{N_B(\Phi_2 - \Phi_1)}{(I_2 - I_1)} = 5V$$

$$\text{So mutual inductance} = \frac{\text{Change in flux linkages on secondary}}{\text{Change in current in primary}}$$

Example 7.3. (Self-induction). A coil of 800 turns is wound on a wooden former and a current of 5A is passed through it to produce a magnetic flux of 200µWb. Calculate the average value of e.m.f. induced in the coil when the current is (a) switched off in 0.08 seconds (1 significant figure) and (b) reversed in 0.2 seconds (1 decimal place).

(a) $\text{Eav} = \dfrac{-N(\Phi_2 - \Phi_1)}{t} = \dfrac{-800 \times (0 - 200 \times 10^{-6})}{0.08}$

$\text{Eav} = 2V$

b) $\text{Eav} = \dfrac{-N(\Phi_2 - \Phi_1)}{t}$ here $\Phi_2 = \Phi_1$ numerically but is in the reverse direction, or $\Phi_2 = -\Phi_1$

$$\text{This Eav} = \frac{2N\Phi_1}{t} = \frac{2 \times 800 \times 200 \times 10^{-6}}{0.2}$$

$$\text{Eav} = 1.6V$$

Example 7.4. (Mutual induction). If the coil of the above example has a secondary coil of 2000 turns wound onto it, find the e.m.f. induced in this second coil when the current of 5A is switched off in 0.08 seconds (1 significant figure.). It can be assumed that all the flux of 200µWb created by the 5A current in the primary links with the secondary coil.

$$\therefore M = \frac{N_B(\Phi_2 - \Phi_1)}{(I_2 - I_1)} = 5V$$

Note. The e.m.f. of the secondary $s\ \dfrac{5}{2} = 2.5$ times the induced e.m.f. in the primary and is

proportional to the turns ratio $\dfrac{2000}{800} = 2.5$. This is the basic principle of the transformer

and the ignition system spark-co l. It shows how a large voltage is induced in a secondary coil by the flux associated with a low-voltage primary coil. For a petrol-engine ignition system, the e.m.f. in the secondary may be close to 8000V compared with 12V applied to the primary coil. This is achieved by using a coil of appropriate turns ratio between the primary and secondary coils, by providing an iron magnetic circuit to concentrate the flux for maximum linkage and by interrupting the primary circuit quickly by an engine-driven cam-operated switch.

Example 7.5. The ignition coil of a petrol engine has an inductance of 4.5H and carries a current of 4A. If, when the distributor points close, the circuit current collapses uniformly to zero in 2ms, find the average e.m.f. induced in the coil (1 decimal place).

$$Eav = \frac{-L(I_2 - I_1)}{t}$$

$$= \frac{-4.5(0 - 4)}{2 \times 10^{-3}}$$

$$Eav = 9000V = 9kV$$

Coupling factor

There is a relationship between the mutual induction of 2 coupled coils and their individual self-inductances, depending upon the magnetic coupling between them. The mutual inductance of the 2 coils is expressed as the change in flux-linkages of one coil to the change in current in the other.

$$M = N_B \frac{d\Phi}{dI_A} \text{ while } M = N_A \frac{d\Phi}{dI_B}$$

$$\text{Hence: } M^2 = N_B \frac{d\Phi}{dI_A} \times N_A \frac{d\Phi}{dI_B}$$

and rearranging so,

$$M^2 = N_A \frac{d\Phi}{dI_A} \times N_B \frac{d\Phi}{dI_B}$$

$$\text{But as } L_A = N_B \frac{d\Phi}{dI_A} \text{ and } L_B = N_B \frac{d\Phi}{dI_B}$$

L_A and L_B will give the same product M^2

$$\text{Thus } M^2 = L_A \times L_B$$

$$\text{or } M = \sqrt{L_A L_B}$$

This is the *maximum* value of mutual inductance available between the 2 coils, but this is difficult to achieve due to magnetic leakage and fringing effects. The magnetic coupling between the coils is also affected by their separation and the angular displacement between them. In general, mutual inductance is given by:

$M = k\sqrt{L_A L_B}$ where k is the coupling factor, where for perfect coupling $k = 1$.

Inductance of 2 coils in series

In our study of mutual inductance, we have only considered the effect of 2 magnetically coupled but electrically separated coils. However, the effects of mutual-inductance and self-inductance can be applied to electrically connected coils. Consider first the effects of 2 coils carrying the same current, and wound so their magnetic fields *assist* one another (figure 7.4a).

From our previous work on mutually coupled coils, we saw that a coil has an e.m.f. induced into it due to (1) the self-inductance of the coil and (2) the mutual inductance of the other coil.

$$\text{i.e. } E_A = \left(-L_A \frac{dI}{dt}\right) + \left(-M\frac{dI}{dt}\right) = -(L_A + M)\frac{dI}{dt}$$

$$\text{and } E_B = \left(-L_B \frac{dI}{dt}\right) + \left(-M\frac{dI}{dt}\right) = -(L_B + M)\frac{dI}{dt}$$

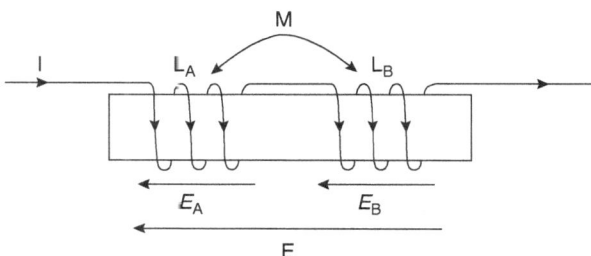

▲ **Figure 7.4a** *Combined Inductance of two coils in series*

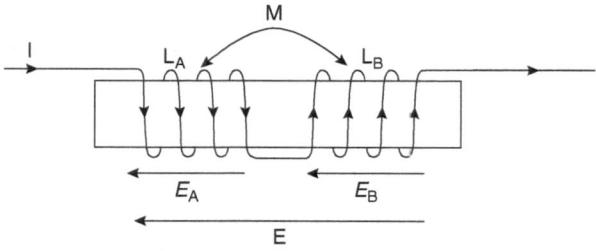

▲ **Figure 7.4b** *Mutual inductance of two coils in series*

However, E_A and E_B are in series assisting each other

$$\therefore E = E_A + E_B$$

$$E = \left[-(L_A + M)\frac{dI}{dt}\right] + \left[-(L_B + M)\frac{dI}{dt}\right]$$

$$E = -(L_A + L_B + 2M)\frac{dI}{dt}$$

So that $E = -L\dfrac{dI}{dt}$ Where L is the total inductance.

$$\therefore L = L_A + L_B + 2M \text{ henries (for coils assisting)}$$

Similarly we may consider 2 coils wound in *opposition* (figure 7.4b).

In this case the induced e.m.f. due to mutual inductance *opposes* that due to self-inductance. Hence, by similar proof to that shown for coils assisting, it is shown that when the coils oppose each other:

$$L = L_A + L_B - 2M \text{ henries.}$$

Example 7.6. Two coils of inductances 10mH and 15mH respectively have a coupling factor of 0.8 between them. What is their combined inductance when they are connected in series (a) assisting (3 significant figures) and (b) opposing (2 significant figures)?

$$M = k\sqrt{L_A L_B} = 0.8\sqrt{10 \times 15} \text{ mH}$$

$$= 9.8\text{mH}$$

(a) $L = L_A + L_B + 2M$

$= 10 + 15 + (2 \times 9.8)$ mH

$L = 44.6$mH

(b) $L = L_A + L_B - 2M$

$= 10 + 15 - (2 \times 9.8)$ mH

$L = 5.4$ mH

Magnetic Induction

It is worth considering the magnetic energy stored within the inductance component of the circuit, or its 'inertia'. If we assume that at time t = 0 a coil of inductance L henries and resistance R is connected across the terminals of a battery of e.m.f. V, the circuit equation will be:

$$V - L\frac{dI}{dt} - IR = 0$$

Or rephrased as: $V = L\dfrac{dI}{dt} + RI.$ (1)

If the total power output of the battery is VI then the total work done by the battery in raising the current in the circuit from zero at time t = 0 to I_T at a later time t = T is:

$$W = \int_0^T VI \, dt.$$ (2)

Putting equation (1) into equation (2).

$$W = L\int_0^T I\frac{dI}{dt}dt + R\int_0^T I^2 dt,$$

giving the result

$$W = \frac{1}{2}LI_T^2 + R\int_0^T I^2 \, dt.$$

The first term is the amount of energy stored in the inductor at time T, while the second term represents the irreversible conversion of electrical energy into heat in the resistor. The energy stored in the inductor can be recovered after the inductor is disconnected from the battery. For example, the energy in an inductance of 3 henries/m with a current of 2A is given by:

$$W = \frac{1}{2} \times I_T^2 = \frac{1}{2} \times 3 \times 2 \times 2 = 6J$$

Dynamic Induction

As mentioned earlier, this condition covers cases where there is relative movement between a magnetic field and a conductor. Obviously this concerns either a stationary conductor and a moving field or a stationary field and a moving conductor, or even both moving together but relatively! To avoid repetition of basic theory, the immediate explanations and diagrams will refer to a fixed field and moving conductor.

The magnetic diagrams in figure 7.5 show a field produced by 2 permanent magnets and a conductor moved so as to cut the field, thus altering the flux-linkages.

Three cases are shown.

For case (a) there is no change of flux-linkages, i.e. no cutting of the magnetic field. The conductor moves at a velocity of v metres/second in the same direction of the lines of flux and no e.m.f. is recorded on the voltmeter. For case (b) the conductor moves at right angles to the field of flux density B teslas and the voltmeter shows a constant deflection. Flux-linkages are considered to change since the flux lines are 'cut' as a conductor passes through. If the conductor is moved from left to right, a polarity is noted, which reverses if the conductor is moved from right to left. Alternatively, if the field is reversed so that the flux lines pass from a bottom N pole to a S pole at the top of the diagram, and the conductor moved left to right, a reversed polarity is again indicated. The investigation will show further deductions. Thus:

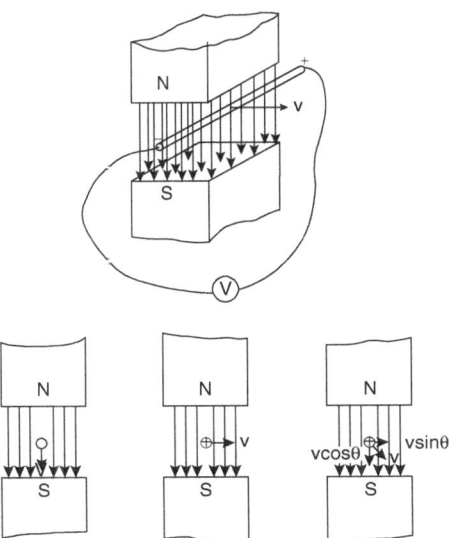

▲ **Figure 7.5** *Movement of a conductor relative to a magnetic field*

The magnitude of the induced e.m.f. varies with the speed of cutting the field or rate of change of flux-linkages. Hence $E \propto v$. Again, if the field cut is varied by altering the flux density, the e.m.f. will vary as B or $E \propto B$. Also, the longer the conductor cutting a field, the greater will be the magnitude of the e.m.f. and $E \propto \ell$. Summarising these 3 conditions we see that $E \propto B\ell v$, where ℓ is the length of the conductor in metres.

Case (c) of the diagram shows the conductor cutting the field at an angle θ. It is an intermediate condition between cases (a) and (b) and is best treated by resolving v into 2 component velocities at right angles to each other. Consider $v \cos \theta$ to be the component velocity in the direction of the flux lines, then $v \sin \theta$ will be the component of velocity at right angles to the field.

In accordance with the reasoning for cases (a) and (b) we see that velocity $v \cos \theta$ is responsible for no induced e.m f. while velocity $v \sin \theta$ is responsible for the induced e.m.f. and $E \propto v \sin \theta$.

$E \propto B\ell v \sin \theta$ is a more general expression than that already deduced and will cover all possible conditions.

For instance, for the condition of case (a) $\theta = 0°$ and as $\sin 0° = 0$ ∴ $B\ell v \sin 0° = 0$ or $E = 0$ as stated.

Again, for case (b), if $\theta = 90°$ then $\sin 90° = 1$ and $B\ell v \sin 90° = B\ell v$ giving $E = B\ell v$.

E.m.f. due to dynamic induction

As explained, the induced e.m.f. is proportional to B, v and the sine of the angle made by the direction of cutting and the field direction. The actual magnitude of such an e.m.f. can be deduced in more definite terms thus:

Consider case (b) of figure 7.5. In 1 second, the area cut by a conductor of length ℓ metres and *moving* at a velocity of v metres/second is ℓv square metres. If the flux density in this area is B teslas, then the flux cut per second by the conductor = $B\ell v$ webers. Using Faraday's law, we see that $B\ell v$ acts as a measure of the magnitude of induced e.m.f. in volts or induced e.m.f. $E = B\ell v$ volts.

If case (c) is considered, the flux cut is proportional to the component of velocity perpendicular to the field or the induced e.m.f. $E = B\ell v \sin \theta$ volts. The above formula is deduced as follows: figure 7.6 shows a conductor Q, carrying a current of ℓ amperes in the direction shown. As before, the flux density of the field is taken as B teslas and the length of the conductor is ℓ metres. A force is exerted on the current-carrying conductor in a magnetic field so the conductor in the diagram experiences a force $BI\ell$ newtons to the left. Accordingly a force of $BI\ell$ newtons must be applied in the opposite direction to oppose movement of the conductor.

Consider the conductor to move from position Q to position P x metres away. The work done by the conductor in moving from Q to P = Force × distance = $BI\ell x$ newton metres or joules.

Let E volts be the e.m.f. induced in the conductor as a result of cutting the magnetic field and t seconds the time taken to do this work.

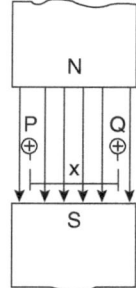
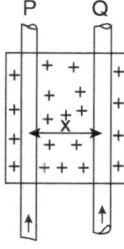

▲ **Figure 7.6** *Relationship of flux to the velocity component perpendicular to the field*

Then mechanical power expended $= \dfrac{Bl\ell x}{t}$ watts and if this appears as electrical power,

it will be EI Watts (e.m.f. × current) or $EI = \dfrac{Bl\ell x}{t}$ and $E = \dfrac{B\ell x}{t}$ volts

As $\dfrac{x}{t} =$ velocity of the cutting v then as before, $E = B\ell v$ volts.

From above, $B\ell x$ is (the flux density × area) or the flux ⊠ cut by a conductor moving from position Q to P in time t seconds, and since

$$E = \frac{B\ell v}{t} \quad \text{then } E = \frac{B \times \text{area}}{t} = \frac{\Phi}{t} \quad \text{or} \quad E = \frac{\Phi}{t}$$

Thus $E(\text{volts}) = \dfrac{\Phi(\text{webers})}{t(\text{seconds})}$ and we have an alternative formula for the e.m.f. generated in a conductor cutting a magnetic field. It is similar to that deduced for the statically induced e.m.f. namely $E = \dfrac{N\Phi}{t}$ where N is the number of turns of the coil and Φ is the change of flux.

Thus the formula for $E = \dfrac{\text{flux-linkages}}{\text{time}}$ or $E = \dfrac{\Phi}{t}$ as the flux-linkages are numerically equal to Φ, there being only one conductor.

Important Note. $\dfrac{\Phi}{t}$ is $\dfrac{\text{Flux cut}}{\text{time}}$ and we have an alternative way of stating Faraday's law, now expressed as: 'The e.m.f. generated in a conductor is proportional to the rate of cutting lines of flux or is proportional to the flux cut/second.' This form of Faraday's law is more applicable to dynamic induction and will be used in connection with the generator, motor and alternator.

Example 7.7. A conductor is moved to cut a magnetic field at right angles. Find the e.m.f. induced in it, if the average density of the field is 0.45 teslas, the length of conductor is 80mm and the speed of cutting is 8.88 metres/second (2 decimal places).

In the Formula $E = B\ell v$ we have

$$E = 0.45 \times 0.08 \times 8.88 = 0.32\text{V}.$$

An alternate solution could be:

Area swept by the conductor/second = 0.08×8.88 m². The flux in this area would be $\boxtimes = 0.45 \times 0.08 \times 8.88$ webers, and e.m.f. = flux cut per second

$$\text{or } E = \frac{0.45 \times 0.08 \times 0.88}{1} = 0.32\text{V}.$$

Example 7.8. A 4-pole generator has a flux of 12mWb/pole. Calculate the value of e.m.f. generated in one of the armature conductors, if the armature is driven at 900 rev/min (2 decimal places).

In 1 revolution a conductor cuts $4 \times 12 \times 10^{-3} = 0.048$Wb (4 poles at 12mWb per pole).

Time of 1 revolution of the armature $= \dfrac{1}{900}$ minutes $= \dfrac{60}{900}$ or $\dfrac{1}{15}$s

∴ Rate of cutting flux $= \dfrac{\text{Flux cut per revolution}}{\text{time taken to complete a revolution}}$

Thus $E = \dfrac{0.048}{\frac{1}{15}} = 0.048 \times 15 = 0.72$ volts/conductor.

THE WEBER. 'The weber is that magnetic flux which, when cut by a conductor in one second, generates in the conductor an e.m.f. of value equal to one volt.'

Alternative ways of defining the weber or SI unit of flux are: 'An e.m.f. of one volt is generated when a conductor cuts flux at the rate of one weber/second', or 'an e.m.f. of one volt is generated when the flux linked with one turn changes at the rate of one weber/second.'

Direction of induced e.m.f. (hand rules)

The direction of the induced e.m.f. is deduced from first principles, with Lenz's law or by application of a rule first enunciated by Professor T. A. Fleming and commonly known as Fleming's Right-Hand Rule. It is noted that there is also a left-hand rule and to avoid confusion, the following is suggested for memorising the appropriate rule. The generator is studied *before* the motor and for the average person, use of the right hand is preferred *before* the left. Hence use the right-hand rule for the generator and the left-hand rule for the motor. A generator is a machine that generates or induces e.m.f.

in armature conductors through electrodynamic induction. The right-hand rule is now explained.

RIGHT-HAND RULE (Fleming's). Consider a conductor in a magnetic field as shown (figure 7.7a).

The magnetic field is in the direction left to right and the conductor moves at right angles and upwards with a velocity of *v* metres/second. The e.m.f. across the ends of the conductor is as shown, i.e. the polarity is such that, if the ends of the conductor are joined externally through an ammeter, current will flow as indicated. Current direction in the conductor is seen and if attention is given to figure 7.7b, it is observed that the field due to the conductor current is *clockwise,* strengthening the field at the top and weakening the field at the bottom. According to Lenz's law, opposition is offered to the motion of the conductor, with field lines concentrating before the conductor. A force opposed to the direction of movement is apparent. The right hand can be drawn and used to find the induced current direction and thus the induced polarity. This is shown (figure 7.8).

To use the rule, place the thumb, index finger and second finger of the right hand at right angles to each other. Point the index finger in the direction of the flux lines and the thumb in the direction of moving the conductor. The current in the conductor, due to the induced e.m.f., will be in the direction indicated by the second finger. For the example considered (figure 7.6) the current will be into the paper. The flux lines/index finger is to the right and the thumb/conductor up.

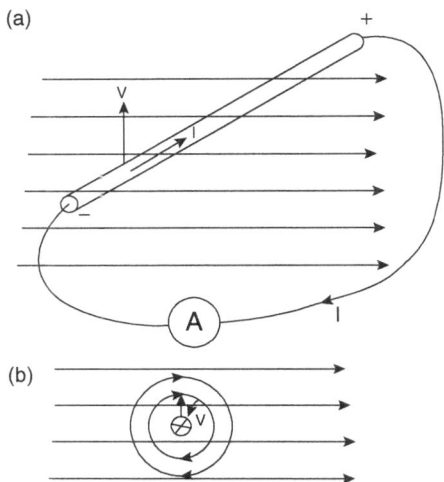

▲ **Figure 7.7** *Direction of induced e.m.f.*

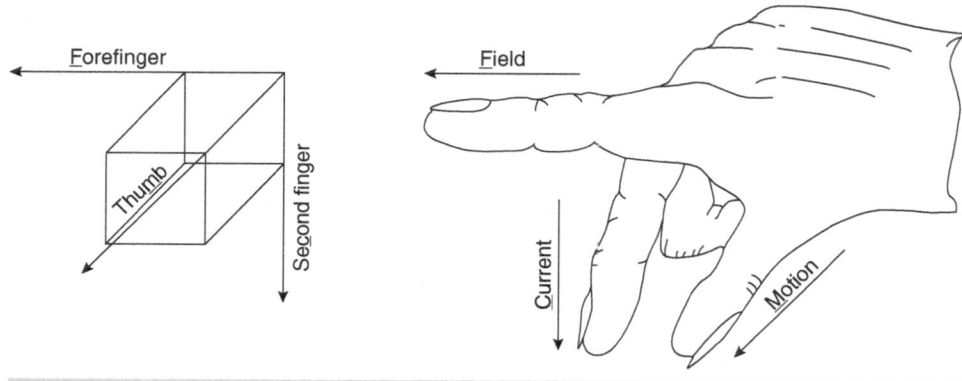

▲ **Figure 7.8** *Fleming's Right- Hand rule*

The Simple Magneto-Dynamo

Once the principles of electromagnetic induction were discovered, it was quickly realised that it was possible to construct a machine converting mechanical energy into electrical energy and generate electricity as a result of being driven by a steam engine or water turbine. The idea of making insulated conductors move through a stationary magnetic field presented no difficulties for a small machine and so the construction of such a magneto-dynamo followed fundamental requirements. A typical machine is illustrated (figure 7.9), and consists of permanent magnets to provide a field and a simple coil mounted on, yet insulated from, a shaft that can be rotated. To allow contact to be made with moving conductors, they are connected to slip-rings also mounted on but insulated from the shaft. Fixed *brushes* in turn contact slip-rings to make sliding connections and allow an external circuit to be completed.

It is seen that the coil consists of 2 'active' conductors: AB and CD, connected in series by the connection BC (which, together with the front connections to the slip-rings, play no part in the generation of e.m.f. but merely carry current to an external circuit). The load resistance of the external circuit is shown concentrated in *R*, connected to machine terminals: X and Y.

Consider the machine operation as follows:

As one conductor AB moves down through the field, the other CD moves up and the induced e.m.f.s will be such that A is +ve relative to B and C is +ve relative to D. The induced current, if allowed to flow, will be as shown by the arrows and, as it is

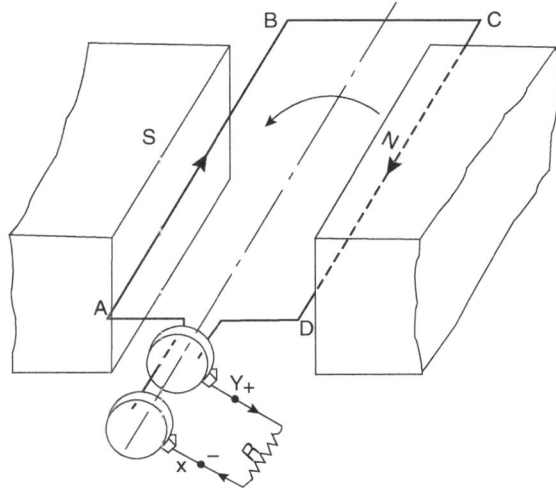

▲ **Figure 7.9** *Simple Magneto-Dynamo*

from terminal Y to terminal X through the external circuit, Y is +ve with respect to X. The student can check the right-hand rule for the polarity of the terminals in each half revolution. It is noted that the right-hand rule, as described, can be applied to conductor AB, the condition being that AB is moving from the top vertical position round past the centre of the magnet pole and then to the bottom vertical position, moving anticlockwise. The position where it moves past the pole at right angles is of first importance, being a condition of maximum e.m.f.

After the coil has rotated a half revolution, conductor DC begins to move down and conductor AB up. The polarity induced is the reverse of that for the first half revolution, D being +ve relative to C and B is +ve relative to A. Terminal X is now the +ve terminal and Y is the negative. An alternating e.m.f. is generated (figure 7.10), which illustrates 4 positions of the coil viewed from the slip-ring end. For position 1, A and D are moving horizontally along the field and no e.m.f. is being generated. A similar condition exists for position 3, but for positions 2 and 4 maximum e.m.f. is being generated, as field-cutting at right angles is taking place. For intermediate positions of e.m.f. generation, as represented by $E \propto B\ell v \sin \theta$, the conductors cut a uniform field at an angle but move at a uniform velocity. Thus e.m.f. generated at any instant is not constant but varies and it is customary to use a small letter for what is termed the *instantaneous* value. The expression $e = B\ell v \sin \theta$ volts gives the magnitude of the voltage generated. If voltage is plotted against revolutions in degrees or radians, a waveform such as that illustrated is obtained.

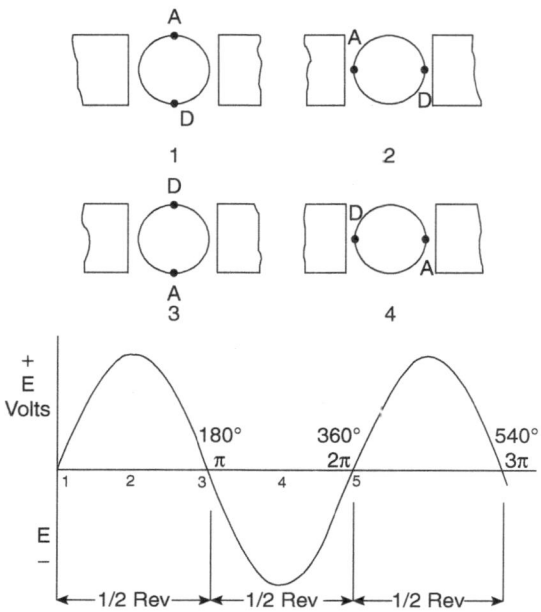

▲ **Figure 7.10** *Generation of an alternating e.m.f.*

The simple D.C. generator

The simple magneto-dynamo machine, as described (figure 7.9), or electrical generator, as it is now called, has a uniform field arranged to be cut by the conductors. It provides an e.m.f. whose magnitude varies sinusoidally, that is, the e.m.f. polarity and value follows a sine wave. A sinusoidal waveform is desired for A.C. working but for a D.C. generator, modifications are needed to achieve a near constant unidirectional voltage magnitude and polarity. It is apparent that a distinction is being made between the generation of D.C. and A.C. and from here on the division between the 2 methods of generating, transmitting and utilising electrical energy will become marked. In this book, both D.C. and A.C. theory is discussed. The main part of electrical theory is concerned with A.C. circuits and machines and if later study difficulties are to be minimised, attention must be given to A.C. theory at the start.

The first of the modifications referred to for the D.C. machine involves introduction of an iron or magnetic material into the armature or moving-coil part of the assembly. The coil made up from insulated wire is wound onto an iron armature carried on the shaft. The magnet system is provided with iron pole-shoes or arc extensions as shown (figure 7.11). As the length of the flux path through air is reduced to 2 small air gaps, the remainder being through the iron of the armature and field system, flux density or *B* value in the air

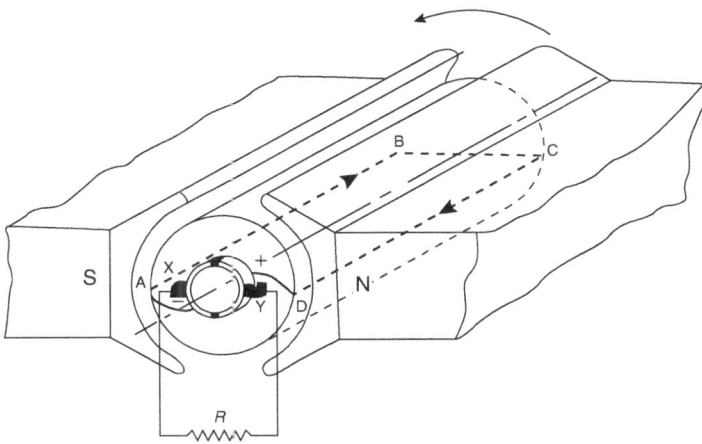

▲ **Figure 7.11** *Simple D.C. generator*

gaps increases and the conductors cut a stronger field. As the air gaps are now small and of constant width, the flux lines will cross them as shown and the field will be uniform over the pole-faces. Moving conductors thus pass from a small arc with substantially no magnetic flux into a large arc of constant flux density. Flux lines are radial in the gaps and cut at right angles for most of the distance under the pole. The e.m.f. waveform is as shown (figure 7.12), i.e. proportional to the flux density through which it passes.

COMMUTATION. To obtain a constant unidirectional e.m.f. or true D.C. generator, the next modifying step is to fit a form of automatic reversing switch or *commutator*, which, even though the moving coil, continues to generate an alternating e.m.f., and ensures that a unidirectional or 'rectified' e.m.f. appears at the machine terminals. Figure 7.11 shows how a commutator is fitted. It consists of a metal slip-ring that is split into 2 parts, each insulated from the other and from the shaft. The ends of the coil are connected to each half or segment of the commutator. The stationary brushes are adjusted that they bridge the gap in the slip-rings at the instant when the e.m.f. induced in the coil has zero value and is due to reverse.

Figure 7.13 shows the side view of the commutator and the reversing action of the switching arrangement is seen more clearly. The diagrams are considered to be complementary to those of figure 7.10 although only conditions for positions 2 and 4 are shown. The obvious position for the brushes is on the 'magnetic neutral axis' and that the brush Y will always be the +ve and brush X the −ve terminal. The new shape of the waveform is shown.

For position 2 it is seen that the −ve end D of conductor CD is connected to the +ve brush Y, while the +ve end A of AB is connected to the −ve brush X. For position 4, when the e.m.f.s have reversed in the conductors of the coil, end D, which is now +ve, is connected

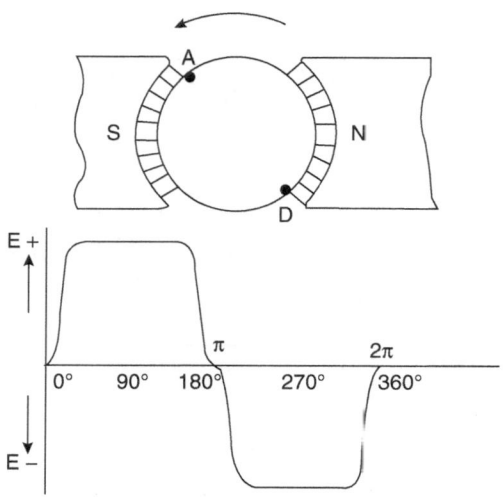

▲ **Figure 7.12** *E.m.f. waveform*

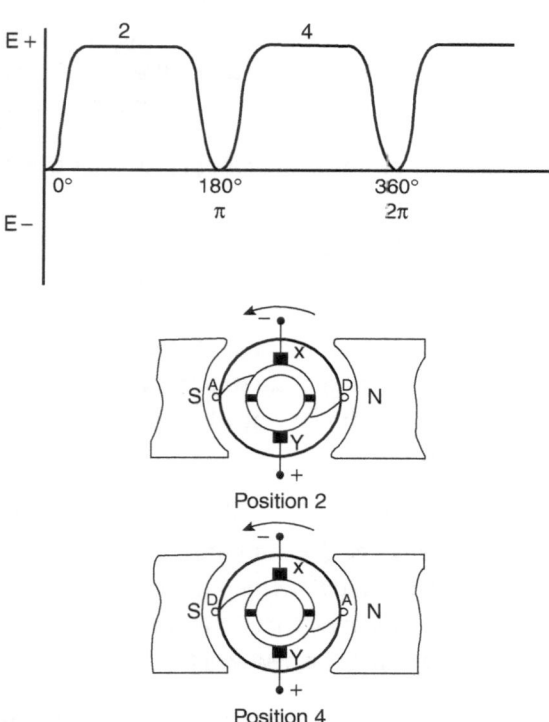

▲ **Figure 7.13** *Reversing action of the switching arrangement*

to the −ve brush X and end A of AB is now −ve and connected to the +ve brush Y. The brush polarity is decided by the direction of current flow in the external circuit. Thus current flows from Y (the +ve brush) to X (the −ve brush) and then to A onto B, etc. The apparent anomaly of +ve end A being connected to a −ve brush and so on is explained. The resulting effect of the commutating action is to produce a pulsating but overall unidirectional e.m.f. at the terminals of the generator.

PRACTICAL REQUIREMENTS. To obtain a more uniform e.m.f., the 2-part commutator and single coil can be repeated to give an arrangement employing a great many segments and a larger number of coils. Each coil consists of a number of turns to give a larger output voltage.

The example shown in figure 7.14 is an armature with 2 coils at right angles. It follows that for this arrangement when coil A develops maximum e.m.f., coil B generates no e.m.f. and when the armature rotates through a quarter of a revolution, the conditions will be reversed, vice versa. The accompanying diagram (figure 7.15) shows the waveforms of the generated e.m.f.s. The generator terminal voltage never falls to zero but 2 disadvantages are evident. First, all the conductors are not used to maximum advantage since only one coil at a time supplies the external circuit. Secondly, but of prime importance, is the new condition of commutation. As the brushes must be placed in a position to contact the coil in which e.m.f. is being generated, it follows that if the generator is on load, i.e. supplying current, at the instant when the connected segments leave the brush, as an e.m.f. still exists and current is flowing, arcing can take place at the brushes. If coil A is considered and figures 7.14 and 7.15 noted, at the instant when the gap between segments is bridged by the brushes, coil A is still cutting the field and coil B has only just entered the field. Thus coil A tends to be short-circuited by a coil in

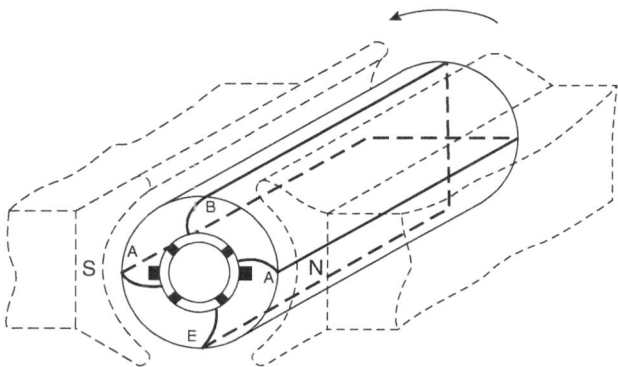

▲ **Figure 7.14** *Armature with 2 coils at right angles*

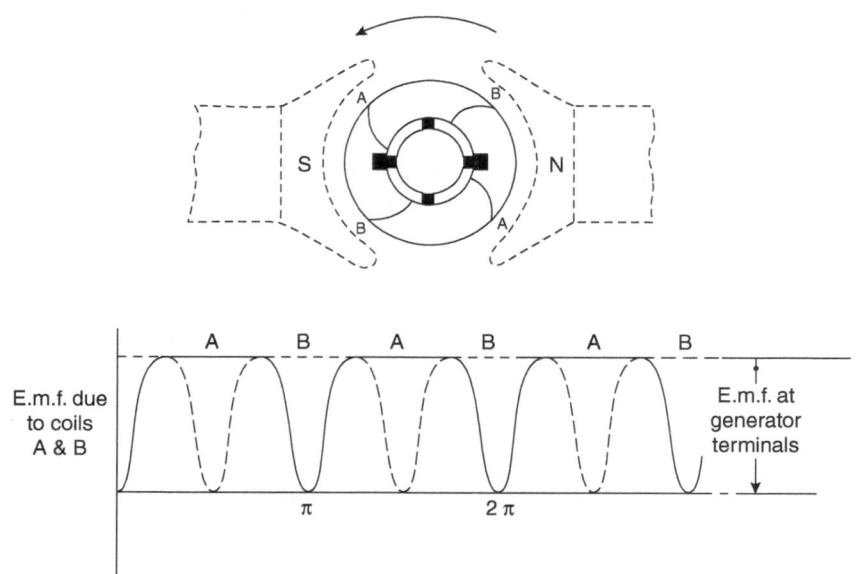

▲ **Figure 7.15** *Variation of E.m.f. at generator terminals*

which the e.m.f. may not have risen to the required value and current will flow instead in coil B. This current is diverted from the load current and adversely affects commutation. If the number of coils is increased, we achieve a smoother output voltage but arcing at the brushes may persist.

THE DRUM WINDING. Figure 7.16 shows the basic arrangement. This armature winding arrangement is accepted as the modern method of connecting active conductors together. The amount of 'copper' on the armature is used to maximum advantage as, except for the overhang at the back of a coil and the front connections to the commutator segments, the winding consists of copper conductors placed to cut magnetic flux and generate e.m.f. The basic winding uses several coils in series between brushes, arranged at constant angles to each other. The resultant e.m.f. is more uniform and larger as many coils in series are used. An example of a simple drum winding is given.

A, B, C, D, etc. are insulated conductors fitted into slots cut into the iron of the armature. There are also 4 commutator segments: Nos 1, 2, 3 and 4. Conductors may be connected to each other and to the segments in several ways and one possible arrangement is shown (figure 7.16a).

With rotation as shown, the e.m.f. in B, C and D are from front to back, while for F, G and H the e.m.f. are from back to front. The full lines show connection to segments and the dotted lines show connections at the back, which constitute the overhang of the coils.

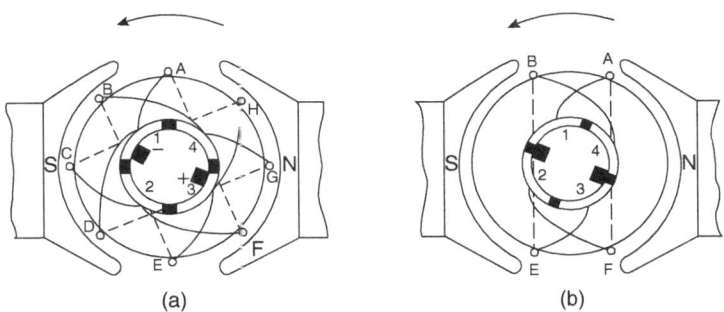

(a)   (b)

▲ **Figure 7.16** *Different possible conductor connection arrangements*

Starting at the −ve brush on segment 1, current enters the armature from the external circuit and divides into 2 parallel paths. One path passes to A and flowing down this comes at the back to F, hence up across to segment 2 and then onto and down C to H, from where it rises up and goes to the brush on segment 3. This would be the +ve brush. The other current path is from segment 1 to conductors D, G, B, E and onto segment 3 or the +ve brush. There are thus 2 circuits in parallel and it will also be noticed that, as a brush passes from one segment to the next, one coil is short-circuited and the brush must be located so as to short the coil at the instant when its e.m.f. and resultant current is zero. Such an instant is shown in figure 7.16b giving correct commutation conditions for the short-circuited coils between segments 1 and 2 and between segments 3 and 4.

Example 7.9. A slow-speed D.C. generator has an armature of diameter 3.0m and active conductors of length 510mm. The average strength of the field in the air gap is 0.8T and the armature speed is 200 rev/min. If the armature has 144 conductors arranged in 8 parallel paths, find the e.m.f. generated at the machine terminals (1 decimal place).

Using the formula $E = B\ell v$ volts so $B = 0.8$ teslas. $\ell = 510 \times 10^{-3}$m and v is obtained thus:

In 1 second the armature revolves $\dfrac{200}{60}$ or $\dfrac{10}{3}$ times. Also in

1 revolution one conductor travels $\pi d = 3.14 \times 3 = 9.42$m

So in 1 second the conductor travels $9.42 \times \dfrac{10}{3} = 31.4$m

E.m.f. generated per conductor $= 0.8 \times 510 \times 10^{-3} \times 31.4\text{m} = 12.8$V

Now there are 144 conductors in 8 parallel paths with conductors in series in each parallel path

$$= \frac{144}{8} = 18$$

Thus the e.m.f. generated in 1 parallel path $= 18 \times 12.8 = 230.4V$

Practice Examples

7.1. Calculate the e.m.f. in mV generated in the axles of a railway train when travelling at 100km/h. The axles are 1.4m in length and the component of the Earth's magnetic field density is 40µT (2 decimal places).

7.2. Find the generated e.m.f./conductor of a 6-pole D.C. generator having a magnetic flux/pole of 64mWb and a speed of 1000 rev/min. If there are 468 conductors, connected in 6 parallel circuits, calculate the total generated e.m.f. of the machine (1 decimal place). Find also the total power developed by the armature when the current in each conductor is 50A (5 significant figures).

7.3. An iron-cored coil of 2000 turns produced a magnetic flux of 30mWb when a current of 10A is flowing from the D.C. supply. Find the average value of induced e.m.f. if the time of opening the supply switch is 0.12s. The residual flux of the iron is 2mWb (1 decimal place).

7.4. A 1-turn armature coil has an axial length of 0.4m and a diameter of 0.2m. It is rotated at a speed of 500 rev/min in a field of uniform flux density of 1.2T. Calculate the magnitude of the e.m.f. induced in the coil (3 decimal places).

7.5. When driven at 1000 rev/min with a flux pole of 20mWb, a D.C. generator has an e.m.f. of 200V. If the speed is increased to 1100 rev/min and at the same time the flux/pole is reduced to 19mWb, what is then the induced e.m.f. (3 significant figures)?

7.6. A coil of 200 turns is rotated at 1200 rev/min between the poles of an electromagnet. The flux density of the field is 0.02T and the axis of rotation is at right angles to the direction of the field. The effective length of the coil is 0.3m and the mean width 0.2m. Assuming that the e.m.f. produced is sinusoidal, calculate (a) the maximum value of e.m.f. and (b) the frequency (1 decimal place).

7.7. A coil of 1200 turns is wound on an iron core and with a certain value of current flowing in the circuit, a flux of 4mWb is produced. When the circuit is opened, the flux falls to its residual value of 1.5mWb in 40ms. Calculate the average value of the induced e.m.f. (2 significant figures).

7.8. The armature of a 4-pole generator rotates at 600 rev/min. The area of each pole-face is 0.09m² and the flux density in the air gap is 0.92T. Find the average e.m.f. induced in each conductor (3 decimal places). If the armature winding is made up of 210 single-turn coils connected so as to provide 4 parallel paths between the brushes, find the generator terminal voltage (2 decimal places).

7.9. A solenoid 1.5m long is wound uniformly with 400 turns and a small 50 turns coil of 10mm diameter is placed inside and at the centre of the solenoid. The axes of the solenoid and the coil are coincident. Calculate (a) the flux in μW linked with the small coil when the solenoid carries a current of 6A (3 decimal places) and (b) the average e.m.f. induced in mV in the small coil when the current in the solenoid is reduced from 6A to zero in 50ms (3 decimal places).

7.10. Two coils A and B having 1000 and 500 turns respectively are magnetically coupled. When a current of 2A is flowing in coil A it produces a flux of 18mWb, of which 80% is linked with coil B. If the current of 2A is reversed uniformly in 0.1s, what will be the average e.m.f. in each coil (3 significant figures)?

8

ELECTROSTATICS AND CAPACITANCE

Have them make an ark of acacia wood – two and a half cubits long a cubit and a half wide, and a cubit and a half high. Overlay it with pure gold, both inside and out, and make a gold moulding around it.

Exodus 25.10–11 (New International Version)

Electric Field

The term electric field has been used already when discussing the P.D. required for electron movement in a circuit. It will be given some attention here, as it is directly associated with electron or current flow.

Elements, such as metals, have the same electrical properties as their atoms. If these atoms are charged and become ions (either through removal of electrons making them +ve, or through an excess of electrons making them −ve), the body of which they are part will be either +ve or −ve. Since atoms seek to remain neutral, ions will try to gain or lose electrons through exchange with neighbouring atoms. The same property exists for charged bodies and if a +ve charged body (deficient in electrons) is placed in contact with a −ve charged body (with excess electrons), electron flow will occur from the −ve body to the +ve body (with conventional current flowing in the opposite direction) until both have the same degree of charge. Before bodies are placed in contact with each other, a force will be detected between them and the adjacent space will show signs of this force. The space within which this force may be detected contains the *electric field*. A P.D. is said to exist between charged bodies, resulting in an electric field. When the bodies are placed in contact, an equalising of charges takes place with the basic requirement that a P.D. must exist between 2 points before a current flows. Earlier studies showed that an appropriate electrical device, for example, a battery or generator, functions by developing an e.m.f., resulting in a P.D. between its terminals and between the 2 bodies. If an e.m.f. or P.D. is maintained continually with a battery, then a continuous current will flow. The main subject of this book deals with *dynamic electricity* and its effects. However, since an e.m.f. or P.D. is only maintained until the current starts to flow and then falls to zero, as the charges on both bodies equalise, then *static electricity or electrostatics* will be the subject of this chapter.

Before commencing electrostatics, students are reminded that the unit of charge carried by the electron is *too small* for practical purposes. Experiment shows the −ve charge of an electron = 16×10^{-19} coulombs, where 1 coulomb = 6.25×10^{18} electron charges. Passage of charges constitutes a current and the practical current unit is the ampere, defined in terms of the coulomb and the time taken for this to pass. Thus if a charge of 1 coulomb takes 1 second to pass through a point in a circuit, the rate of flow of electricity is 1 coulomb per second and the current is 1 ampere.

Thus 1 ampere (1A) = 1 coulomb/second.

Both the coulomb and the ampere are used as practical engineering units and defined earlier. The properties of conductors and insulators were also described earlier and explained in terms of electron theory where:

A *conductor* is a material like metal, carbon and certain liquids, which contains mobile electrons that move under the influence of an applied P.D. and allow free passage of current.

An *insulator* is a material with few free electrons, such as rubber, glass, mica and most oils, in which electrons are bound strongly to a nucleus. As little electron movement occurs, current flow is negligible.

Electrostatics

Mention has already been made of static charges known from ancient times because of their attractive and repulsive effects. The Greeks knew that rubbed amber attracted light bodies such as cork and fibrous material. Amber is said to be 'charged' with electricity and the phenomena discussed here deal with the presence of electric charges at rest, i.e. electrostatics.

Experiments show that the easiest method of generating static electricity is by rubbing or friction. Thus a glass rod rubbed with silk is electrified to attract pieces of paper, but if a similarly treated glass rod is suspended by a thread, and brought near the original charged glass rod, a repulsive effect is noted. An ebonite rod rubbed with fur also becomes charged and, if brought near a suspended charged glass rod, attraction is now noted.

Summarising, glass and ebonite acquire charges of 2 types: positive (+ve) and negative (−ve), and like charges *repel* while unlike charges *attract*. Allocation of charge type, +ve charge to the glass and −ve charge to the ebonite, is arbitrary, but all uncharged bodies consist of +ve and −ve charges, which neutralise each other. If these charges are separated by an applied force, their presence is detectable and if they move from one body to another their movement is explained by the passage of current. It is noted that these assumptions agree with electron theory and it is apparent that a −ve charged body has an excess of electrons and a +ve charged body a lack of electrons. In the uncharged state, atoms of a material are neutral, i.e. charges due to electrons and protons exactly balance.

THE ELECTROSCOPE. Deductions made from experiments in electrostatics are key to theory, and for demonstrations a simple charge detector is needed. Such a detector, the gold-leaf electroscope, is often used for investigations, consisting of 2 leaves of gold foil attached to a metal rod, which is held in a glass jar from which it is insulated (figure 8.1). A metal disc may be fitted to the rod and the container may be a metal box-like frame with glass sides. Electroscope leaves hang down when no charges are present, but if charge is imparted to the instrument the leaves diverge.

Assume a +ve charge is given to the electroscope by stroking the disc with a glass rod charged by rubbing with silk. The +ve charge imparted to the disc spreads over

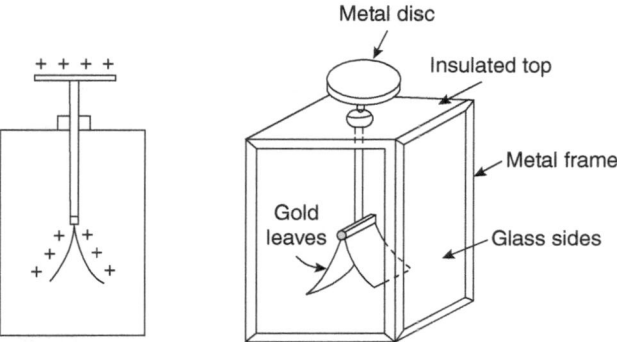

▲ **Figure 8.1** *An electroscope*

the insulated metal and the leaves, which, having the same charges, repel and diverge (figure 8.1). If an ebonite rod, −ve charged from rubbing with fur, is brought near the electroscope, the leaves converge. The explanation is that the +ve charges on the electroscope are attracted by the −ve charged rod and rise up to concentrate in the area of the disc. The charge on the leaves thus diminishes and the repulsion force between them falls. In the same way, a +ve charged glass rod brought near the instrument causes the leaves to diverge further as the +ve charges present are repelled down towards the leaves. Charge density in this region increases and increased divergence is observed. The instrument is not used outside a laboratory, but is useful for demonstrations.

Potential difference (P.D.)

When 2 bodies charged as described are brought into contact, a small current flows between them while the charges equalise, during which a P.D. exists between them. For the bodies to maintain their charges they must be insulated from earth, i.e. mounted on insulating rods. With the electroscope charge is given to it is with respect to earth; the gold leaves, rod and disc being one body and the mass of earth being the other. Hence leaves are charged +ve to earth if a +ve charge is given to the electroscope. Similarly a −ve charge given to the electroscope makes the leaves −ve with respect to earth. It is seen that if 2 bodies are charged +ve and −ve, they are at a potential to each other, i.e. a P.D. exists between them and they are also at a potential to earth. One body is +ve to earth and the other −ve to earth, the earth mass considered to be at zero potential.

Electrostatic charging

Frictional effects have been mentioned and in practical engineering, it is the most dangerous cause of electric charging. Build-up of charges must be considered and precautions taken in the artificial silk, paper, rubber, cable-making and associated industries to discharge coils of material after processing. Such processes involve kneading, rolling, drawing, etc. and friction may generate large voltages, which could be dangerous to those handling the material. The electrostatic charging of aircraft and motor-vehicles is a well-known hazard. In the case of the former, because of the large voltage levels, a means of earthing is needed before people can even leave aircraft. Electrically conducting rubber tyres have been developed to this end. For motor-cars the problem is not as important, as the resulting charges are small. However, precautions are needed for special load-carrying road vehicles such as petrol-tankers, which must be 'earthed' before loading or unloading of fuel commences. The action of the 'lightning conductor' will be mentioned shortly, but its use is concerned with the +ve and −ve charging of clouds, arising from atmospheric activity.

CHARGING BY INDUCTION. Imagine an electroscope to be uncharged and a +ve charged body brought near to the disc. Mobile metal electrons are attracted to the disc and +ve charged atoms or ions are displaced to the leaves (figure 8.2a). If the disc is touched with the charged rod (figure 8.2b), electrons from the main earth mass flow up and neutralise ions or +ve charges so the leaves collapse or converge. If then the rod is taken off and the adjacent +ve charged body removed, leaves diverge slightly (figure 8.2c). The −ve charges, previously held by the adjacent +ve charged body, spread over the electroscope and charges of the same polarity spread down to the leaves.

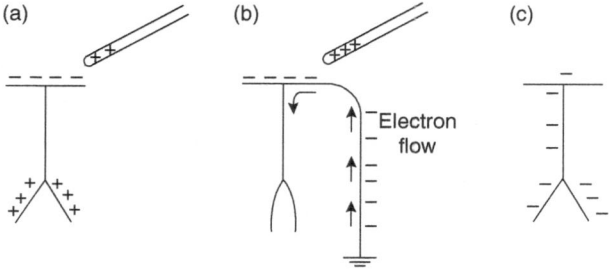

▲ **Figure 8.2** *Charging by induction*

Charging by induction produces a charge of opposite polarity, for example, the inducing charge was +ve so a −ve charge results on the instrument. If a −ve inducing charge is used, a +ve charge results instead.

Distribution of charge

The statements set out below are the results of experiments with a charged electroscope and a proof plane – a small metal disc fitted with an insulated handle. A proof plane is placed in contact with the body being investigated and electrified to the same polarity.

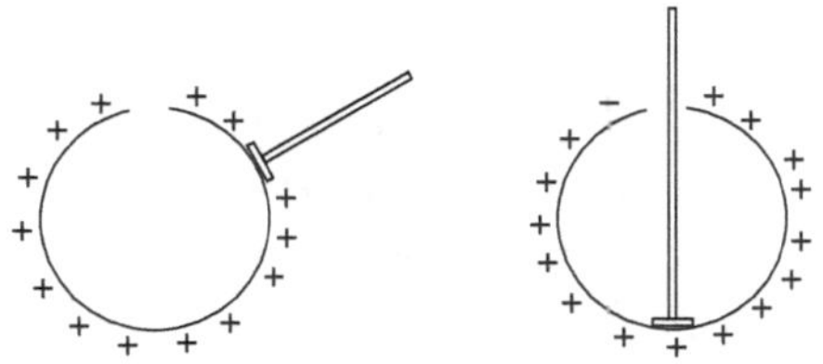

▲ **Figure 8.3** *Charge distribution*

If brought near a charged electroscope, movement of the leaves with the appropriate interpretation enables conclusions to be made. The following are deductions resulting from such investigations (figure 8.3).

1. A hollow body only charges on the outside. So a proof plane contacted with the outer surface and presented to a charged electroscope shows a deflection. If contacted with the inside, it shows no deflection.

2. If a sphere is charged, charge spreads uniformly over its surface and the *surface charge density* is uniform. If a charged body is non-spherical, charge concentration is greatest in the region where the radius of curvature is the *smallest*. Figure 8.4 depicts this, with the charge distribution or surface charge density represented by a dotted envelope.

3. If a charged body has a sharp point, charge concentrates at the point and surface density may be so great that dust or particles in the air coming in contact with the body are charged and repel. On moving away, each particle takes a small portion of the original charge and the effect of a point on a charged body is to discharge the body.

Let us now consider the action of a lightning conductor. Take a +ve charged cloud near a high building fitted with a lightning conductor, consisting of a copper rod and a well-earthed conductor. The building and rod acquire a −ve charge by induction, with air particles becoming −ve by contact, which move towards the cloud seeking to neutralise it. Alternatively, the space between the cloud and building being charged by the −ve particles has its insulating effect lowered until a breakdown occurs and a spark (lightning discharge) passes between the cloud and earth. A current flows during the discharge as electrons pass from earth to the cloud, which is discharged safely. Current will select the path of least resistance, and is conducted along the most suitable path, the lightning rod, and damage is avoided.

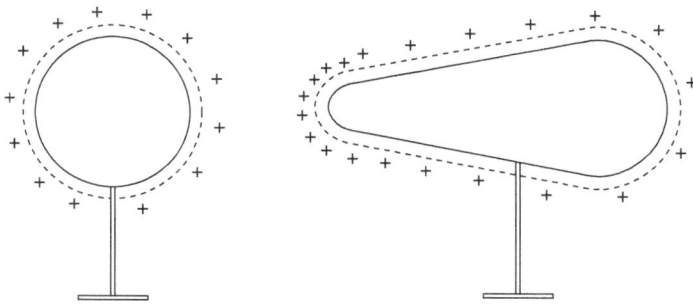

▲ **Figure 8.4** *Charge density and surface geometry*

Electrostatic fields of force

When 2 bodies are charged, a force of attraction or repulsion is produced depending on the charges' polarities. The size of this force relative to the charges can be investigated if the existence of *lines of flux* are assumed. A force is exerted on a small +ve charged body placed near a large +ve charged body. An electric field of force exists in the space around a large charged body. If a small body is small enough to constitute a point with +ve charge then, if free to move, it will travel in a fixed direction in the electric field and the path traced by it will represent a line of flux. Thus the large +ve charged body will have many lines of flux passing out from it. The similarity with the representation used for magnetic fields is noted. For electrostatic fields (figure 8.5), one key fact is observed. Each flux line terminates at the *surface* of a charged body and doesn't pass through the body to form a closed path, as is the case for magnetic flux lines. The medium through which the electric flux lines pass is called the *dielectric* and lines terminate at the *surface* of another body where balancing charges of opposite polarity appear. Thus figure 8.5a shows a +ve charged metal sphere in the centre of a room. Lines leave the surface perpendicularly, in all directions and planes, terminating on earthed walls, floor

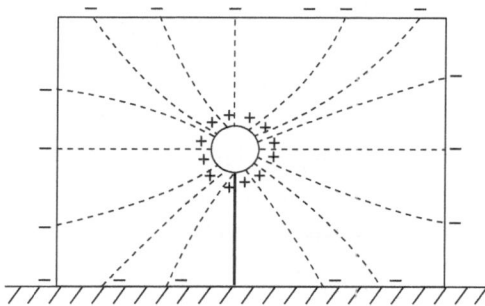

▲ **Figure 8.5a** *Electrostatic fields of force*

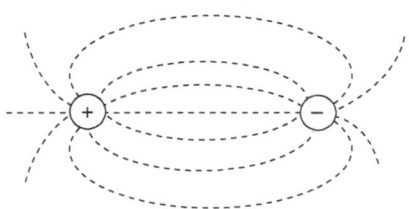

▲ **Figure 8.5b** *Electrostatic fields of force*

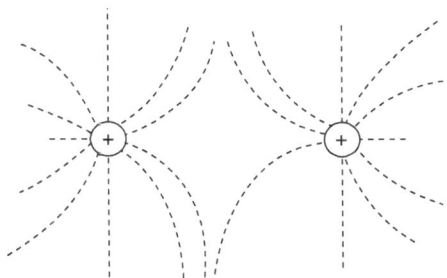

▲ **Figure 8.5c** *Electrostatic fields of force*

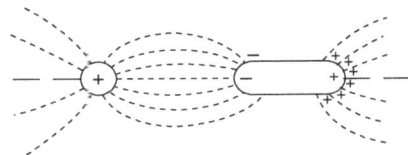

▲ **Figure 8.5d** *Electrostatic fields of force*

and ceiling, the earth mass being −ve to the charged body. Figure 8.5b shows the field arrangement associated with 2 bodies when charged in opposition, and figure 8.5c shows the field with like charges on the bodies. Figure 8.5d shows the arrangement when an uncharged body is placed in the field and how induced polarities result. As for the magnetic field, flux lines are imagined to behave like elastic threads, which contract if permitted.

Electrostatic flux

As for the magnetic field, we also introduce a term called 'flux' for the electric field. The symbol Ψ (Greek letter psi) and the number of electrostatic lines of flux passing through a medium is called the flux. As the practical charge unit is the coulomb, to establish an electrostatic unit we consider one line of flux to originate from 1 coulomb. So a charge of 10 coulombs produces 10 lines of flux, or Ψ = 10 coulombs. As most practical electrostatic work uses *capacitors* made from flat plate-like conductors, adjacent and parallel to each other, and as the medium between the plates or dielectric carries the flux, it is convenient to introduce the term *electric flux density* – symbol D, with $D = \dfrac{\Psi}{A}$ where A is the area of the dielectric in square metres. Then $D = \dfrac{\Psi}{A}$ or $\dfrac{Q}{A}$ coulomb per square metre.

Electric potential

The basic idea of electric potential was introduced earlier. If 2 bodies are charged and connected then as a current flowed while the charges equalise, there must have been a difference of potential between them. Earth's mass is taken to be at zero potential so if a body charged with Q coulombs of electricity is connected to earth, a current flows. Current will be from the body to earth if it has +ve charge, and from earth to the body if it has a −ve charge. Work is done in this process, so if we consider 1 joule of work done while 1 coulomb is passed, the P.D. of the body's electric potential will be 1 volt, where voltage measures the P.D. between the body and earth.

In a practical capacitor, plates are charged with respect to each other. If a +ve unit charge is placed in the field between charged capacitor plates, a force will push the +ve charge towards the −ve plate and is a measure of the *intensity* or *strength* of the field. The symbol for electric field strength or field intensity is E and the force is measured in newtons. So a charge of Q coulombs placed in an electric field E will experience a force of $F = QE$ newtons or $E = \dfrac{F}{Q}$ newtons per coulomb.

Alternatively field strength is the electric force or electrical potential gradient, measured in volts per metre.

Thus, $E = \dfrac{V}{d}$ where d is the distance between the plates in metres.

Both expressions for electric field intensity and potential gradient or electric force are numerically equal. If practical units are substituted, the same amount of work is done by 1 newton of force acting through 1 metre distance between the plates as done by 1 coulomb conveyed by a pressure of 1 volt.

$$\text{Thus, since } E = \frac{F}{Q} \text{ or } \frac{V}{d} \text{ then } \frac{F}{Q} = \frac{V}{d} = \text{ or } Fd = VQ$$

Note.

1 joule = 1 volt × 1 coulomb, or 1 joule = 1 newton × 1 metre.

Capacitance

The capacitor

Several references have already been made to the electrical capacitor (or condenser, as it was called). In its simplest form, it consists of 2 metal plates separated from each other by an air gap. As will be seen, the plates' area, separation, and type of dielectric insulating medium all affect a capacitor's performance, but the key fact learned from such experiments is that capacitors store electricity. Thus if the plates are connected to a supply source through a sensitive milliammeter, a current will pass at the instant the switch closes. The current quickly falls to zero, as the P.D. between the plates rises, as indicated by a voltmeter. The form of current fall-off is a topic itself, but a capacitor will attain a 'charged' state. If the supply is disconnected and the plates are shorted together, a discharge current flows in the opposite direction, and although initially large, soon decays to zero. The initial voltage, although showing the value of the charging supply, will also decay to zero.

Experiments with a simple capacitor establish a basic relationship between the quantity of electricity Q that can be stored and the charging voltage V. The former is proportional to the latter or $Q \propto V$. As this is directly proportional, a constant is introduced to give the expression:

$$Q = CV$$

C is termed the *capacitance* and is a unit defined in terms of unit quantity and unit voltage. The unit of capacitance is the *farad* and a capacitor has a capacitance of 1 farad if 1 coulomb of electricity is stored when 1 volt is applied across the plates.

Example 8.1. Find how many electrons are displaced when a P.D. of 500V exists between the plates of a 4μF capacitor (3 significant figures).

$$\text{Since } Q = CV$$

then $Q = 4 \times 10^{-6} \times 500 = 2 \times 10^{-3}$ coulombs, but 1 coulomb $= 6.3 \times 10^{18}$ electrons

$$\therefore \text{ No of electrons } = 12.6 \times 10^{15}.$$

Capacitor systems

Capacitors can be connected in series or parallel and the student should compare the expressions giving equivalent capacitance values with those giving equivalent resistance values, for comparable arrangements.

SERIES CONNECTION. The arrangement is shown in figure 8.6.

Let capacitors have values: C_1, C_2 and C_3 farads respectively, and the applied voltage V be dropped as shown.

Then as $V = V_1 + V_2 + V_3$ and,

Since $V_1 = \dfrac{Q_1}{C_1}$ $V_2 = \dfrac{Q_2}{C_2}$ and $V_3 = \dfrac{Q_3}{C_3}$

we can write:

$$V = \frac{Q_1}{C_1} + \frac{Q_2}{C_2} + \frac{Q_3}{C_3}$$

If C is taken to be the equivalent capacitance of the arrangement then:

$$V = \frac{Q}{C}$$

$$\text{or } \frac{Q}{C} = \frac{Q_1}{C_1} + \frac{Q_2}{C_2} - \frac{Q_3}{C_3}$$

but the same current flows through each capacitor for the same time. $\therefore Q = Q_1 = Q_2 = Q_3$ and the above can be simplified to:

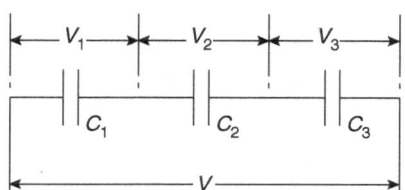

▲ Figure 8.6 *Capacitors in series*

$$\frac{1}{C} = \frac{1}{C_1} + \frac{1}{C_2} + \frac{1}{C_3} \text{ etc.}$$

PARALLEL CONNECTION. The arrangement is shown in figure 8.7 with the same voltage applied to each capacitor.

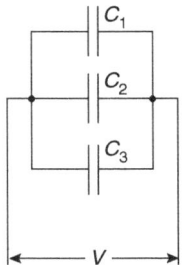

▲ **Figure 8.7** *Capacitors in parallel*

Then for each capacitor $Q_1 = C_1 V$ $Q_2 = C_2 V$ and $Q_3 = C_3 V$

If the total quantity of charge $= Q$ then,

$$Q = C_1 V + C_2 V + C_3 V = V(C_1 + C_2 + C_3)$$

$$\text{or } \frac{Q}{V} = C_1 + C_2 + C_3$$

If C is the equivalent capacitance of the arrangement, then

$$Q = CV \text{ or } CV = V(C_1 + C_2 + C_3)$$

$$\text{Such that } C = C_1 + C_2 + C_3.$$

Example 8.2. If 2 capacitors of values 100µF and 50µF respectively are connected (a) in series (4 significant figures) and (b) in parallel (3 significant figures) across a steady applied voltage of 1000V, calculate the joint capacitance.

(a) Series. Joint capacitance C is given by:

$$\frac{1}{C} = \frac{1}{100} + \frac{1}{50}$$

$$C = \frac{100}{3} = 33.33 \text{ µF.}$$

(b) Parallel. Joint capacitance is given by $C = 100 + 50$ or $C = 150$ µF.

Capacitor current

From the relation $Q = CV$, the following can be deduced.

$$\text{Since } Q = It \text{ then } It = CV$$

$$\text{or } I = C\frac{V}{t}$$

The expression shows that current only flows when the voltage across a capacitor changes, as $\dfrac{V}{t}$ represents a rate of change of voltage. The current at any instant can be found if the rate of change of the voltage is known at that instant. If, however, the rate of change is uniform for a period of time, then a constant current will flow, as illustrated by the next example.

Example 8.3. The P.D. across the plates of a 50µF capacitor varies thus:

From time $t = 0$ to $t = 1$ms, V rises uniformly from 0–200V

From time $t = 1$ to $t = 3$ms, V is constant at 200V

From time $t = 3$ to $t = 5$ms, V falls uniformly from 200–0V

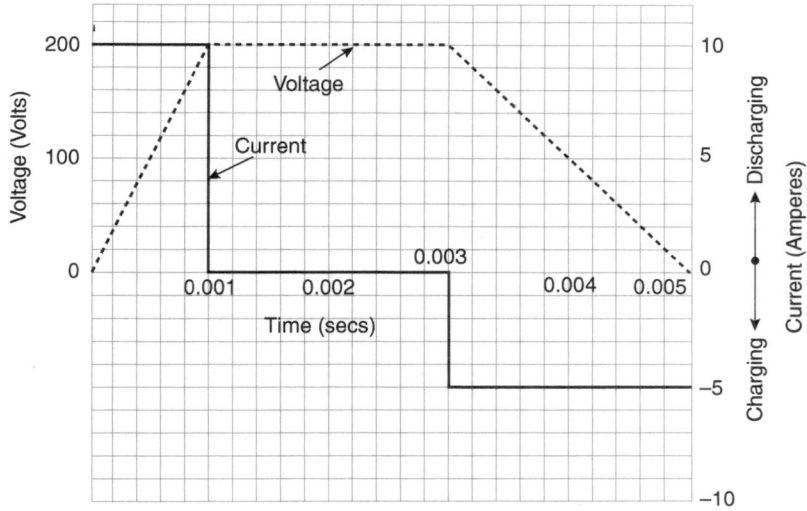

▲ **Figure 8.8** *Voltage and current vs time*

Illustrate the voltage variations on a graph and deduce the shape of the current wave during the period of 5ms.

Since $Q=CV=It$ then $It=CV$ or $I=C\dfrac{V}{t}$

(a) $I = 50 \times 10^{-6} \times \dfrac{(200-0)}{0.001} = 10A\,(\text{charging})$

(b) $I = 50 \times 10^{-6} \times \dfrac{(200-200)}{0.002} = 0A$

(c) $I = 50 \times 10^{-6} \times \dfrac{(0-200)}{0.002} = -5A\,(\text{discharging})$

The required graphs are shown in figure 8.8.

Energy stored in an electric field or dielectric

Consider the voltage to rise uniformly across the capacitor plates, to a value of V volts, in a time t seconds. The average P.D. value will be $\dfrac{V}{2}$ volts and the charging current constant equal to I amperes. The average power supplied during the charging period will be $\dfrac{V}{2} \times I$ watts and the energy fed in will be $\dfrac{V}{2} \times I \times t$ joules.

This energy is not converted into heat, as a capacitor has no resistance, but it does work in establishing the electric field. It is this energy that is stored, and then recovered when the field collapses as the capacitor discharges. Thus:

$$\text{Energy stored} = \frac{V}{2}It \text{ joules or} = \frac{V}{2}Q \text{ joules}$$

$$\text{or alternatively, } W = \frac{1}{2}CV^2 \text{ joules.}$$

Example 8.4. Consider the capacitor arrangement of Example 8.2 and calculate the total energy stored for a steady applied voltage of 1000V for (a) series (2 decimal places) and (b) parallel (2 significant figures) connections.

(a) Series

Energy stored is given by $W = \frac{1}{2}CV^2$ joules

$$= \frac{1}{2} \times 33.33 \times 10^{-6} \times 1000^2$$

$$= 16.67J.$$

(b) Parallel

Energy stored or $W = \frac{1}{2} \times 150 \times 10^{-6} \times 1000^2$

$$= 75J.$$

Note. The equivalent capacitance for each arrangement is used for C in the energy expression.

Relative permittivity

If a parallel-plate capacitor is made so one plate is connected to earth and the other to an electroscope, the effect of altering the dielectric can be noted. With air as the insulating medium between the plates, a capacitor is charged to a given value as indicated by the amount of divergence of an electroscope's leaves. If a sheet of insulating material, for example, a slab of paraffin wax, is placed in the air gap, a *converging* effect is produced, indicating that charge *appears* to have reduced. However, if the insulation is removed, the leaves again diverge to the original extent. A correct assumption is that the capacitance has increased, i.e. the capacitor now accepts a greater charge for the same leaf divergence. This is confirmed if a capacitor is charged to an amount giving the original leaves divergence and the wax insulation is removed. The leaves will then diverge to a greater extent, showing a larger charge has been imparted. The experiment shows that capacitance can vary with the type of dielectric, a property termed its *permittivity*. Permittivity is likened to the permeability of a magnetic material. Relative permittivity can be defined as the ratio of the capacitor's capacitance with the material taken as the dielectric, compared with the capacitance of the same capacitor with air as the dielectric. Another term for relative permittivity is the *dielectric constant*. The symbol used is $\in_r$ (the Greek letter – small epsilon).

Typical values for relative permittivity are as follows: air 1.0006, paraffin wax 2.2, mica 4.5 to 8 and glass 4 to 10.

Absolute permittivity

Absolute permittivity can also be considered as $\in_r$ was introduced relative to air. As for a magnetic circuit, we can write:

Absolute permittivity = relative permittivity × permittivity of free space

So $\in = \in_r \times \in_0$

Permittivity of free space

As comparisons have been made with the magnetic circuit, permittivity is best understood by comparing it with permeability, its magnetic equivalent.

$$\text{Permeability is defined as the ratio} = \frac{\text{Flux density}}{\text{Magnetising force}}$$

$$\text{or } \mu = \frac{B}{H}$$

$$\text{Similarly, permittivity} = \frac{\text{Electric flux density}}{\text{Electric force}} \quad \text{or } \in = \frac{D}{E}$$

For air, permittivity is measured to be

$$\frac{1}{4\pi \times 9 \times 10^9} \text{farads per metre}$$

i.e. $\in_0 = 8.85 \times 10^{-12}$ farads per metre.

Although other alternatives are derived to estimate $\in_0$, the following deduction gives the required answer. If unty is taken for the area dimensions A and d the plate spacing:

$$\text{Then since } \epsilon_0 = \frac{D}{E} = \frac{Q/A}{V/d} = \frac{Q}{V} \times \frac{d}{A}$$

$$\text{or } \epsilon_0 = \frac{CV}{V} \times \frac{d}{A} = \frac{Cd}{A}$$

When d and A are equal to 1, ϵ_0 = the capacitance value of the arrangement. For a vacuum, the capacitance value of the standard capacitor, using unity for A and d, is measured to be 8.85×10^{-12} SI units.

$$\text{or } \epsilon_0 = \frac{1}{4\pi \times 9 \times 10^9} \text{ also expressed as,}$$

$$8.85 \times 10^{-12} \text{ farads per metre.}$$

Note. Although a vacuum is mentioned for the above capacitor arrangement, air can be taken as the dielectric, as the variation is small enough to be neglected.

Capacitance of a parallel-plate capacitor

Consider the area of the plates to be A square metres and their spacing d metres, i.e. the thickness of the dielectric. Applied voltage is taken as V volts, resulting in a charge of Q coulombs. As charge Q is assumed to be uniformly distributed over the whole area of the plates, electric flux density D will be $\frac{Q}{A}$.

The electric force or potential gradient E in the dielectric is $\frac{V}{d}$ volts per metre and permittivity ϵ (by definition) $= \frac{D}{E}$.

$$\text{Thus } \epsilon = \frac{D}{E} \text{ or } \epsilon = \frac{Q/A}{V/d} = \frac{Qd}{VA}.$$

When $\epsilon = \frac{CVd}{VA} = \frac{Cd}{A}$ or $C = \frac{\epsilon A}{d}$ but $\epsilon = \epsilon_0 \epsilon_r$

$$\text{So } C = \frac{\epsilon_0 \epsilon_r A}{d} \text{ farads.}$$

Example 8.5. A capacitor consists cf 2 parallel metal plates, each 300mm × 300mm, separated by a sheet of polythene 2.5mm thick, with relative permittivity 2.3. Calculate the energy stored in the capacitor when connected to a D.C. supply of 150V (3 significant figures).

$$C = \frac{8.85 \times 10^{-12} \times 2.3 \times (300 \times 10^{-3})^2}{2.5 \times 10^{-3}} \text{ farads}$$

$$C = 733 \times 10^{-12} \text{F}$$

$$C = 733 \text{p}^{\bar{}}$$

$$\text{Energy stored} = \frac{1}{2}CV^2 \text{ joules}$$

$$= \tfrac{1}{2} \times 733 \times 10^{-12} \times 150^2$$

$$= 8.25 \times 10^{-6} \text{ joules}$$

$$= 8.25 \mu J$$

Example 8.6. A capacitor of 5μF charged to a P.D. of 100V is connected in *parallel* with an identical uncharged capacitor. What quantity of electricity flows into the second capacitor and to what voltage will it be charged?

Consider the first capacitor designated A, then as $Q = C_A V$, $Q = 5 \times 10^{-6} \times 100 = 5 \times 10^{-4}$ coulombs.

When capacitor B is connected across A, charge passes from A to B until the potential of each is the same. The capacitor arrangement is considered as a parallel connection or the joint capacitance is the same as that of 1 unit of 10μF.

$$\text{Applying the formula } Q = CV$$

$$\text{Then } V = \frac{Q}{C} = \frac{5 \times 10^{-4}}{10 \times 10^{-6}}$$

$$\text{o}^{\bar{}} V = 50 \text{ volts, the final voltage.}$$

This may be deduced from the fact that the capacitors are similar so charge passes from A to B until the potential of each is the same.

Transient effects in D.C. circuits

We have considered the factors affecting capacitance and charge on capacitor plates, without reference to real circuits. We will return to the effects of capacitance on A.C. circuits later but here it is appropriate to consider the application of a D.C. voltage to a capacitive circuit.

In a resistive D.C. circuit, current rises to its final steady state and the P.D. across it settles to a steady value almost immediately. Similarly, on opening a circuit, current and P.D. fall almost immediately to zero. However, when a capacitor and resistor, or an inductor and resistor, are connected in series to a D.C. supply, the final steady state of current and P.D. is *not* achieved immediately. The change takes a short period of time depending on the circuit component values. Such changes of current and P.D. up to the final steady state conditions are called *transient* values, which grow or decay in an *exponential* way.

Capacitor in a D.C. circuit

Consider figure 8.9, an unchanged capacitor in series with a resistor R connected to a D.C. supply V.

$$V = P.D. \text{ across } R + P.D. \text{ across } C$$

$$V = iR + v$$

$$\therefore i = \frac{V - v}{R}$$

At the instant of switch-on, the instantaneous voltage value across the capacitor (v) is zero, and the current (i) is maximum, limited only by resistance R.

At switch-on $i = \dfrac{V}{R}$, but this current is the charging capacitor current and, as the capacitor charges, v increases, which in turn decreases i. The rate of charge therefore decreases until, when the capacitor is fully charged, current is zero. However zero current is achieved at infinite time but, for practical purposes, it is assumed this occurs in a time equal to 5 times the initial rate of charge. If the initial rate of charge is maintained, the capacitor will have fully charged in a definite time depending upon the circuit components. This time is the *time constant* and given the symbol τ (Greek letter TAU)

Thus $\tau = CR$ seconds

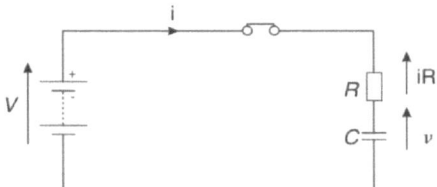

▲ **Figure 8.9** *A capacitor in a D.C. circuit*

The rate of charge decreases, a graph of capacitor voltage against time is in a non-linear manner, following an exponential law (figure 8.10).

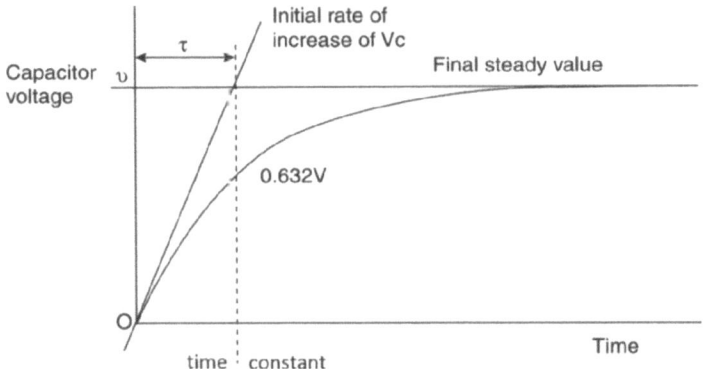

▲ **Figure 8.10** *Capacitor voltage vs time*

The equation for this line is:

$$v = V(1 - e^{-t/\tau}) \text{ volts}$$

Where v = Instantaneous P.D. across capacitor

V = Applied circuit voltage

t = Time from switch-on

τ = CR = Time constant

It is noted the capacitor voltage will be 0.632 of the final steady value after a time equal to one time constant. Similarly, during a second time constant, voltage only increases by 63.2% of the remaining voltage, and after each subsequent time interval, voltage increases similarly. As current i is a maximum value when capacitor voltage v is zero, and voltage increases exponentially, current will decrease exponentially as shown in figure 8.11.

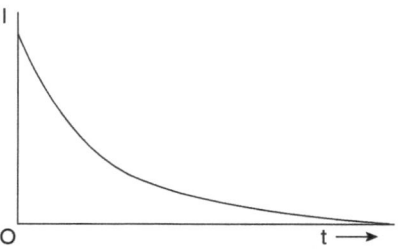

▲ **Figure 8.11** *Current vs time*

The equation for this curve is:

$$i = \text{I}e^{-t/\tau} \quad \text{or} \quad i = \frac{V}{R}e^{-t/\tau}$$

Consider now a fully charged capacitor C in figure 8.12 discharged through resistor R. Capacitor C acts as a supply source and as it discharges the P.D. across it falls and again the curve is exponential.

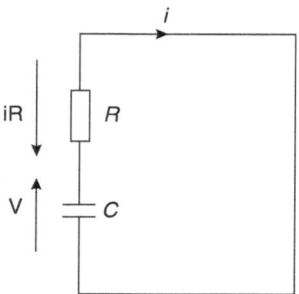

▲ **Figure 8.12** *Circuit for a discharging capacitor*

As the current is limited by the resistance $\left(i = \dfrac{V}{R}\right)$, at the start of discharge the current is maximum $\left(i = \dfrac{V}{R}\right)$, so the current curve also follows a falling exponential curve.

Hence on discharge:

$$v = Ve^{-t/\tau}$$

$$\text{and } i = \text{I}e^{-t/\tau}$$

Example 8.7. A 40kΩ resistor and a 20µF capacitor are connected in series to a 200V D.C. supply. Find the circuit current (3 significant figures) and the P.D. across the capacitor after 0.2 seconds (2 decimal places) from switch-on.

Time constant $\tau = CR = 20 \times 10^{-6} \times 40 \times 10^3$

$$= 0.8s.$$

$$i = \frac{V}{R}e^{-t/\tau} = \frac{200}{40 \times 10^3}e^{-0.2/0.8} = 5 \times 10^{-3}\, e^{-0.25}$$

$$\therefore \text{ After 0.2s } i = 3.89mA$$

$$v = V(1 - e^{-t/\tau}) = 200(1 - e^{-0.2/0.8})$$

$$\therefore \text{ After 0.2s } v = 44.24V$$

Example 8.8. A 20µF capacitor, fully charged to a voltage of 300V, is discharged through a 1MΩ resistor. Find the time taken for the capacitor voltage to fall to 60V (1 decimal place).

$$\tau = CR = 20 \times 10^{-6} \times 1 \times 10^6$$

$$= 20s$$

$$= 20s$$

$$v = Ve^{-t/\tau}$$

$$60 = 300e^{-t/20}$$

$$0.2 = e^{-t/20}$$

$$\frac{-t}{20} = -1.61$$

$$t = 32.2s.$$

Practice Examples

8.1. Two capacitors of 0.02µF and 0.04µF are connected in series across a 100V D.C. supply. Find the voltage drop across each unit (1 decimal place).

8.2. For the circuit shown, calculate the effective capacitance between A and B. The capacitance values shown are in microfarads (1 significant figure).

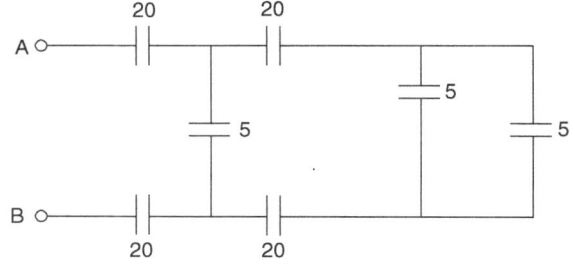

8.3. A variable capacitor having a capacitance of 1000μF is charged to a P.D. of 100V. The plates of the capacitor are then separated by means of an insulated rod, so that the capacitance is reduced to 300μF. Find, by calculation, by how much the capacitance changes (1 decimal place).

8.4. A plate capacitor consists of a total of 19 metal-foil plates each 2580mm² and separated by mica 0.1mm thick. Find the capacitance of the assembly if the relative permittivity of mica is 7 (3 significant figures).

8.5. A P.D. of 10kV is applied to the terminals of a capacitor consisting of 2 circular plates, each having an area of 10 000mm², separated by a dielectric 1mm thick. If the capacitance is 3×10^{-4}μF, calculate the electric flux density (1 significant figure) and the permittivity of the dielectric (2 decimal places).

8.6. A capacitor consists of 2 parallel metal plates, each 200mm by 300mm, separated by a sheet of polythene 3.5mm thick, having a relative permittivity of 3.0. Calculate the energy stored in the capacitor when connected to a D.C. supply of 300V (4 significant figures).

8.7. Calculate the capacitance value of a capacitor that has 10 parallel plates separated by insulating material 0.3mm thick. The area of one side of each plate is 1500mm² and the relative permittivity of the dielectric is 4 (2 significant figures).

8.8. Two capacitors A and B having capacitances of 20μF and 30μF respectively are connected in series to a 600V D.C. supply. Determine the P.D. across each capacitor (3 significant figures). If a third capacitor C is connected in *parallel* with A and it is then found that the P.D. across B is 400V, calculate the value of C (2 significant figures) and the energy stored in it (1 decimal place).

8.9. A D.C. voltage of 500V is applied to a 40μF capacitor. Find the value of the charging current at the instants when the voltage is varying as follows:

Time $\left(\dfrac{1}{1000} \text{ sec}\right)$	0–1	1–2	2–3	3–4	4–5
Voltage values	0–100	100–150	160 constant	150–50	50–0

8.10. The capacitance of a single-phase concentric cable has a value of $C = 0.289$μF. The diameter of the inner core is 12mm and the insulation has a radial thickness of 10mm. Calculate the permittivity of the insulating material (2 decimal places).

9

BASIC ALTERNATING CURRENT (A.C.) THEORY

The invention of the wheel was perhaps rather obvious; but the invention of an invisible wheel, made of nothing but a magnetic field, was far from obvious, and that is what we owe to Nikola Tesla.

Reginald Kapp

Let us revise briefly the fundamental principles of Chapter 7. Figures 9.1a and 9.1b show a simple A.C. generator where a coil rotates in a uniform magnetic field. The sides of the coil, i.e. the conductors, cut the magnetic flux and an e.m.f. is induced, which from first principles is $e = B\ell v$ volts. The letter e, for the induced e.m.f., is introduced because this value varies from instant to instant. Even though the coil rotates at a uniform velocity v, the cutting rate is not constant, but depends upon the *angle* at which the conductors cut the magnetic flux. The velocity is resolved into a cutting component ($v \sin \theta$) and a non-cutting component ($v \cos \theta$). Only the cutting velocity component is responsible for e.m.f. and creates a general expression for the e.m.f. at any instant: $e = B\ell v \sin \theta$ volts.

The A.C. Waveform

In the expression $e = B\ell v \sin \theta$, as for an alternator, B, ℓ and v are assumed constant and equal to K. The expression thus becomes $e = K \sin \theta$ and a value for K is obtained if we consider the instant when the coil's sides cut the field at right angles. Velocity

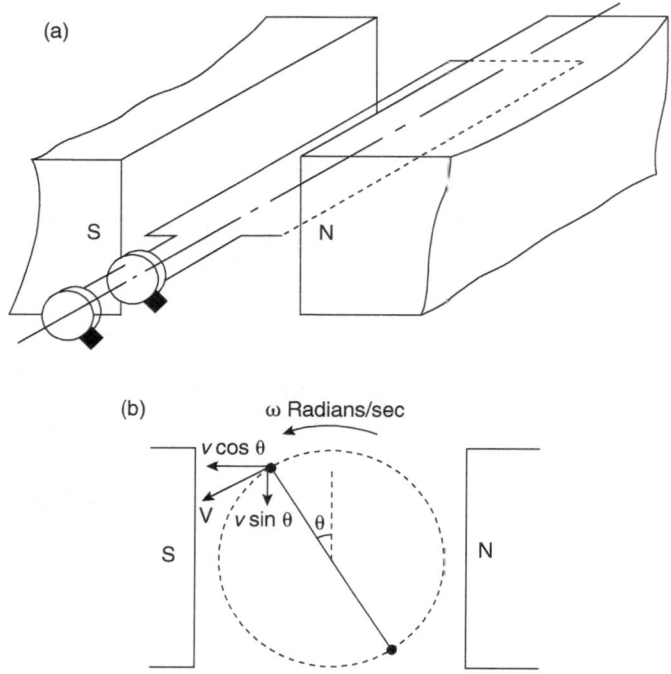

▲ **Figure 9.1** *The A.C. waveform*

component $v \sin \theta$ generates a maximum e.m.f. designated E_m. At this instant $e = E_m$ and we write $E_m = K \sin \theta$. But as $\sin \theta = \sin 90 = 1$. ∴ $E_m = K$.

Substituting back in the original expression:

$$e = E_m \sin \theta$$

This is not just clever manipulation of an equation but is important as it shows that the generated e.m.f. varies *sinusoidally*, where e is the *instantaneous* value and E_m the *maximum* value.

If we examine the waveform plotted to a time or angle base, it must be remembered from vector treatment that a sine wave is deduced from a rotating vector's vertical component – for electrical work a rotating vector is called a *phasor*. If the phasor's length represents E_m, then for any angle θ, the instantaneous value is the vertical projection used as an ordinate for the waveform, when plotted to an angle or time base. Figure 9.2 illustrates this process and the method is summarised: Draw a circle of radius equal to the maximum wave's value. Start from the horizontal, and move the phasor through a known angle and project the vertical value onto an angle or time scale. Choose suitable

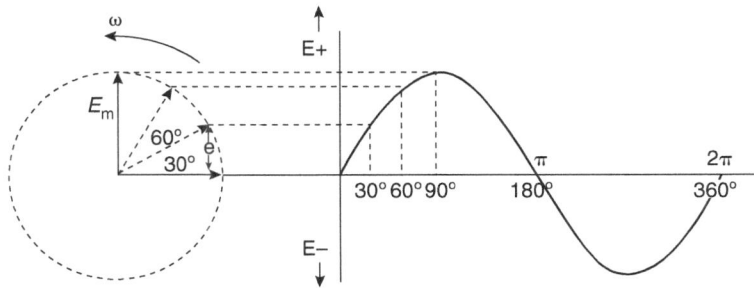

▲ **Figure 9.2** *Phasor diagram*

scales to avoid distorting the wave's sinusoidal shape. The connection between the construction and representation of a sinusoidally induced e.m.f. is seen in the various triangles:

$$\frac{e}{E_m} = \sin \theta \text{ so } e = E_m \sin \theta.$$

The expression is the same as that deduced previously but can be modified for phasor representation.

In accordance with accepted procedure, assume the phasor E_m rotates from the zero or horizontal position *anticlockwise* with an angular velocity of ω radians/second (ω is a Greek letter – small omega). So $\theta = \omega t$ where t is time in seconds and the equation written as $e = E_m \sin \omega t$.

Figure 9.3 shows some of the terms used in A.C. theory. *Periodic time* = the time for 1 complete cycle. The *frequency f* of a wave = the number of complete cycles in a time interval of 1 second. In accordance with SI recommendations, the *hertz* (Hz) is used

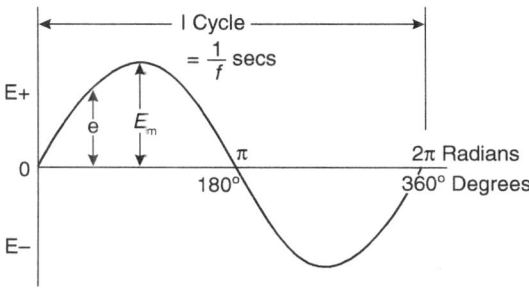

▲ **Figure 9.3** *Frequency representation of a wave*

for frequency measurement. Current marine practice is usually to have either 50Hz or 60Hz A.C. systems. The maximum value reached by a wave is the *peak* value or *amplitude* and at any instant is termed the *instantaneous* value and denoted by a small letter *e*. We observe that sinusoidal current conditions also occur and that the expression for a current following a sine wave law can also be written as $i = I_m \sin \omega t$.

Following introduction of the sine wave, derived from a phasor, and the generation of a sinusoidal e.m.f. by a rotating coil, the treatment can be combined by the following. The phasor is assumed to rotate at a constant angular velocity of ω radians/second and the waveform, if at a frequency of f hertz, will cover in 1 second an angle of 360f degrees or 2πf radians. The phasor meanwhile will pass through ω radians in 1 second so ω can be equated to $2\pi f°$ or $360f°$. The earlier expressions can now be written in their most useful form, namely:

$$e = E_m \sin 2\pi ft.$$

This is the first fundamental A.C. theory formula. It is important to note that if $\dfrac{22}{7}$ or 3.14 is substituted for π, the angle will be in radians but can be converted into degrees by multiplying by 57.3. A simpler method is to substitute 180° for π, converting into degrees directly.

Example 9.1. Find the *instantaneous* value of a 50Hz sinusoidal e.m.f. wave, maximum value 100V, at a time 0.003s after the zero value (1 decimal place).

Substituting in $e = E_m \sin 2\pi ft$ we have:

$$e = 100 \sin (2 \times 180 \times 50 \times 0.003)$$

$$\text{or } e = 100 \sin 54°$$

$$\text{and } e = 80.9\text{V}.$$

Important Note. A problem can occur if the instantaneous value is given and the *time* is required. Attention must be given to the following example, which illustrates this point.

Example 9.2. Find the *first* time after zero when the instantaneous value of a sinusoidal current wave is 6.8A. The maximum value is 12A and the frequency 50Hz. Find also the *second* time after zero.

$$\text{Here } i = I_m \sin 2\pi ft$$

$$\text{or } 6.8 = 12 \sin (2 \times 180 \times 50 \times t)$$

$$\text{Thus } \frac{6.8}{12} = \sin(180 \times 100 \times t)$$

and $0.566 = \sin(18 \times 10^3 \times t)$. Solution of this equation can only be made by use of electronic calculator or sine tables, from which an angle can be found whose sine equals 0.566.

Thus let θ = this angle, then $\sin\theta = 0.566$. So $\theta = \sin^{-1}(0.566)$, hence $\theta = 34°30' = 34.5°$

$$\therefore \sin\theta = \sin(18 \times 10^3 \times t) \text{ or } 18 \times 10^3 \times t = 34.5$$

$$\text{and } t = \frac{34.5}{18 \times 10^3} = 1.9\text{ms}$$

The second time value is also required, and occurs $\frac{1}{2}$ cycle later. Subtracting the interval from the zero value is needed to attain a second instantaneous height of 6.8A.

$$\text{Thus time for } \frac{1}{2} \text{ cycle} = \frac{1}{100} \text{ seconds} = 0.01\text{s (as } T = 1/f)$$

So second time required $= 0.01 - 0.0019 = 0.0081$s or 8.1ms.

Representation of Sinusoidal Alternating Quantities

Earlier it was shown that an alternating voltage or current can be respectively represented by:

$$e = E_m \sin 2\pi ft \quad \text{or} \quad i = I_m \sin 2\pi ft.$$

This notation method conveys all that is required about the quantity, i.e. it follows a form whose amplitude, frequency and instantaneous value at any particular time can be found. This method of notation is called a trigonometrical representation.

Trigonometrical representation

This is useful for 2 alternating quantities, but not necessarily simultaneous quantities. A 50Hz alternating voltage can create an alternating current in a circuit to alternate at

50Hz. The current need not, however, be *in-phase* with the voltage, which may reach its maximum value a little time before the current reaches its maximum value. The voltage is said to *lead* the current or the current to *lag* behind the voltage. There is a *phase difference* between the 2 quantities or waveforms. Such a phase difference is shown by inclusion of a phase angle in radians. Thus if 2 current waveforms are represented by:

$$i_1 = I_{1m} \sin 2\pi ft \quad \text{and} \quad i_2 = I_{2m} \sin \left(2\pi ft + \frac{\pi}{3}\right)$$

the second waveform will lead the first by $\frac{\pi}{3}$ radians or $\frac{180}{3} = 60°$. A third waveform written $i_3 = I_{3m} \sin \left(2\pi ft - \frac{\pi}{6}\right)$ will lag the first or reference waveform by $\frac{180}{6} = 30°$.

This mathematical trigonometrical form of representation is used for trigonometrical operations, such as multiplication, division, expansion, etc. and these will be illustrated in A.C. theory later.

Phasor representation

This is used for A.C. quantities such as current, voltage, flux, etc. and was introduced with vectors. Since voltages or currents are quantities whose magnitudes and directions are known, they can be described by rotating vectors. As 'phase' is often involved it is customary to represent these with phasors. Thus a voltage phasor drawn to scale represents the magnitude of a voltage and the direction in which it acts is shown by an arrow. This technique was introduced earlier in this chapter, and we will now consider the standard phasor operation methods.

The relation between the graphical deduction of a waveform from a phasor has been covered. For most practical A.C. work waveforms and instantaneous values are of little importance compared with its magnitude and phase, so using only a phasor as shown

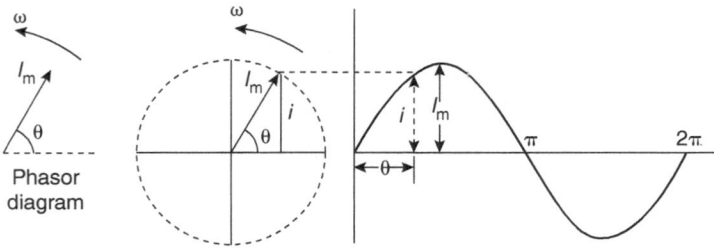

▲ **Figure 9.4** *Phasor for wave representation*

makes representation simpler (figure 9.4). Further simplification can result from a correct use of *phasor diagrams* to illustrate A.C. circuit relationships, especially if more than one current and/or voltage is considered at one time.

Note. For most practical electrical engineering work *r.m.s.* (root mean square) values are used. The full meaning of this term will be considered later in this chapter, but here it is convenient to make a phasor equal to this value rather than the maximum value with phase difference shown by phasors. Consider two 50Hz sinusoidal voltages represented E_{1m} and E_{2m}. The phase angle φ is known, voltages being of the same frequency but out-of-phase. Voltages are written $e. = E_{1m} \sin \omega t$ and $e_2 = E_{2m} \sin (\omega - \phi)$ where the angle ϕ is say 60° or $\left(\dfrac{\pi}{3}\right)$ radians and,

$$e_2 = E_{2m}\sin\left(\omega - \frac{\pi}{3}\right).$$

Waveforms are drawn graphically with the first leading the second. If the instant when the first goes through its zero value is the start of the angle or time scale, and the first wave considered the reference, then the phasor diagram can be drawn (figure 9.5).

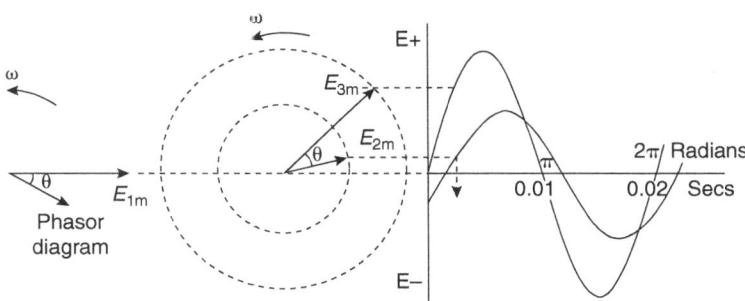

▲ **Figure 9.5** *Phasor diagrams*

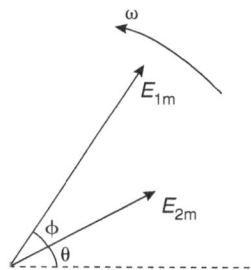

▲ **Figure 9.6** *Changes over time in the phasor diagram*

Here we depict the 2 phasors and their relation to each other where the first phasor is the reference and the second lags it by an angle ϕ.

Direction of phasor rotation is anticlockwise so E_{2m} is *behind* E_{1m} by angle ϕ. If an instant θ degrees later in time is considered the diagram can be redrawn (figure 9.6). The horizontal is taken as the zero time or reference axis.

Addition and Subtraction of Alternating Quantities

When 2, or more, sinusoidal voltages or currents act in a circuit the resultant is obtained by either of the following: (1) trigonometrical methods or (2) phasor methods.

(1) TRIGONOMETRICAL METHODS. These methods require knowledge of trigonometrical identities and follows established procedures. Examples of this approach will be given.

(2) PHASOR METHODS. The resultant of 2 or more phasors may be obtained (a) *graphically* or (b) *mathematically*.

(a) The *graphical method* is performed by setting out phasors to scale at the given angles, completing the parallelogram or polygon and measuring the resultant. Figure 9.7 shows the method employed with phasor addition shown. To subtract a phasor, reverse its direction and proceed as before.

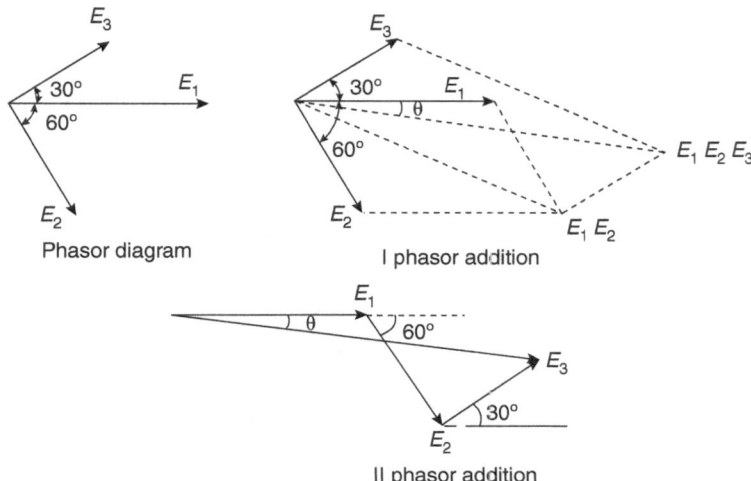

▲ **Figure 9.7** *Combining alternating phasor quantities*

In the diagram phasor addition s shown: (I) by completing the parallelogram to obtain the resultant of 2 phasors and using this resultant with the third phasor to obtain the final resultant, (II) by completing the polygon as shown. Both methods are tedious and unfortunately errors accumulate.

If the resultant of 2 waveforms is needed, either of 2 procedures can be taken.

The first method uses the fact that the sum of 2 sine waves with the same frequency is itself a sine wave. Thus any instantaneous value of the resultant wave is the *sum* of the individual instantaneous values taken from the 2 waves. Each wave is drawn graphically in accordance with the method already outlined, care being taken to displace one from the other by a given phase angle. By adding instantaneous values as shown (figure 9.8b) the *resultant* instantaneous value will give a point on the resultant wave, for example, $e = e_1 + e_2$. Other points are similarly obtained and a smooth curve drawn between points.

Note. One waveform is chosen as the reference, the other and the resultant are drawn to its base and zero value.

An alternative procedure for obtaining the resultant waveform is as follows. As individual voltages or currents are treated as phasors, the resultant of any 2 (or more) values is obtained as described. If a para lelogram is completed, as in figure 9.8b, the resultant E_m gives the maximum value of the resultant wave. Using E_m as the radius of the largest circle, the resultant waveform is deduced as before. For example, if $E_{1m} = 10V$, $E_{2m} = 6V$ and the phase difference is 60°, the resultant E_m will be 14V, E_m will also lag 21°45′ behind E_{1m}. The same procedure for magnitude and phase angle will give the resultant r.m.s. value, if r.m.s. values are used for component values instead of maximum values.

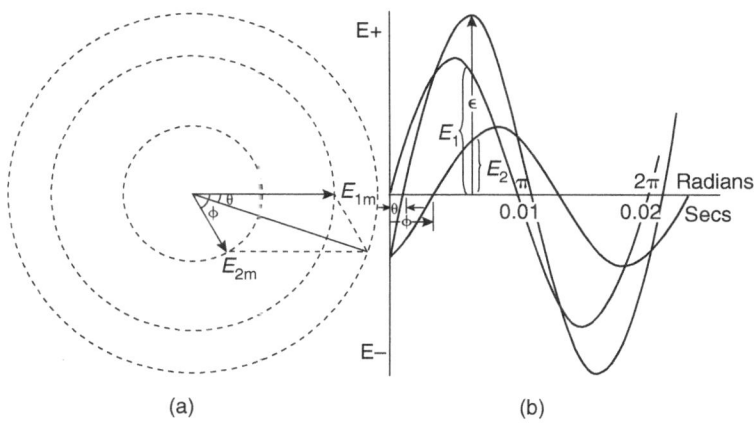

(a)   (b)

▲ **Figure 9.8** *Resulting phasor quantities*

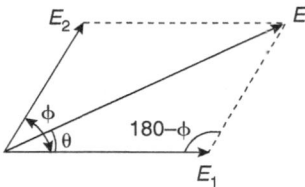

▲ **Figure 9.9** *Phasor parallelogram*

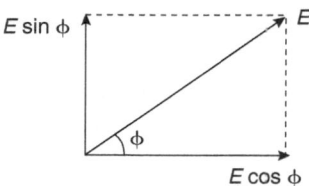

▲ **Figure 9.10** *E.m.f. phasor horizontal and vertical components*

The second method of obtaining a resultant waveform is a quicker method if component waveforms are not required. The resultant phasor is obtained graphically or by one of the following mathematical methods.

(b) The *mathematical method* is performed in 2 ways: (i) by use of the Cosine Rule or (ii) by resolving into horizontal and vertical components.

(i) The Cosine Rule can be used if the resultant of only 2 phasors is needed. Consider figure 9.9. The resultant E is obtained from $E = \sqrt{E_1^2 + E_2^2 + 2E_1E_2 \cos \phi}$ where E_1 and E_2 are the given phasors and ϕ the angle between them. The phase angle θ of the resultant is obtained from the Sine Rule:

$$\frac{E}{\sin(180 - \phi)} = \frac{E_2}{\sin\theta}$$

For more than 2 phasors, the resultant is combined with a third phasor etc. The next method is suggested for summation of more than 2 phasors as it is quicker and is the method used with series and parallel A.C. circuits.

(ii) Horizontal and vertical components. Any phasor can be split into 2 components at right angles to each other, which combined *together* produces the same effect as the original phasor. Thus in figure 9.10 the e.m.f. phasor E is split into a horizontal and a vertical component. If E lies at an angle ϕ to the horizontal, a horizontal component $E \cos \phi$ and a vertical component: $E \cos (90 - \phi)$ or $E \sin \phi$.

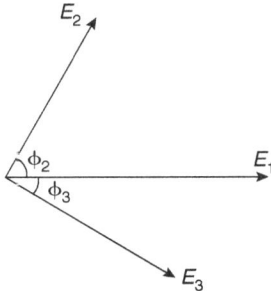

▲ **Figure 9.11** *Adding phasor quantities*

If all the phasors to be added are as shown (figure 9.11) the resultant is obtained.

The sum of the horizontal components is:

$$E_H = E_1 \cos \phi_1 + E_2 \cos \phi_2 + E_3 \cos \phi_3.$$

Similarly, the sum of the vertical components will be:

$$E_v = E_1 \sin \phi_1 + E_2 \sin \phi_2 - E_3 \sin \phi_3.$$

Note. Due allowance must be made for the *signs*. If vertical components are considered to be +ve when acting up, then $E_3 \sin \phi_3$ must be *subtracted* from the sum as it acts in a downward or −ve direction.

The resultant E is obtained from $E = \sqrt{E_H^2 + E_v^2}$ and ϕ, the angle at which it acts is found from the sine, cosine or tangent values. So $\cos \phi = \dfrac{E_H}{E}$.

The method is illustrated by the following example.

Example 9.3. Find the resultant of the following currents where:

$$i_1 = 5 \sin \omega t$$

$$i_2 = 4 \sin \left(\omega t + \frac{\pi}{3} \right)$$

$$i_3 = 3 \sin \left(\omega t - \frac{\pi}{6} \right)$$

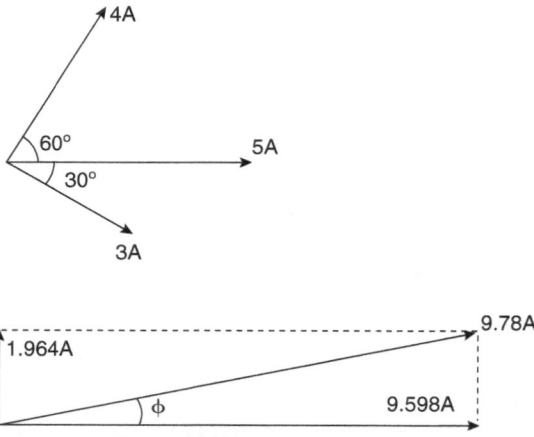

▲ **Figure 9.12** *Phasor trigonometrical method*

Express the resultant in a trigonometrical form, i.e. in the same form as the individual quantities.

Figure 9.12 should be considered for this solution.

$$I_H = 5 \cos 0 + 4 \cos 60 + 3 \cos 30$$

$$= (5 \times 1) + (4 \times \frac{1}{2}) + (3 + 0.866)$$

$$= 9.598A$$

And $I_v = 5 \sin 0 + 4 \sin 60 - 3 \sin 30$

$$= (5 \times 0) + (4 \times 0.866) - \left(3 \times \frac{1}{2}\right)$$

$$= 1.964A$$

From which $I = \sqrt{I_H^2 + I_v^2} = \sqrt{9.598^2 + 1.964^2}$

$$= 9.78A$$

$$\cos \phi = \frac{I_H}{I} = \frac{9.598}{9.78} = 0.98 \ \phi = \sin^{-1}(0.98)$$

$$\phi = 12° \text{ (approx.)} = \frac{\pi}{15} \text{ radians}$$

So $i = 9.78 \sin\left(\omega t + \frac{\pi}{15}\right)$.

Note. The following points are again made in this example. (1) In line with mathematical practice, phasors drawn to the right and drawn up are given +ve signs, while those drawn to the left and drawn dcwn are −ve. In the example all the I_H components are +ve. Consider the phasor diagram. For I_V 3 sin 30° acts down and is subtracted from 4 sin 60°, which acts up. (2) The resulting sign of I_V indicates whether the resultant I is in the first or fourth quadrant, i.e. whether it lags or leads the reference (which here is the horizontal). In the solution ϕ is about 12° so $\left(\dfrac{180}{12} = 15\right) = \dfrac{\pi}{15}$ radians and the resultant is written as shown.

In the treatment introducing the mathematical method and in these examples (figures 9.9–9.12) the suffix m is omitted from the e.m.f. symbol E, illustrating that the method is equally applicable to maximum values *and* r.m.s. values. The meaning of r.m.s. values will be considered next. As these are commonly used in A.C. work, it is vital to appreciate that phasor representations, applications and solutions will follow the r.m.s. convention.

Root Mean Square and Average Values

R.m.s. or effective value

The magnitude of an A.C. current varies from instant to instant and the power dissipated in a resistor varies accordingly. Energy dissipated over time will manifest itself as heat. A resulting steady temperature rise will occur due to constant power dissipation, i.e. due to passage of a constant current will give the same heating effect in the same time. From a heating aspect, an A.C. current value will have an equivalent value of D.C. current if the heating effect is proportional to 'current squared'. Since $P = I^2R$, the magnitude of this equivalent value can be found.

Let I amperes be the equivalent D.C. current with the same *average* heating effect as the A.C. current of *varying* instantaneous value i amperes.

For the D.C. condition:

$$\text{Energy expended} = \left(\text{current}^2 \times \text{resistance}\right) \times \text{time}$$

$$= I^2Rt$$

For the A.C. condition:

$$\text{Energy expended} = \left(\text{mean or average of } i^2R\right) \times \text{time}$$

$$= \left(\text{mean or average } i^2\right) \times R \times t$$

As both energy conditions are considered equal then I^2Rt = (mean or average of i^2) Rt

$$\text{so } I = \sqrt{\text{mean or average of } i^2}$$

The effective or r.m.s. value of an alternating current is the square root of the mean or average of the squares of the instantaneous values. This is true for the shape of *any* half cycle.

Note. The above deduction shows where the term 'root mean square' comes from, as the alternative to 'effective'. In practical electrical engineering the term r.m.s. is used.

The following definition is useful.

The r.m.s. or effective value of an alternating current or voltage is the value of direct current or voltage that, passed through or applied to a given resistor for a time of 1 cycle, produces the same amount of heat as the alternating current or voltage.

It is noted in this definition, voltage is mentioned although the r.m.s. value of a voltage wave wasn't *specifically* mentioned. The heating effect on a resistor of value R ohms is taken as the basis of discussion, but an alternative to $P = I^2R$ is $P = \dfrac{V^2}{R}$, so an r.m.s. value of alternating voltage wave of instantaneous value v can be deduced similarly.

The waveform's r.m.s. value can be graphically obtained (figure 9.13). The current or voltage instantaneous values are plotted to a time or angle base with suitable scales. The base of 1 half cycle is subdivided into equal divisions with mid-ordinates i_1, i_2, i_3, etc. up to i_n. These are measured to scale and substituted in the expression:

$$I = \sqrt{\frac{i_1^2 + i_2^2 + i_3^2 + \cdots i_n^2}{n}}$$

Only a half cycle is considered, as the next half cycle is similar to the first, but −ve. As the *square* of the current ordinates is required, +ve values result and the r.m.s. average taken over a complete cycle will be the same as for a half cycle.

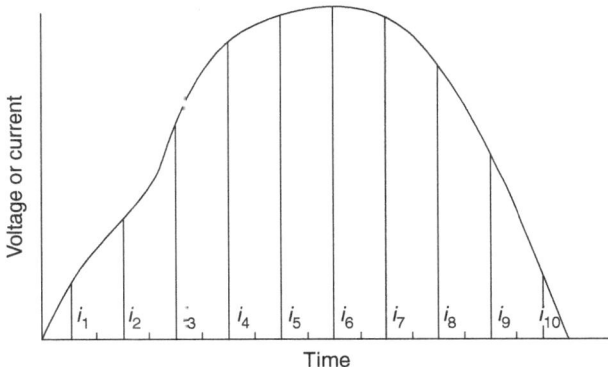

Time

▲ **Figure 9.13** *Voltage or current vs time*

It is noted that the mid-ordinate rule is applied to the *ordinates squared* and *not* to the ordinates directly.

Example 9.4. The following alternating voltage measurements result at intervals over 1 half cycle:

Time (t ms)	0	0.45	0.95	1.5	2.1	2.5	3.1	3.9	4.5	5.0
Voltage (V volts)	0	20	36	40	37.5	33	32	31	20	0

Calculate the voltage's r.m.s. value (1 decimal place) and the wave's frequency.

The solution is shown (figure 9.14), where if the waveform is plotted to scale and dividing the base into 10 equal parts, mid-ordinates can be drawn and measured.

Mean or average of $v_2 = \dfrac{9427}{10} = 942.7$

R.m.s. value $= \sqrt{942.7} = 30.7V$

Time for a half cycle $= \dfrac{5}{1000}$ th seconds, so time for a whole cycle $= \dfrac{2 \times 5}{1000}$

$= \dfrac{1}{100}$ second or 100Hz.

Time, $\dfrac{1}{1000}$ th sec

▲ **Figure 9.14** *Voltage vs time*

For a sine wave the r.m.s. value is shown mathematically = 0.707 of the maximum value. The most direct approach involves calculus but a graphical method illustrates the relationship.

Consider an A.C. current of sinusoidal waveform with a maximum or peak value of 4 amperes. The current squared is found to be half the maximum height, i.e. half of 16 = 8 as shown (figure 9.14). The number of amperes of continuous current that gives the same heating effect is $\sqrt{8}$ = 2.828 amperes and is the square root of the mean of the squares of the current, i.e. the true r.m.s. value.

The ratio of r.m.s. to maximum value $= \dfrac{2.828}{4}$ or $\dfrac{\text{r.m.s value}}{\text{Maximum value}} = \dfrac{2.828}{4} = \dfrac{0.707}{1}$,

this ratio is true for any A.C. current or sinusoidal voltage.

The r.m.s. value is always used in electrical engineering. Sine-wave working is assumed and any departure from it will be stated. If a supply voltage is stated as 220V, this will be the r.m.s. value and the voltage varies over a cycle between zero and a peak of $\dfrac{220}{0.707} = 311.2\text{V}$.

Table 9.1

$v_1 = 12$	$v_1^2 = 144$
$v_2 = 32$	$v_2^2 = 1024$
$v_3 = 39.5$	$v_3^2 = 1560$
$v_4 = 39.5$	$v_4^2 = 1560$
$v_5 = 35$	$v_5^2 = 1225$
$v_6 = 32.5$	$v_6^2 = 1056$
$v_7 = 31.5$	$v_7^2 = 992$
$v_8 = 31$	$v_8^2 = 961$
$v_9 = 28$	$v_9^2 = 784$
$v_{10} = 11$	$v_{10}^2 = 121$
	Total 9427

Average value

This is the true average value as understood mathematically of a half cycle, as that of a full A.C. cycle is zero. A waveform's average value is of interest for devices that don't depend on the effect of current squared, such as a rectifier. Rectifiers were developed to convert A.C. into D.C. current, without rotating machinery, and are found in various forms. It is common for semiconductor devices to meet the necessary rectification requirements, but certain industrial and military devices, for example, valve radio sets, with greater immunity to EMP than semiconductor devices, gas or vapour-filled vacuum valves, convert A.C. mains voltage to a direct voltage and may operate other valves.

In the marine industry it is still common to find gas or vapour type rectifiers, the most common of which is the 'mercury-arc' rectifier seen in ship dockyards and which provide

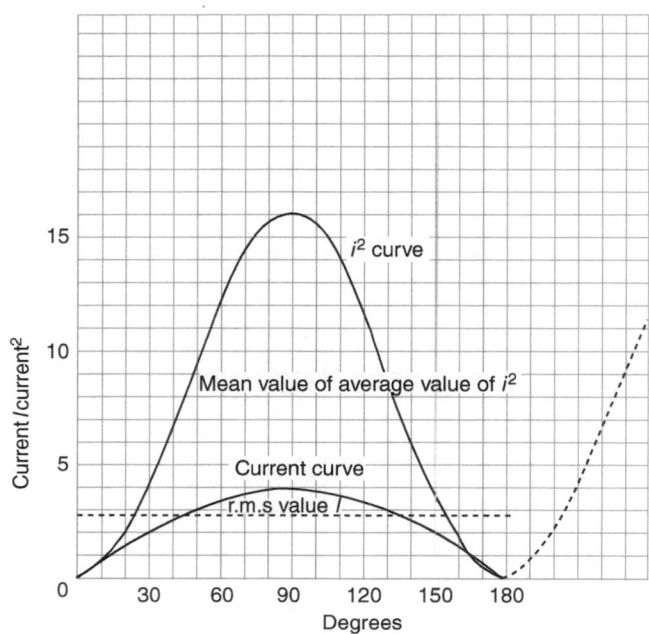

▲ **Figure 9.15** *Current characteristics*

a 'shore supply' to D.C. ships. Metal and semiconductor rectifiers are, however, used for A.C. in ships, large enough to supply D.C. current for 'degaussing' of induced magnetic signature, yet small enough to build into moving-coil indicating instruments, enabling them to be used in A.C. circuits.

D.C. current or voltage has a value equal to the *average* value of the A.C. waveform rectified and for a non-sinusoidal wave is obtained graphically, as follows. Referring to the waveform in figure 9.13, let I_{av} equal the average value, so

$$I_{av} = \frac{i_1 + i_2 + i_3 \cdots i_n}{n}$$

Example 9.5. Consider the same waveform and ordinates that are the subject of Example 9.4. Find the average value.

Here $I_{av} = \dfrac{12 + 32 + 39.5 + 39.5 + 35 + 32.5 + 31.5 + 31 + 28 + 11}{10}$

$= \dfrac{292}{10} = 29.2A$

or Average value $= 29.2A$.

For a sine wave the average value is 0.6365 of the maximum value. This is proven mathematically or graphically. As for the r.m.s. value, the most direct calculation method requires calculus but a graphical method can be checked by plotting a sine wave to a time or angle base. Subdivide the base into equal divisions, erect the mid-ordinates and again obtain the average value using the mid-ordinate rule or substitute in the expression:

$$I_{av} = \frac{i_1 + i_2 + i_3 \cdots i_n}{n}$$

The ratio of average to maximum value is $\frac{2}{\pi}$ or $\frac{\text{Average value}}{\text{Maximum value}} = \frac{2}{3.14} = 0.6365.$

This ratio is true for any A.C. current or sinusoidal voltage.

Form factor

This factor, when given a numerical value, states how near a waveform approaches the theoretical ideal sine wave. For any waveform it is defined as the ratio of the r.m.s. to the average value.

$$\text{Thus form factor} = \frac{\text{r.m.s. value}}{\text{Average value}}$$

For a sine wave, the form factor s 1.11, obtained from:

$$\text{Form factor} = \frac{0.707 \text{ Maximum value}}{0.6365 \text{ Maximum value}} = \frac{0.707}{0.6365} = 1.11.$$

Example 9.6. For the Example 9.5 already considered, in obtaining the r.m.s. and average values, find the form factor. The form factor will be $\frac{30.7}{29.2} = 1.05.$

Peak factor

The term 'peak factor' is encountered when dealing with A.C. waveforms, and defined as the ratio of the maximum value to the r.m.s. value. Thus:

$$\text{Peak factor} = \frac{\text{Maximum value}}{\text{r.m.s. value}}$$

For a sine wave the peak factor will be

$$\frac{1}{0.707} = 1.4 \,.$$

Practice Examples

9.1. Three circuits carrying currents: I_1, I_2 and I_3 are joined in parallel. $I_1 = 4A$, $I_2 = 6A$ *lagging* I_1 by 30° and $I_3 = 2A$ leading I_1 by 90°. Find by a phasor construction drawn to scale the resultant current and its phase angle with reference to current I_1 (2 decimal places).

9.2. A sinusoidal, 25Hz A.C. voltage has a maximum value of 282.8V. Find the time interval, after the zero value, when the voltage wave reaches (a) its first (3 decimal places) and (b) its second instantaneous value of 200V (2 decimal places).

9.3. A sinusoidal e.m.f. of 100V maximum value is connected in series with an e.m.f. of 80V maximum value, lagging 60° behind the 100V e.m.f. Determine the maximum value of the resultant voltage (1 decimal place) and its phase angle with respect to the 100V e.m.f. (4 decimal places).

9.4. The table below gives instantaneous A.C. current values that vary smoothly over 1 half cycle.

Time (milliseconds)	0	1	2	3	4	5	6	7
Current (amperes)	0	0.4	0.75	1.1	1.4	1.7	1.9	2.0
Time (milliseconds)	8	9	10					
Current (amperes)	1.8	1.3	0					

Plot the curve of current and find its r.m.s. value (3 decimal places). Calculate the power dissipated when the above current flows through a resistance of 8Ω (2 decimal places).

9.5. Three currents of peak values 10A, 17.32A and 20A respectively flow in a common conductor. The 17.32A current lags the 10A current by 90° and leads the 20A current by 60°. Draw a phasor diagram and find the resultant current value (2 decimal places) giving its phase relation with respect to the 10A current.

9.6. An alternating sine-wave voltage, having a peak value of 340V, is applied to the ends of a 24Ω resistor. Calculate the r.m.s. value of the current in the resistor (2 decimal places).

9.7. Represent by phasors the following e.m.f.s:

$$e_1 = 100 \sin \omega t,$$

$$e_2 = 50 \cos \omega t, \; e_3 = 75 \sin\left(\omega t + \frac{\pi}{3}\right), \text{ and}$$

$$e_4 = 125 \cos\left(\omega t - \frac{2\pi}{3}\right).$$

Determine by calculation the values of E and θ if:

$$e_1 + e_2 + e_3 + e_4 = E \sin(\omega + \theta).$$

9.8. Two alternators are coupled together to allow the phase angle between their generated e.m.f.s to vary. If the machines are connected in *series* and generate 100V and 200V respectively, find the total output voltage when the phase difference is: (a) zero, (b) 60°, (c) 90°, (d) 120° and (e) 180° (all 1 decimal place).

9.9. A stepped A.C. current wave has the following values over equal intervals of time.

Value (amperes)	4	6	6	4	2	0	0	-2	-4
Time interval (seconds)	0–1	1–2	2–3	3–4	4–5	5–6	6–7	7–8	8–9 etc.

Plot the waveform and find what D.C. current value gives the same heating effect (2 decimal places).

9.10. The 50Hz induced e.m.f.'s. in 4 separate coils of an A.C. generator are each of maximum value 4V and successively 10° out-of-phase. If these coils are connected in *series*, find by calculation and phasor construction the resultant maximum value, expressed in the form $e = E_m \sin(\omega t + \theta)$, where θ is the angle of phase difference with respect to the first coil.

10

THE SERIES A.C. CIRCUIT

We now know a thousand ways not to build a light bulb.

Thomas Edison

The approach we will take in teaching the subject of series A.C. circuits will be to introduce the essential fundamentals first and then to add details as required. At this stage several new terms are mentioned and their relationships stated. It will be helpful to memorise these terms.

Impedance

For the A.C. circuit, conditions are comparable with those for Ohm's law discussed with the D.C. circuit. We will modify Ohm's law as it applies to the A.C. circuit; where current is directly proportional to applied voltage and *inversely* proportional to the 'opposition' or resistance of the circuit to the flow of current. This A.C. opposition is termed the circuit *Impedance* (symbol – Z; unit – Ohm) but is due to more than a circuit's simple D.C. ohmic resistance R. The difference between Z and R is now considered.

For an A.C. circuit, the current flowing is given by:

$$\text{Current } I = \frac{\text{Applied voltage}}{\text{Impedance}}$$

$$\text{or } I \text{ (amperes)} = \frac{V \text{ (volts)}}{Z \text{ (ohms)}}$$

Note the 3 variations of the relationship. Thus:

$$I = \frac{V}{Z} \quad \text{or} \quad V = IZ \quad \text{or} \quad Z = \frac{V}{I}$$

For the D.C. circuit, it is known that $I = \frac{V}{R}$ where R is the circuit's ohmic resistance. If a straight conducting wire is connected to a D.C. supply of V volts, the measured current I will be $\frac{V}{R}$ amperes. However, if the same wire is then connected to an A.C. voltage supply, the measured current I will be $\frac{V}{Z}$ amperes and of the same magnitude as for the D.C. supply. In this case Z and R are equal as the impedance is due to resistance only. The circuit is said to be 'resistive' or 'non-inductive'.

If the wire is then wound into a coil or solenoid and the same voltage V applied, the current will be *smaller*, i.e. the new impedance is greater than the ohmic resistance. Likewise if an iron core is inserted into the solenoid, the impedance will increase and it should be clear that the impedance is made up of ohmic resistance and an extra current-limiting quantity. This extra quantity is the *Reactance* (symbol – X; unit – Ohm). It is noted that the impedance is *not* given by a straightforward arithmetical summation of resistance R and reactance X but by a right-angled relationship where:

$$\text{Impedance} = \sqrt{\text{Resistance}^2 + \text{Reactance}^2}$$

$$\text{or } Z \text{ (ohms)} = \sqrt{R^2(\text{ohms}) + X^2(\text{ohms})}$$

The relationship between R, X and Z is represented by an 'Impedance Triangle' shown in figure 10.1.

The angle ϕ is the 'phase angle' and $\cos \phi$ is a measure of a circuit's *Power Factor*. The reactance X for a wound coil is of only one particular form, namely *inductive reactance*. If a coil with its associated magnetic field is used in an A.C. circuit, then its inductive reactance must be known, which requires knowledge of the *inductance* and the supply frequency. The term 'inductive reactance' is associated with an *inductor* coil – usually iron-cored. Alternative terms for an inductor are: *reactor, choke* or *coil*, where all of these devices store energy in their magnetic fields. The term reactor is often used for a large coil passing heavy currents strengthened to withstand the greater than usual electromagnetic forces.

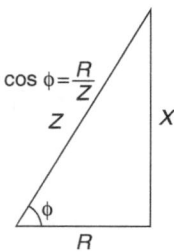

In contrast to inductive reactance there is also a *capacitive reactance* (symbol – X; unit – Ohm), a term associated with a capacitor. For comparison, a capacitor stores energy in its electric field, while a resistor does not store energy but instead dissipates it as heat. The capacitor and capacitive reactance will be considered later, but immediate focus is now given to inductance and inductive reactance.

Example 10.1. An inductor has an ohmic resistance of 3Ω and an inductive reactance of 4Ω. If it is connected to a 20V A.C. supply, find the current that flows (1 significant figure) and the power factor (1 decimal place) at which the coil operates. Note the diagram (figure 10.2).

Here $R = 3\Omega$ and $X = 4\Omega$. Since $Z = \sqrt{R^2 + X^2}$

Then $Z = \sqrt{3^2 + 4^2} = 5\Omega$

Current $I = \dfrac{V}{Z} = \dfrac{20}{5} = 4A$

The circuit power factor is given by $\cos\phi = \dfrac{R}{Z}$ (from the impedance triangle). Thus $\cos\phi = \dfrac{3}{5} = 0.6$ (lagging). The term 'lagging' is associated with an A.C. circuit containing inductive reactance.

Figure 10.2 represents the circuit. All the ohmic resistance is considered to be concentrated in a resistor R and the reactance in an inductor X, even though together they constitute the impedance Z of the inductor. The dotted rectangle represents the inductor and is omitted in future diagrams.

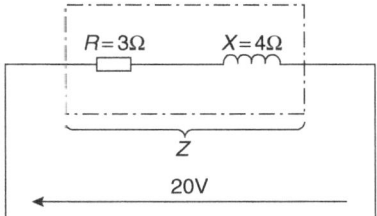

▲ **Figure 10.2** *Impedance circuit*

Inductance

Basic electromagnetic induction theory covered in Chapter 7 showed that if magnetic flux linked with a circuit changes, an e.m.f. is induced in the circuit. Faraday's law shows that the induced e.m.f. is proportional to the rate of change of flux-linkages and this e.m.f. only exists while this flux-linkage change occurs. In an A.C. circuit, current changes continually, so the associated flux of a coil carrying an A.C. current also changes continually; flux-linkages change and an induced e.m.f. is continually generated. By Lenz's law this is a 'back e.m.f.' tending to *oppose* the change causing it.

When the current in an inductive circuit is made to change, due to the inductance, the current value will, at the instant of change, be controlled by more than just voltage and resistance. During the change or *transient* conditions, a back e.m.f. is generated and new conditions of voltage balance occur. For the A.C. circuit, current varies sinusoidally and changes constantly, so inductance provides a vital and continuous effect.

Inductive reactance

Imagine an inductor with no resistance and only inductance of value L henries with an alternating voltage of *V* volts applied giving a current of *I* amperes. Figure 10.3 represents the current and voltage conditions.

Assume the current of *I* amperes (r.m.s. value) to be sinusoidal. As induced e.m.f. $= \dfrac{LI}{t}$ or $= L \times$ (rate of change of current), then at point 'a' the current value is zero, but increasing at its maximum rate, as the slope or gradient of the waveform is steepest here. Maximum induced e.m.f. occurs at this instant 'a' and as this is a 'back e.m.f.', by Lenz's law it opposes the supply voltage. Voltage waveform conditions are illustrated

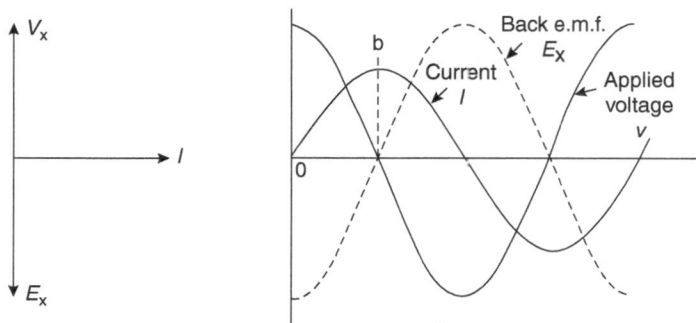

▲ **Figure 10.3** *Inductive reactance*

and the corresponding phasor diagrams are shown with *I* as the reference phasor. At point 'b' on the current waveform, no e.m.f. is induced as current is maximum and *not* changing at this instant. Thus there is a 90° phase difference between current and induced voltage (E_x) and a further 90° difference between the current and the applied voltage (V_x). *Note.* This condition applies to a circuit with inductance only.

Referring to figure 10.3, it is seen that, as current rises to its maximum value I_m in the first quarter cycle, flux-linkages LI_m are created $\left(\text{since } L = \dfrac{N\Phi}{I} \text{ or } N\Phi = LI\right)$. As current falls to zero in the second quarter cycle, linkages are destroyed. For the next half cycle the same number of linkages are created and destroyed. The change of flux-linkages in 1 cycle = $4LI_m$ and the change of flux-linkages in 1 second = $4fLI_m$ (where *f* is the frequency). The average value of induced e.m.f. = rate of change of flux-linkages.

$$\text{So average e.m.f.} = \frac{\text{Flux-linkages}}{\text{time}} = \frac{4fLI_m}{1} \text{ volts}$$

$$\text{Hence back e.m.f. } E_{Xav} = 4fLI_m \text{ volts.}$$

The supply voltage is equal yet opposite, its value being V_x (r.m.s.) or V_{Xav} (average). As r.m.s. values are preferred, the following conversion is needed.

$$\text{Since } V_{Xav} = \frac{2}{\pi}V_{Xm} \quad \text{and} \quad V_{Xav} = 4fLI_m$$

$$\text{then } \frac{2}{\pi}V_{Xm} = 4fLI_m \quad \text{or} \quad V_{Xm} = 2\pi fLI_m$$

giving $0.707\, V_x m = 2\pi fL \times 0.707 I_m$ or $V_x = 2\pi fLI.$

So the voltage drop in an inductor is proportional to the current and a constant that involves the circuit inductance and the supply frequency. This constant is given the name 'reactance' and as it is for an inductive circuit, we represent it with the symbol X and the suffix L.

Thus $X_L = 2\pi fL$ ohms, which for a purely inductive circuit $V_X = IX_L$ where $X_L = 2\pi fL$.

Inductive reactance is measured in ohms and is proportional to both frequency and inductance. As resistance has been omitted, the phase relationship between reactive voltage drop IX_L and current is fixed at 90° so these quantities are in *quadrature* with respect to each other.

1. Circuit with pure resistance

The circuit conditions are illustrated (figure 10.4).

Assume a sinusoidal voltage of value $v = V_m \sin \omega t$ applied to a purely non-inductive resistor of R ohms. The applied voltage overcomes the ohmic voltage drop at each instant. We can write $i = \dfrac{v}{R}$ and since maximum current occurs when the voltage is a maximum: $I_m = \dfrac{V_m}{R}$.

$$\text{Since } \frac{v}{R} = \frac{V_m}{R}\sin \omega t \quad \therefore i = I_m \sin \omega t.$$

The circuit current is also sinusoidal and in-phase with the applied voltage. The phasor diagram is drawn as shown.

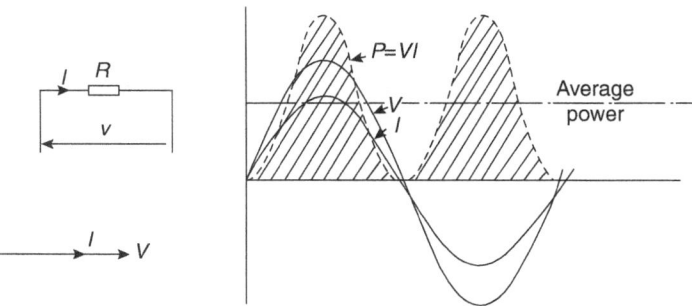

▲ **Figure 10.4** *Circuit with pure resistance*

R.m.s. values are deduced. Power at any instant is given by $p = vi$ or $p = V_m \sin \omega t \times I_m \sin \omega t = V_m I_m \sin^2 \omega t$ and using standard trigonometric identities:

$$= V_m I_m \frac{(1 - \cos 2\omega t)}{2} \quad \ldots *$$

Average power P = Average value of $\left\{ \dfrac{V_m I_m}{2} - \dfrac{V_m I_m \cos 2\omega t}{2} \right\}$

$$= \frac{V_m I_m}{2} - 0,$$

as the average value of a cosine wave across its cycle is zero. So $P = V_m / 2I_m$.

Hence $P = \dfrac{V_m}{\sqrt{2}} \times \dfrac{I_m}{\sqrt{2}} = VI$ or $P = VI$ (watts).

From the expression marked thus*, the power wave is a *periodic* quantity, always +ve and at **twice** the supply frequency. This is confirmed if the power wave is plotted by obtaining values of v and i for various instants in time and multiplying these together to give p, the power value at that instant. The resulting power wave is fully displaced above the horizontal and its maximum value is equal to $V_m I_m$. Being symmetrical, its average value is obtained from the distance its axis is displaced from the horizontal. This will be $\dfrac{V_m I_m}{2}$ and is a measure of average power.

$$\text{Thus: } P = \frac{V_m I_m}{2} = \frac{V_m}{\sqrt{2}} \times \frac{I_m}{\sqrt{2}} \text{ or } P = VI \text{ (watts)}.$$

For a resistive circuit, power equals the product of voltage and current, but this is only true for non-inductive circuits. Generally, if an attempt is made to correlate power with voltage and current (or volt amperes of a circuit), the product of V and I must be multiplied by a so-called *power factor*.

For the condition of a purely resistive circuit, if we write $P = VI \times$ power factor, it is clear that the power factor will be 1, since the power factor is related to the ratio of resistance and impedance of a circuit and is obtained from $\dfrac{R}{Z}$. From a general circuit's impedance triangle: $\dfrac{R}{Z}$ is the cosine of the phase angle ϕ between circuit voltage and current with

$\cos \phi = \dfrac{R}{Z} =$ the power factor.

The assumption made is that $P = VI \times$ power factor and can be rewritten as $P = VI \cos \phi$.

Further as this circuit is resistive $Z = R$ so $\cos \phi = \dfrac{R}{Z}$ and $\dfrac{R}{Z} = 1$, giving the condition of unity power factor as stated.

Note. The following deduction is also of value and identical to the D.C. circuit.

As $P = VI \cos \phi$ we can write $P = VI\dfrac{R}{Z} = \dfrac{V}{Z}IR = I \times I \times R$ or $P = I^2R$ (watts).

Example 10.2. An electric fire rated at 2kW power is connected to a 220V supply. Find the current that will flow and the resistance value of the fire element (1 decimal place).

An electric fire consists of a heating element that is purely resistive, so the circuit operates with unity power factor. Thus $\cos \phi = 1$ or the general expression $P = VI \cos \phi$ becomes $P = VI$

$$\text{Therefore } I = \frac{P}{220} = \frac{2000}{220} = 9.1\text{A}$$

$$\text{Again } Z = \frac{V}{I} = \frac{220}{9.1} = 24.2\Omega$$

$$\text{Here } Z = R \quad \therefore \ R = 24.2\Omega$$

$$\text{Alternatively } P = I^2R \ \therefore R = \frac{P}{I^2} = \frac{2000}{9.1^2}$$

$$\text{Hence } R = \frac{2000}{82.81} = 24.2\Omega.$$

2. Circuit with pure inductance

This condition was introduced under inductance and reactance, but as further deductions are necessary, the circuit is illustrated (figure 10.5) for a coil with no resistance and an inductance of L henries. Assume a sinusoidal current given by $i = I_m \sin \omega t$ flows through a coil. As i varies sinusoidally, the magnetic field also varies and a sinusoidal self-induced back e.m.f. is set up *opposing* the applied voltage at each instant. Treatment of the A.C. circuit with inductance shows this 'back e.m.f.' is regarded as equivalent to a voltage drop caused by the current – a property termed inductive reactance (symbol – X_L; unit – Ohm). Thus we have $E_X = V_X = I \times X_L$. X_L is shown equal to $2\pi fL$. The associated phasor diagram is now considered with the waveform. E_X is the

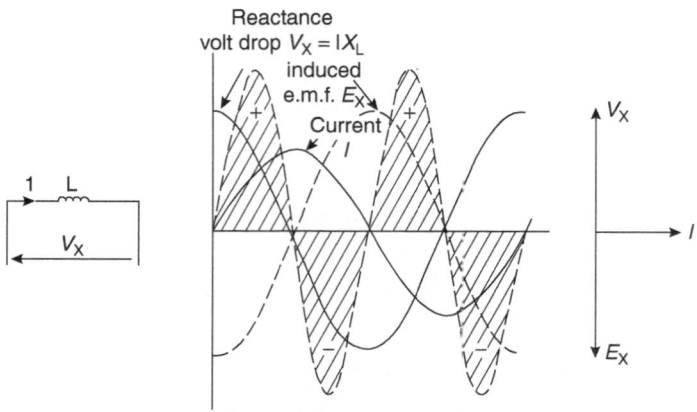

▲ Figure 10.5 *Circuit with pure inductance*

e.m.f. of self-inductance, displaced 90° behind the current I. V_X is the supply voltage and, being 90° ahead of the current, is 180° out-of-phase with E_X. V_X is thus always equal and opposite to E_X, illustrated by the waveforms and shown thus:

The self-induced e.m.f. is written mathematically as $e = -L\dfrac{di}{dt}$. By Lenz's law, as it at all times *opposes* the supply voltage, so we can write $v = L\dfrac{di}{dt}$. Thus $e = -L\dfrac{di}{dt} = -\dfrac{Ld(I_m \sin\omega t)}{dt} = \omega L I_m \cos\omega t$ or $e = \omega L I_m \left(\sin\omega t - \dfrac{\pi}{2}\right)$. Similarly v is deduced as $v = \omega L I_m \left(\sin\omega t + \dfrac{\pi}{2}\right)$.

Note. v is 180° ahead or in anti-phase with e.

As ωL is the reactance X_L, where $\omega = 2\pi f$, $e = E_m \sin\left(\omega t - \dfrac{\pi}{2}\right)$ and $v = V_m \sin\left(\omega t + \dfrac{\pi}{2}\right)$ giving the 90° phase displacement between current and voltage waves as shown. *Note.* $V_m = X_L I_m$. The relationship $X_L = 2\pi f L$ is a fundamental one and deduced earlier.

The power condition is deduced thus: Power at any instant is given by $p = vi$

$$\text{or } p = V_m \sin\left(\omega t + \dfrac{\pi}{2}\right) I_m \sin\omega t$$

$$= V_m I_m \sin\omega t \cos\omega t$$

$$= V_m I_m \dfrac{\sin 2\omega t}{2}$$

Thus instantaneous power $p = \dfrac{V_m}{\sqrt{2}} \times \dfrac{I_m}{\sqrt{2}} \sin 2\omega t$

$$= VI \sin 2\omega t \cdots *$$

Average power P = Average of value of $VI \sin 2\omega t = 0$. As the average value of a sine wave is zero, a different result to a purely resistive circuit.

From the expression marked *, it is seen that the power wave is a sine wave of *twice* the supply frequency. The waveform is symmetrically disposed about the horizontal and the average value is zero. This is confirmed if the power wave is plotted from values obtained from the voltage and current waves. As the axis of the power wave lies along the horizontal, the average power used must be zero as the +ve halves of the power wave are exactly equal to the −ve halves. This indicates that if power is taken from the supply to establish a magnetic field associated with a coil, it is subsequently returned to the supply when the magnet c field collapses.

If the general expression $P = VI \times$ power factor is adopted for a circuit, the power factor here must equal zero as $P = 0$. If used in the form $P = VI \cos \phi$ then $\cos \phi = 0$ since $\cos \phi = \dfrac{R}{Z}$ if $R = 0$ so $\cos \phi = \dfrac{0}{Z} = 0$.

In summary, a circuit with inductance only and no resistance is purely imaginary and has a zero power-factor working condition.

Example 10.3. A 220V, 50Hz supply is applied to an inductor of negligible resistance and the circuit current measured to be 2.5A. Find the coil inductance and the power dissipated (2 decimal places).

$$\text{As } Z = \frac{V}{I} \text{ then } Z = \frac{220}{2.5} = 88\Omega$$

$$\text{Now } R = 0. \therefore X_L = Z \text{ or } X_L = 88\Omega$$

$$\text{In addition as } X_L = 2\pi f L$$

$$\text{so } L = \frac{X_L}{2\pi f} = \frac{88}{2 \times 3.14 \times 50}$$

$$\text{or } L = 0.28\text{H}$$

Also as

$$R = 0 \text{ so } \cos\phi = \frac{0}{88} = 0$$

$$\therefore P = 220{\times}2.5{\times}0 = 0$$

Alternatively since $P = I^2R$

then $P = 2.5^2 \times 0 = 0$

3. Circuit with resistance and inductance in series

Consider a pure resistance and a pure inductance in series as shown (figure 10.6). It is noted that for a practical inductor, resistance and inductance are physically inseparable, and for illustrative purposes they are shown as 2 separate components R and L.

As the circuit conditions for both resistance and inductance have been introduced, if a current I flows, 2 voltage drops $V_R = IR$ and $V_{XL} = IX_I$ exist. Both of these form the applied circuit voltage and we assume that the applied voltage V consists of 2 components. One component V_R is the voltage needed to overcome the resistance voltage drop of the circuit and the other component V_{XL} is the voltage needed to overcome the reactance voltage drop or oppose the self-induced back e.m.f. As these components are at right angles to each other, as shown by consideration of circuit conditions 1 and 2, then the applied voltage is the resultant of the 2 components. The relationships being discussed is illustrated (figure 10.7) showing the relevant waveforms and appropriate phasors.

For the phasor diagram, current is common to both components (being a series circuit) and is used as the reference phasor. The resistance voltage drop $V_R = IR$ is in-phase with the current and drawn horizontally. The reactance voltage drop $V_{XL} = IX_L$ is at right angles to the current and is drawn vertically. The e.m.f. of self-inductance is also shown but is omitted as it serves no useful purpose on the phasor diagram.

The resultant of V_R and V_{XL} is V the applied voltage, and the current will lag V by an angle ϕ, which is the circuit's phase angle.

Simplification of a phasor diagram allows an appropriate 'voltage triangle' to be found, which if modified gives the 'impedance triangle' (figure 10.8).

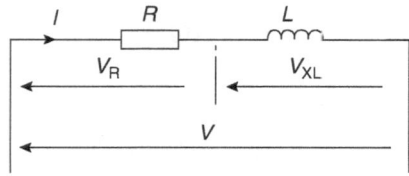

▲ **Figure 10.6** *Circuit with resistance and inductance in series*

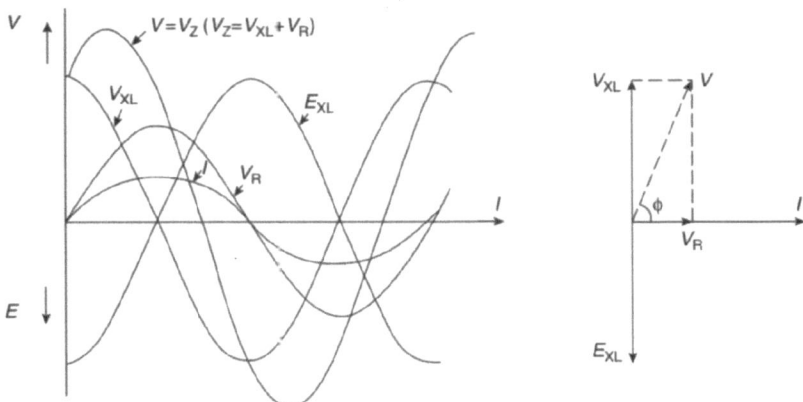

▲ **Figure 10.7** *Related waveforms and phasor relationships*

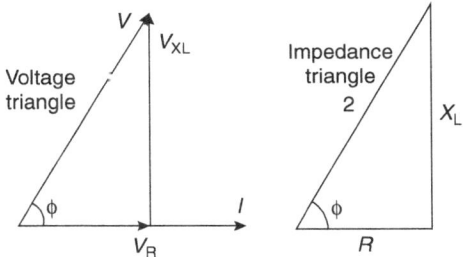

▲ **Figure 10.8** *Voltage and impedance triangles*

From the voltage triangle $V = \sqrt{V_R^2 + V_{XL}^2}$ and $\cos\phi = \dfrac{V_R}{V}$ Since $V_R = IR$ and $V_{XL} = IX_L$ this can be written:

$$V = \sqrt{(IR)^2 + (IX_L)^2} = I\sqrt{R^2 + X_L^2}$$

If Z is the circuit's equivalent impedance then $V = IZ$ or $Z = \dfrac{V}{I}$. So $IZ = I\sqrt{R^2 + X_L^2}$ and we have the impedance triangle relationship of:

$$Z = \sqrt{R^2 + X_L^2} \quad \text{and} \quad \cos\phi = \frac{R}{Z}$$

The power condition for the R, L series arrangement is as follows:

Figure 10.9 shows the basic waveforms of v and i, redrawn to allow the power wave to be obtained.

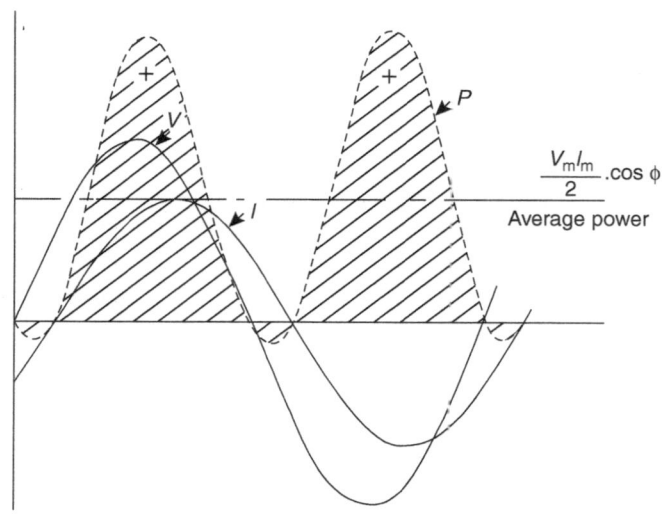

▲ **Figure 10.9** *Average power*

Let $v = V_m \sin \omega t$ be the applied voltage, and

$i = I_m \sin (\omega t - \phi)$ the circuit current lagging the voltage by an angle ϕ.

Then the instantaneous power $p = vi$

$= V_m \sin \omega t \times I_m \sin (\omega t - \phi)$

or $p = V_m I_m \sin \omega t \sin (\omega t - \phi)$

$= V_m I_m \left\{ \dfrac{\cos\phi - \cos(2\omega t - \phi)}{2} \right\}$. Again using standard trigonometric identities:

$= \dfrac{V_m}{\sqrt{2}} \times \dfrac{I_m}{\sqrt{2}} \{\cos\phi - \cos(2\ t - \phi)\}$ ...*

Thus $p = VI \cos\phi - VI \cos(2\omega t - \phi)$.

The Average power P = Average of $VI \cos\phi$ – Average of $VI \cos(2\omega t - \phi)$

Whence $P = VI \cos\phi - 0$ as the average of a cosine wave is 0.

Note. $VI \cos\phi$ is a constant quantity, thus its *average* value is apparent.

In the expression $P = VI \cos\phi$, the power factor is already present, varying between limits of 0 and 1, to enable the extreme conditions of pure R or L to be satisfied. If we examine the power wave we see how power factor or $\cos\phi$ is involved with actual displacement of a power wave axis above the horizontal. The expression marked * shows the power wave is per odic and at twice the supply frequency. It consists of +ve and −ve areas, showing that some power is returned to the supply, the amount depending on the operating circuit's power factor. The greater +ve net result of the power wave area, the greater the circuit's power consumption and the nearer unity the power-factor condition.

Note. The only component responsible for power consumption is the resistance and the deductions made previously are repeated here.

Thus since $P = I^2R$ and $I = \dfrac{V}{Z}$ we write:

$$P = I \times I \times R = \frac{V}{Z}IR \quad \text{or} \quad P = VI\frac{R}{Z} = VI \cos \phi.$$

Power factor is considered further later, but it is pointed out that the product VI is often referred to as the circuit's 'volt amperes' and suggests 'apparent power'. P, we know, is the 'true or active component of power', so we have the relation:

True power = 'Apparent power' $\times \cos \phi$ hence the name 'power factor' for $\cos \phi$.

Example 10.4. A circuit has a resistance of 25Ω and an inductance of 0.3H. If it is connected to a 230V, 50Hz supply, find the circuit current (2 decimal places), the power factor (3 decimal places) and the power dissipation (3 significant figures).

$$X_L = 2\pi fL = 2 \times \pi \times 50 \times 0.3 = 94.2\Omega$$

$$Z = \sqrt{25^2 + 94.2^2} = 97.5\Omega$$

$$I = \frac{230}{97.5} = 2.36A$$

$$\text{Power factor} = \cos \phi = \frac{R}{Z} = \frac{25}{97.5} = 0.256 \text{ (lagging)}$$

$$P = VI \cos \phi = 230 \times 2.36 \times 0.256 = 139W$$

$$\text{or } P = I^2R = 2.36^2 \times 25 = 139W.$$

Note the word 'lagging' introduced after the power-factor figure, indicating a current 'lags' or 'leads' the voltage. The latter is possible, but it is assumed that inductive circuits always operate with a **lagging** power factor. The term is used for a circuit current with respect to the applied voltage, i.e. current lags voltage.

Example 10.5. A coil of wire dissipates 256W when a D.C. current of 8A flows. If the coil is connected to an alternating applied voltage of 120V, the same current flows. Find the resistance (1 significant figure) and impedance of the coil (2 significant figures) and the power dissipated on A.C. (3 significant figures).

D.C. condition. As I = 8A and P = 256W, applied voltage $= \dfrac{256}{8} = 32V$. Coil resistance must be $\dfrac{32}{8} = 4\Omega$

A.C. condition. As I = 8A and applied voltage is 120V then coil impedance $= \dfrac{120}{8} = 15\Omega$

Resistance is, as for the D.C. case = 4Ω

Power dissipated = I^2R or $8^2 \times 4 = 256W$, as for the D.C. condition.

The last part is solved by $P = VI \cos \phi = 120 \times 8 \times \cos \phi$ and $\cos \phi$ obtained from $\dfrac{R}{Z}$

So $\cos \phi = \dfrac{4}{15} = 0.266 \text{ (lagging)}$ and $P = 960 \times 0.266 = 255.4W$

Capacitance

A complete treatment of the capacitor and its property of capacitance was discussed in Chapter 8. However, the capacitor is associated with A.C. circuits so it is necessary to revise capacitance briefly, before proceeding with A.C. theory.

If 2 plate conductors, arranged as plates, are separated by insulation and connected to a D.C. voltage, then at the instant of connection, a current flows. This current is of maximum value at the instant of switching on but dies away to zero. This is termed a 'charging' current and explained by considering the insulation to be in a state of electrical stress likened to a 'back e.m.f.', which builds up in a capacitor to oppose the supply voltage. Once a capacitor is charged and the voltage built up, its presence is apparent if the supply voltage is lowered, where the back e.m.f. causes a current to flow in the reverse direction, i.e. it is a 'discharging' current.

As an alternating voltage varies all the time (rising or falling), it follows that if it is applied to a capacitor or a capacitive circuit, an *exchange* of A.C. current will result. As the voltage across the capacitor plates rises, a charging current results and as the voltage falls, a discharging current results. Current magnitude depends on the circuit's capacitive reactance (symbol – X_c; unit – Ohm), a term corresponding to the inductive reactance with units as discussed previously.

Note. Current only flows if the voltage across the plates is changing (see the expression $Q = C/V$ where Q is the charge, V the voltage and C the capacitance). The change in P.D. across the plates from 0 to V volts occurs when a switch is closed. If the P.D. increases by v volts in t seconds and i is the average charging current, we can write:

$$q = Cv \text{ or it} = Cv \text{ or } i = \frac{Cv}{t}$$

For very short instants in time, the above will become:

$$i = \frac{Cdv}{dt}$$

Capacitive reactance

The action of a capacitor when an alternating voltage is applied across its plates is now considered (figure 10.10). As voltage rises from 0 to V_m in a quarter cycle, the charge on the plates rises to CV_m, as $Q = CV$. During the next quarter cycle, the charge falls to zero. For the next quarter cycle, the charge falls further to $-CV_m$ and returns to zero for the last quarter of a cycle. The total change of charge for a complete cycle is thus $4CV_m$ and this occurs f times a second. The average current during this time is $\dfrac{Q}{t}$

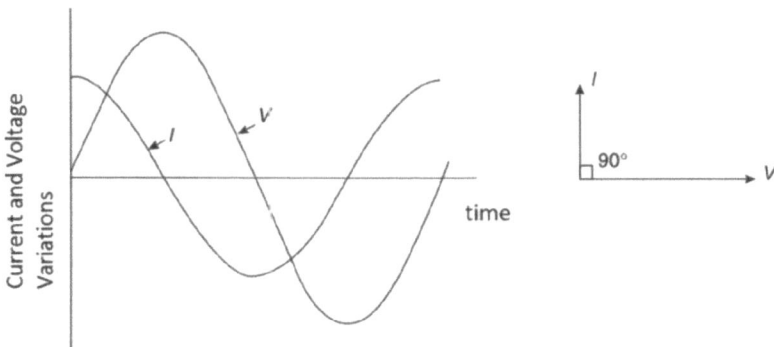

▲ **Figure 10.10** *Capacitive reactance I and V vs time*

$$\text{or } I_{av} = \frac{4fCV_m}{t} = 4fCV_m \text{ since } t \text{ is 1 second.}$$

For a sine wave $I_m = \frac{\pi}{2} \times I_{av}$. Hence $I_m = \frac{\pi}{2} \times 4fCV_m$

$= 2\pi fCV_m$ or in r.m.s. values $0.707\, I_m = 2\pi fC \times 0.707\, V_m$ and $I = 2\pi fCV$.

Hence $\dfrac{V}{I} = \dfrac{1}{2\pi fC} = X_C$. X_C is the capacitive reactance and the expression is in step with that developed for inductive reactance X_L.

Thus $X_C = \dfrac{1}{2\pi fC}$ mega ohms or $\dfrac{10^6}{2\pi fC}$ ohms with C in microfarads.

X_C is usually measured in ohms. It is noted that current leads voltage by 90° as maximum current occurs at the instant of maximum rate of charge of voltage. If a phasor diagram is drawn, the current phasor I is 90° ahead of the applied voltage phasor V.

These conclusions can be shown thus:

Let $v = V_m \sin \omega t$ be the sinusoidal voltage applied across a capacitor's plates.

Since $i = C\dfrac{dv}{dt}$ then $i = \dfrac{Cd(V_m \sin \omega t)}{dt}$ or $i = C\omega V_m \cos \omega t$

$= C\omega V_m \sin\left(\omega t + \dfrac{\pi}{2}\right)$. Capacitor current is in quadrature (in step) with the voltage, and is also sinusoidal. If $\dfrac{1}{\omega C}$ is the capacitance reactance and equal to X_C then

$i = \dfrac{V_m}{X_c}\sin\left(\omega t + \dfrac{\pi}{2}\right)$ and i becomes a maximum when the wave becomes a maximum

or $\sin\left(\omega t + \dfrac{\pi}{2}\right) = 1$.

Thus $I_m = \dfrac{V_m}{X_c}$ or $0.707\, I_m = \dfrac{0.707V_m}{X_c}$ giving $I = \dfrac{V}{X_c}$

Summarising $V = IX_C$ (as for the inductive circuit), except that $X_C = \dfrac{1}{\omega C} = \dfrac{1}{2\pi fC}$.

The current leads voltage by 90° and unlike the inductive circuit, where it was said that an inductor without resistance was not possible, a capacitor with negligible resistance and thus circuit condition 4, as set out below, can exist and is practical.

4. Circuit with pure capacitance

The circuit diagram is shown (figure 10.11) and the waveforms and phasor diagram repeated, so the power condition may be considered.

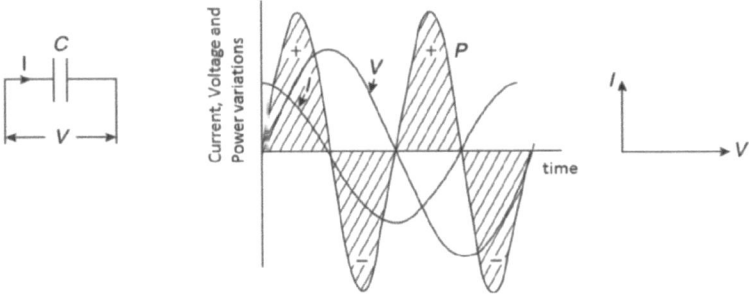

▲ **Figure 10.11** *Pure capacitive circuit*

The power at any instant $p = vi$

$$\text{or } p = V_m \sin \omega t \times I_m \sin\left(\omega t + \frac{\pi}{2}\right)$$

$$= V_m I_m \sin \omega t \cos \omega t$$

$$V_m I_m \frac{\sin 2\omega t}{2} = \frac{V_m}{\sqrt{2}} \times \frac{I_m}{\sqrt{2}} \sin 2\omega t \text{ again using standard trigonometric relationships.}$$

$$= VI \sin 2\omega t. *$$

Thus average power $P = 0$ since the average of a sine wave is zero. The expression marked thus * shows the power wave is periodic and at twice the signal frequency. The diagram also shows power to be +ve at the times when voltage is increasing and energy is put into a capacitor's electrostatic field. When voltage decreases power is shown as −ve, i.e. energy is recovered from the field as the capacitor discharges so no power is wasted. The power wave is symmetrical about the axis and the circuit power factor is zero. Thus if the expression $P = VI \cos \phi$ or $P = VI \times$ power factor is applied to this condition, $\cos \phi = 0$, as from $\cos \phi = \dfrac{R}{Z} = \dfrac{0}{Z} = 0$, there being no resistance.

Example 10.6. A capacitor of value 200μF is connected across a 220V, 50Hz supply mains. Find the current (1 decimal place) that would be recorded and the circuit impedance (1 decimal place).

$$\text{Here } X_C = \frac{1}{2\pi f C} = \frac{1}{2\pi \times 50 \times 200 \times 10^{-6}} \Omega$$

$$X_C = 15.92\Omega$$

$$\text{Current is given by } \frac{V}{X_C} \text{ or } I = \frac{220}{15.92} = 13.8\text{A}$$

Since there is no circuit resistance, impedance is made up of reactance only or $Z = X_C = 15.92\Omega$.

5. Circuit with resistance and capacitance

Figure 10.12 illustrates the circuit conditions and the technique used is similar to that used for the inductive circuit of condition 3.

The applied voltage V is resolved into 2 components V_R and V_X. One component V_R overcomes the resistance voltage drop due to the passage of current I, and the other component V_{XC} maintains the charging current of the capacitor and is at all times equalled and sustained by the internal stress voltage. As seen from condition 4, there is a 90° phase displacement between V_{XC} and I. If current is used as the reference for the waveform and phasor diagram, as it is common to R and C (being a series circuit), then the conditions shown can be deduced. If the voltage triangle (shown heavy) is extracted, the impedance triangle and relationships can be found:

$$V = \sqrt{V_R^2 + V_{XC}^2} = \sqrt{(IR)^2 + (IX_c)^2} = I\sqrt{R^2 + X_C^2}$$

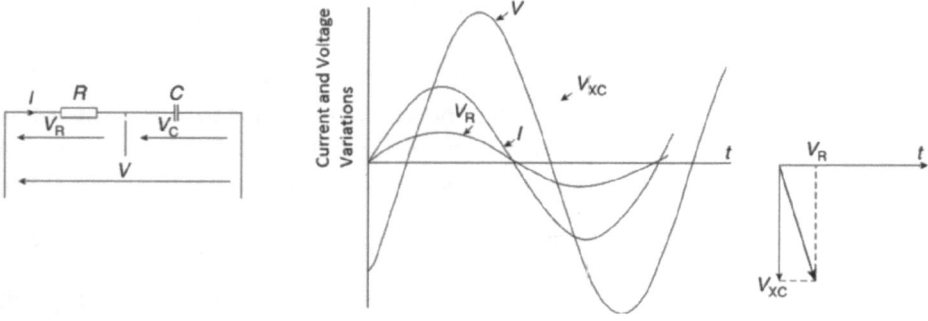

▲ **Figure 10.12** *Circuit with resistance and capacitance*

If Z is the circuit impedance then $\dfrac{V}{I} = Z = \sqrt{R^2 + Xc^2}$

or $Z = \sqrt{R^2 + \left(\dfrac{1}{2\pi fC}\right)^2}$ As before $\cos\phi = \dfrac{R}{Z}$

The power relation follows the form already used several times.

Thus power at instant $p = vi$

where $i = I_m \sin \omega t$ and $v = V_m \sin(\omega t - \phi)$

then $p = Vm\,Im \sin \omega t \sin(\omega t - \phi)$

$$= V_m I_m \left\{ \frac{\cos\phi - \cos(2\omega t - \phi)}{2} \right\}$$

or $p = \dfrac{V_m}{\sqrt{2}} \times \dfrac{I_m}{\sqrt{2}} \{\cos\phi - \cos(2\omega t - \phi)\}$ using standard trigonometric relationships.

$$= VI\cos\phi - VI\cos(2\omega t - \phi)$$

Average power P = Average of $VI \cos\phi$ – Average of $VI \cos(2\omega t - \phi)$ or $P = VI \cos\phi - 0$, since the average of a cosine wave is zero.

Thus $P = VI \cos\phi$ – an expression encountered previously. If the power wave is plotted as before, it will be as shown in figure 10.13.

The power wave is seen as before to be periodic at twice the signal frequency and consist of +ve and –ve sections. The average value is found from the amount by which

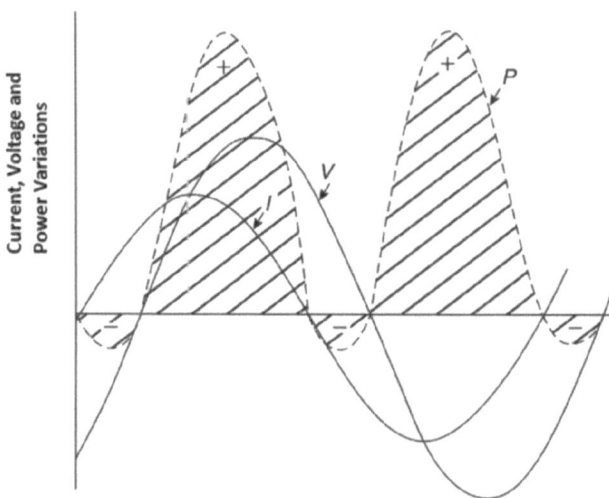

▲ **Figure 10.13** *Average power*

the axis is displaced above the horizontal and this displacement varies with a circuit's power factor. Conditions 1 and 4 are also covered. If $X_C = 0$ and the circuit is purely resistive cos $\phi = 1$ and the wave is fully displaced above the horizontal. If $R = 0$ and the circuit is purely capacitive then cos $\phi = 0$ and the wave is symmetrical about the horizontal giving $P = 0$. As for the inductive circuit, the only component responsible for dissipation of power is resistance.

As before $P = I^2 R$ or $P = I \times I \times R = VI\,\dfrac{R}{Z}$

and $P = VI\,\dfrac{R}{Z}$ or $P = VI \cos \phi$ as deduced.

Example 10.7. A 500W, 100V bulb is connected across 250V, 50Hz mains A.C. supply. Find the value of the capacitor required to be connected in *series* (3 significant figures).

Current taken by bulb is $\dfrac{500}{100} = 5A$.

Resistance of lamp $= \dfrac{100}{5} = 20\Omega$

On 250V, impedance of the circuit will be $\dfrac{250}{5} = 50\Omega$

$$\text{Thus } X_C = \sqrt{50^2 - 20^2}$$

$$= 45.8\Omega$$

$$\text{Again } X_C = \frac{1}{2\pi f C}$$

$$\text{or } C = \frac{1}{2\pi f X_C}$$

$$\therefore C = \frac{1}{2\pi \times 50 \times 45.8}F$$

$$\text{giving } C = 69.5\mu F.$$

The Series Circuit

From the various circuit conditions considered, cases 1 to 5, we see the general techniques used for a series circuit. The phasor diagram is easily drawn with current used as the reference phasor. From this diagram, circuit relationships and expressions are deduced.

Inductive impedances in series

Figure 10.14 shows the circuit arrangement and the relevant phasor diagram.

Impedances A and B consist of both resistances and reactances: R_A, R_B, X_A and X_B ohms respectively, connected in series. From the phasor diagram we deduce an expression for the total circuit impedance Z noting that it is **not** equal to $Z_A + Z_B$.

Using the diagram we have:

$$V = \sqrt{V_R^2 + V_X^2} = \sqrt{(V_{RA} + V_{RB})^2 + (V_{XA} + V_{XB})^2}$$

$$= \sqrt{(IR_A + IR_B)^2 + (IX_A + IX_B)^2}$$

$$\text{or } V = I\sqrt{(R_A + R_b)^2 + (X_A + X_B)^2}$$

If Z is the equivalent circuit impedance then:

$$Z = \frac{V}{I} = \sqrt{(R_A + R_B)^2 + (X_A + X_B)^2}$$

or summarising, for more than 2 inductive impedances:

$$Z = \sqrt{(R_A + R_B + R_c \cdots)^2 + (X_A + X_B + X_c \cdots)^2}$$

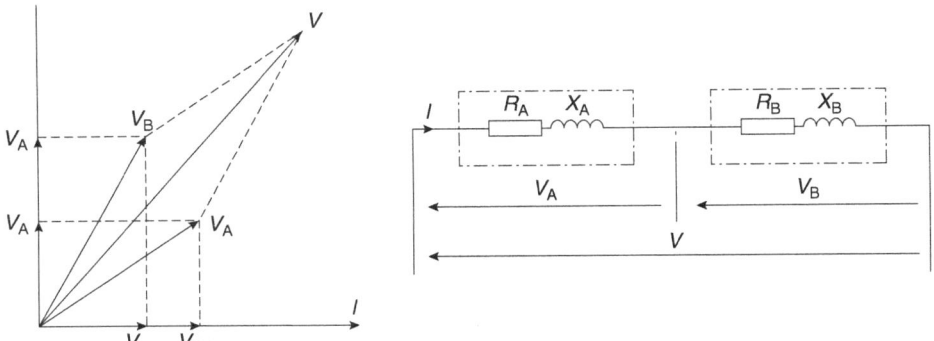

▲ **Figure 10.14** *V vs I characteristics and impedance circuit diagram*

Also the power factor is given by:

$$\cos \phi = \frac{V_R}{V} = \frac{IR}{IZ} = \frac{R}{Z} = \frac{(R_A + R_B + ...)}{Z}$$

The example below shows how simply this expression can be adapted for practical use.

Example 10.8. Two coils A and B are connected in series to 50Hz mains A.C. supply. The current is 1A and the voltage across each coil is measured to be 45V and 70V respectively. When the coils are connected to a D.C. supply, the current is also 1A, but the voltages across the coils are now 20V and 40V respectively. Find the impedance, reactance and resistance of each coil, the total circuit impedance, the applied A.C. voltage and the power factor of the complete circuit.

On D.C.   On A.C.

$$R_A = \frac{20}{1} = 20\Omega \qquad\qquad Z_A = \frac{45}{1} 45\Omega$$

$$R_B = \frac{40}{1} = 40\Omega \qquad\qquad Z_B = \frac{70}{1} = 70\Omega$$

Then $X_A = \sqrt{45^2 - 20^2} = 40.3\Omega$

Also $X_B = \sqrt{70^2 - 40^2} = 57.4\Omega$

Total $R = 20 + 40 = 60\Omega$

Total $X = 40.3 + 57.4 = 97.7\Omega$

Total impedance $Z = \sqrt{60^2 + 97.7^2} = 114\Omega$

Applied voltage $= 114 \times 1 = 114V$

Circuit power factor $= \frac{R}{Z} = \frac{60}{114} = 0.53$ (lagging)

Inductive and capacitive impedances in series

Figure 10.15 shows the following arrangement:

From the phasor diagram we deduce the expression for the total circuit impedance Z. It is noted that although V_{RA} and V_{RB} are in-phase and can be added, V_{XA} and V_{XB} are in

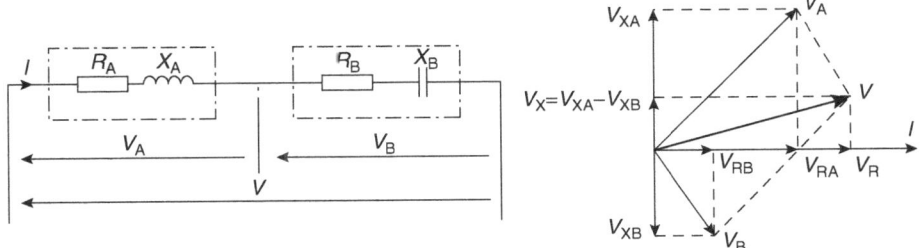

▲ **Figure 10.15** *Inductive and capacitive impedance in series with phasor diagram*

anti-phase and the resultant of the vertical phasors are obtained by *subtraction*. Thus from the resulting final diagram (shown heavy):

$$V = \sqrt{V_R^2 + V_X^2} = \sqrt{(V_{RA} + V_{RB})^2 + (V_{XA} - V_{XB})^2}$$

$$= \sqrt{(IR_A + IR_B)^2 + (IX_A - IX_B)^2}$$

$$= I\sqrt{(R_A + R_B)^2 + (X_A - X_B)^2}$$

If Z is the equivalent circuit impedance then:

$$Z = \frac{V}{I} = \sqrt{(R_A + R_B)^2 + (X_A - X_B)^2}$$

Summarising $Z = \sqrt{(R_A + R_B)^2 + (X_A - X_B)^2}$

Also for the circuit, the power factor

$$\cos \phi = \frac{V_R}{V} = \frac{IR}{IZ} = \frac{R}{Z} \text{ or, } \cos \phi = \frac{R_A + R_B}{Z}$$

The above 2 circuit conditions give rise to the general series circuit set out in the next section.

The general series circuit

From work already done on circuit theory, a fundamental expression is deduced from the phasor diagram (figure 10.16).

Since it is a series circuit, current is common and is used as the reference phasor. Note this circuit condition is similar to that of inductive and capacitive impedances in series,

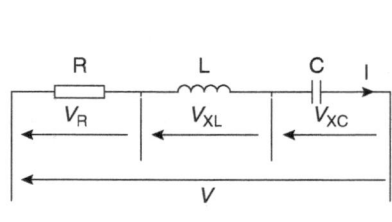
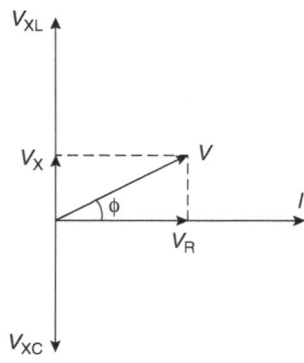

▲ **Figure 10.16** *The general series circuit*

except that all the circuit resistance is considered within one resistor R. Then for the phasor diagram:

$V_R = IR$ and is in-phase with the current

$V_{XL} = IX_L$ and is 90° ahead of I

$V_{XC} = IX_C$ and is 90° behind I

V_{XL} and V_{XC} are 180° out-of-phase or anti-phase and a phasor difference can be obtained where $V_x = V_{XL} - V_{XC}$. Here V_{XL} is assumed greater than V_{XC}.

Further deduction from the diagram is possible, thus:

$$V = \sqrt{V_R^2 + V_X^2} = \sqrt{(V_R^2 + (V_{XL} - V_{XC})^2}$$

$$= \sqrt{(IR)^2 + (IX_L - IX_C)^2}$$

$$= I\sqrt{(R)^2 + (X_L - X_C)^2} \text{ or } \frac{V}{I} = \sqrt{(R)^2 + (X_L - X_C)^2}$$

If Z is taken as the equivalent impedance of the circuit then:

$$Z = \frac{V}{I}$$

$$\text{or } \frac{V}{I} = Z = \sqrt{(R)^2 + (X_L - X_C)^2}$$

$$\text{Thus } Z = \sqrt{R^2 + \left(2\pi fL - \frac{1}{2\pi fC}\right)^2}$$

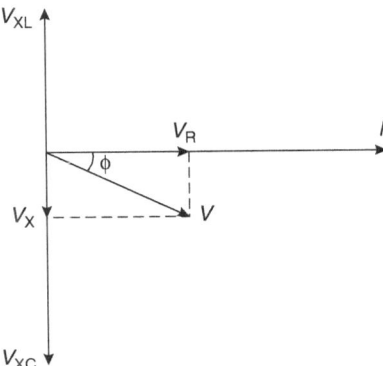

▲ **Figure 10.17** *Capacitive reactance dominating over inductive reactance*

Example 10.9. A series circuit is made up of an inductor of resistance 20Ω and inductance 0.08H, connected in series with a 100μF capacitor. If the circuit is connected across 200V, 50Hz mains A.C. supply, find (a) the circuit current (1 decimal place) and (b) its power factor (2 decimal places).

$$\text{Here } X_L = 2\pi fL = 2\pi \times 50 \times 0.08 = 25.2\Omega$$

$$X_C = \frac{1}{2\pi fC} = \frac{1}{2\pi \times 50 \times 100 \times 10^{-6}} = 31.75\Omega$$

Resultant reactance $= X = X_L - X_C = 25.2 - 31.75$

$$= -6.55\Omega$$

The −ve sign denotes that the capacitive reactance predominates and the phasor diagram will be as shown (figure 10.17).

From the diagram as before $V = \sqrt{V_R^2 + V_X^2}$

$$\text{or } Z = \sqrt{R^2 + X^2}$$

$$\therefore Z = \sqrt{20^2 + 6.55^2} = 21\Omega$$

$$\text{The circuit current} = \frac{200}{21} = 9.5\text{A}.$$

The power factor is given by $\cos \phi = \dfrac{R}{Z} = \dfrac{20}{21} = 0.95$ (leading), i.e. the current leads the voltage as the circuit capacitive reactance dominates.

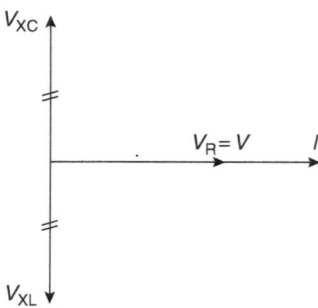

SERIES RESONANCE. From the phasor diagram (figure. 10.18) a state can occur when V_{XL} and V_{XC} are equal, and also V_R is equal to the supply voltage V. An examination of the general series circuit expression

$$Z = \sqrt{R^2 + \left\{2\pi fL - \frac{1}{2\pi fC}\right\}^2}$$

shows that an unusual condition arises when $2\pi fL = \dfrac{1}{2\pi fC}$ in magnitude, i.e. the capacitive reactance equals the inductive reactance. Under this condition $Z = R$ and the circuit is said to be in a 'state of resonance'. The current passed will be limited by the value of R only and although large voltages may be present across components L and C, their effect on the supply voltage V is not evident. Series resonance is used to advantage for practical purposes, especially in radio.

Since inductive and capacitive reactances vary with frequency, a point exists where the reactances are equal at a particular frequency known as the resonant frequency.

$$\text{Since } X_L = X_C$$

$$2\pi fL = \frac{1}{2\pi fC}$$

$$\therefore \text{ Resonant frequency } f = \frac{1}{2\pi\sqrt{LC}}$$

Example 10.10. A 4µF capacitor is connected in series with a coil of inductance 39.6mH and resistance 40Ω to a 200V mains A.C. supply. Calculate (a) the frequency when the current is a maximum value (1 significant figure) and (b) the P.D. across the capacitor at this frequency (1 decimal place).

When current is maximum, Z is minimum.

i.e. $Z = R$ and circuit is in its resonant condition

$\therefore$ Maximum current $I = \dfrac{V}{R} = \dfrac{200}{40}$

$I = 5A$ at resonance

(a) Resonant frequency $f = \dfrac{1}{2\pi\sqrt{LC}}$

$$= \dfrac{1}{2\pi\sqrt{39.6 \times 10^{-3} \times 4 \times 10^{-6}}}$$

$$F = 399.89 \text{ Hz (say 400Hz)}$$

$$X_c = \dfrac{1}{2 \text{ fC}} = \dfrac{10^6}{2 \times 400 \times 4} \qquad X_c = 99.5\Omega$$

(b) P.D. across capacitor $V_{Xc} = I X_c = 5 \times 99.5$

$$= 497.5 \text{ volts}$$

Example 10.11. A coil of unknown inductance and resistance is connected in series with a 25Ω, non-inductive resistor across 250V, 50Hz mains A.C. supply. The P.D. across the resistor is 150V and across the coil 180V. Calculate the resistance and inductance of the coil and also find its power factor.

The circuit and phasor diagrams are shown (figure 10.19).

This example is important as it involves basic fundamentals but has a simple solution. The phasor diagram is first explained with the various voltage drops considered separately. V_R is the voltage drop across resistor $R = IR$. V_C is the voltage drop across the coil and is the resultant of 2 voltage drops: V_R across the resistance of the coil $= IR$ and V_L across the reactance of the coil $= IX_L$. V_R is in-phase with current and V_L is 90° *ahead* of the current. From the phasor diagram it is clear that V is the resultant of V_C and V_R and that the expression given for simple phasor summation can be applied here.

$$\text{Thus } V = \sqrt{V_R^2 + V_C^2 + 2V_R V_C \cos\phi_L}$$

$$\text{or } 250^2 = 150^2 + 180^2 + 2 \times 150 \times 180 \times \cos\phi_L$$

$$\therefore 62\,500 = 22\,500 + 32\,400 + 54\,000 \cos\phi_L$$

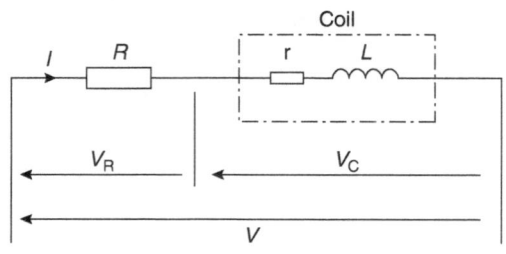

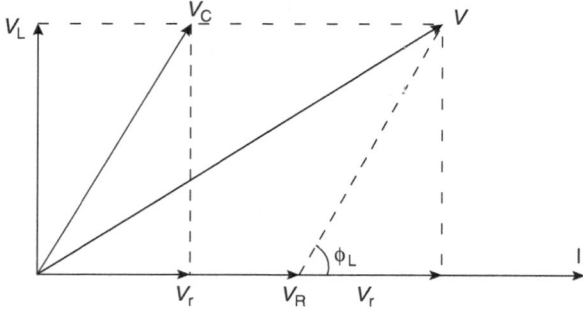

▲ **Figure 10.19** *Circuit and phasor diagram*

$$\text{or } 54\,000 \cos \phi_L = 62\,500 - 54\,900$$

$$\cos \phi_L = \frac{7600}{54\,000} = 0.141 \,(\textbf{lagging})$$

$$\text{The current flowing } = \frac{150}{25} = 6\text{A}$$

$$\text{The impedance of the coil} = \frac{180}{6} = 30\Omega$$

$$\text{Resistance of coil } = Z \cos \phi_L = 30 \times 0.141 = 4.23\Omega$$

$$\text{Reactance of coil} = \sqrt{30^2 - 4.23^2} = 29.7\Omega$$

$$\text{Inductance of coil} = \frac{29.7}{2 \times \pi \times 50} = 0.0945\text{H}$$

$$\text{Power factor of coil} = \cos \phi_L = 0.141 \text{ (lagging)}.$$

Example 10.12. A coil of resistance 10Ω and inductance 0.1H is connected in series with a capacitor of capacitance 150μF, across a 200V 50Hz mains A.C. supply. Calculate (a) the inductive reactance, (b) the capacitive reactance, (c) the circuit impedance, (d) the circuit current, (e) the circuit power factor, (f) the voltage drop across the coil, (g) the voltage drop across the capacitor (all 1 decimal place).

(a) Inductive reactance $= 2\pi fL = 2 \times \pi \times 50 \times 0.1 = 31.4Ω$

(b) Capacitive reactance $= \dfrac{1}{2\pi fC} = \dfrac{10^6}{2 \times \pi \times 50 \times 150} = 21.2Ω$

(c) Resultant reactance $= 31.4 - 21.2 = 10.2Ω$ (inductive)

Impedance $= \sqrt{R^2 + X^2} = \sqrt{10^2 + 10.2^2} = 14.3Ω$

(d) Circuit current $= \dfrac{200}{14.28} = 14.0A$

(e) Power factor $= \dfrac{10}{14.28} = 0.7$ (lagging) – As the circuit reactance is overall inductive

(f) Impedance of coil $= \sqrt{10^2 + 31.4^2} = 33.0Ω$

Voltage drop across coil $= 14 \times 33 = 462V$

(g) Voltage drop across capacitor $= 14 \times 21.2 = 296.8 = 297.0V$.

Note. Although resonance is *not* occurring here, the condition is very close to this and large voltages can build up across circuit components. Thus the fact that the voltages across the coil and capacitor are larger than the supply voltage agrees with theory.

Practice Examples

10.1. A circuit has a resistance of 3Ω and an inductance of 0.01H. The voltage across its ends is 60V and the A.C. frequency is 50Hz. Calculate (a) the impedance (2 decimal places), (b) the power factor (2 decimal places) and (c) the power absorbed (1 decimal place).

10.2. A 100W lamp for a 100V supply is placed across a 220V, 50Hz supply. What value of resistance must be placed in series with it so that it will work under its proper conditions (3 significant figures)? If a coil is used instead of the resistor and if the coil resistance is small compared to its reactance, what is the coil inductance (3 decimal places)? What is the total power absorbed in each case (2 significant figures)?

10.3. An inductive load takes a current of 15A from a 240V, 50Hz supply and the power absorbed is 2.5kW. Calculate (a) the load's power factor (3 decimal places), (b) the load's resistance (1 decimal place), reactance (2 significant figures) and impedance (1 decimal place). Draw a phasor diagram showing the voltage drops and the current components.

10.4. Two inductive circuits A and B are connected in series across 230V, 50Hz mains. The resistance values are A 120Ω and B 100Ω. The inductance values are A 250mH and B 400mH. Calculate (a) the current, (b) the phase difference between the supply voltage and current, (c) the voltages across A and B and (d) the phase difference between these voltages (all 3 decimal places).

10.5. Two coils are connected in series. When 2A D.C. is passed through a circuit, the voltage drop across the coils is 20V and 30V respectively. When passing 2A A.C. at 40Hz, the voltage drop across the coils is 140V and 100V respectively. If the 2 coils in series are connected to a 230V, 50Hz mains A.C. supply, find the current flowing (2 decimal places).

10.6. A simple transmission line has a resistance of 1Ω and a reactance at normal frequency of 2.5Ω. It supplies a factory with 750kW, 0.8pf (lagging) at a voltage of 3.3kV. Determine the voltage at the generator (3 significant figures) and its power factor (3 decimal places). Find also the generator output (3 significant figures) and draw the phasor diagram.

10.7. A non-inductive resistor of 8Ω is connected in series with an inductive load and the combination placed across a 100V supply. A voltmeter (drawing negligible current) is connected across the load and then across the resistor and indicates 48V and 64V respectively. Calculate (a) the power absorbed by the load (1 decimal place), (b) the power absorbed by the resistor (3 significant figures), (c) the total power taken from the supply (1 decimal place) and (d) the power factors of the load and whole circuit (2 decimal places).

10.8. A circuit, consisting of a resistor and a capacitor connected in series across a 200V, 40Hz mains A.C. supply, takes a current of 6.66A. When the frequency is increased to 50Hz and the voltage maintained at 200V, the current becomes 8A. Calculate the values of resistance (2 decimal places) and capacitance (3 significant figures) and sketch the phasor diagram (not to scale) for either frequency.

10.9. A coil, having an inductance of 0.5H and a resistance of 60Ω, is connected in series with a 10μF capacitor. This combination is now connected across a sinusoidal supply and it is found that at resonance, the P.D. across the capacitor is 100V. Calculate the circuit current flowing under this condition (3 decimal places). Sketch the phasor diagram (not to scale).

10.10. A coil has a resistance of 400Ω and, when connected to a 60Hz main A.C. supply, an impedance of 438Ω. If the coil is then connected in series with a 40μF capacitor and a P.D. of 200V, 50Hz is applied to the circuit, find the current (3 decimal places), the P.D. across the capacitor (1 decimal place) and the P.D. across the coil (3 significant figures).

A.C. PARALLEL CIRCUITS AND SYSTEMS

More than the diamond Koh-i-noor, which glitters among their crown jewels, they prize the dull pebble which is wiser than a man, whose poles turn themselves to the poles of the world, and whose axis is parallel to the axis of the world.

Ralph Waldo Emerson

A.C. Circuits

Power in the A.C. circuit

From the various circuit conditions considered in Chapter 10, for various series combinations of resistors, and inductive and capacitive reactance, it was seen that the current flowing was sinusoidal and displaced from the applied sinusoidal voltage by an angle φ, termed the *phase angle* of the circuit. The general expressions were:

For voltage $v = V_m \sin \omega t$ and for current $i = I_m \sin(\omega t - \phi)$ with a *lagging* phase angle assumed for convenience.

The instantaneous power $p = vi = V_m I_m \sin \omega t \sin(\omega t - \phi)$

or $p = V_m I_m \left\{ \dfrac{\cos\phi - \cos(2\omega t - \phi)}{2} \right\} = VI \cos\phi - VI \cos(2\omega t - \phi)$

and Average power $P = VI \cos\phi - 0$ or $P = VI \cos\phi$

VI is often called a circuit's *Apparent Power* and P the *Active Power.*

So active power = apparent power × power factor

Cos ϕ is the 'power factor' as it is the factor by which the apparent power must be multiplied to obtain the active power value expended in a circuit.

So power factor $= \dfrac{\text{active power}}{\text{apparent power}}$ or $\cos\phi = \dfrac{P}{VI} \dfrac{I^2 R}{|Z|} = \dfrac{R}{Z}$ as deduced earlier.

Furthermore, from the active power equation: P (watts) $= VI$ (volt amperes) $\times \cos\phi$

Wattage is given by volt amperes × power factor and expressed by $W = \text{VA} \cos\phi$ or kW $= \text{kVA} \cos\phi$.

Note. The term kVA is an accepted rating of A.C. generators, motors or transformers but it *does not* indicate the power rating. More information is required before this can be deduced and the power factor must be specified. The volt amperes or VA of a circuit term was retained from the earliest electrical engineering days, before standardised terms and symbols were introduced. VA or kV ratings are used internationally for A.C. circuits and machines.

Active and reactive components

These terms are used in connection with current but under certain conditions may apply to voltage and power. Consider the phasor diagram (figure 11.1) for a simple A.C. circuit with current lagging voltage by an angle ϕ. The current I splits into its 2 quadrature components I_a and I_r as shown so $I_a = I \cos\phi$ and $I_r = I \sin\phi$.

As $I \cos\phi$ is a current *in-phase* with the voltage V and as $VI \cos\phi$ is the measure of power expended in a circuit, then $I \cos\phi$ is the current component responsible for power dissipation and $I \cos\phi$ is called the *active* power, and is the working component of current. $I_r = I \sin\phi$, being always at right angles to the voltage, is responsible for no power and is called the *reactive*, wattless or idle current component. Example 11.1 illustrates these terms.

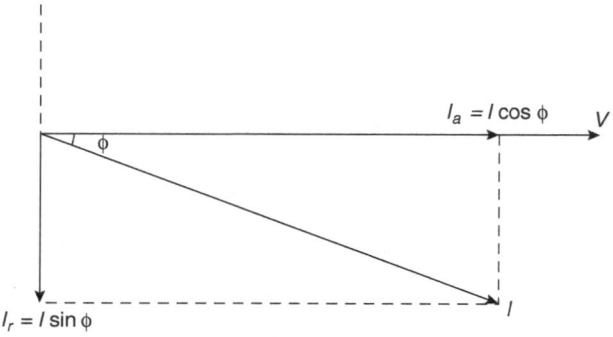

▲ Figure 11.1 *Active and reactive components*

Example 11.1. A single-phase A.C. motor of 15kW and 90% efficiency runs from a 400V single-phase supply. Find the current taken from the mains supply if the motor operates at 0.8 power factor (lagging). What is the value of the active current, the reactive current and the motor rating in volt amperes (1 decimal place).

Motor power output = 15kW = 15 × 1000 watts

$$\text{Motor power input} = \frac{15 \times 1000}{90} \times 100 \text{ watts}$$

$$= 16\,667\text{W or 16.7kW}$$

$$\text{Volt ampere rating} = \frac{16.67}{0.8} = 20.84\text{kVA}$$

The line current is obtained by dividing the volt ampere value by the supply voltage.

$$\text{Thus } I = \frac{20.84 \times 1000}{400} = 52.1\text{A}$$

Active component of current $I_a = I \cos \phi = 52.1 \times 0.8 = 41.7\text{A}$.

Reactive component of current $= I_r = I \sin \phi = 52.1 \times 0.6 = 31.3\text{A}$.

It is pointed out that the relation for sin ϕ = 0.6 when cos ϕ = 0.8 refers to a right-angled triangle of sides: 10, 8 and 6. For example, cos ϕ is frequently given as 0.707 or sometimes 0.7, referring to a right-angled isosceles triangle and sin ϕ in this case is also 0.707 or 0.7 (approx.).

The parallel circuit

The parallel circuit is treated separately to show students that the procedure is different to that for the series circuit. Nevertheless the method employed follows the technique of phasor summation, i.e. resolving into horizontal and vertical components or in terms of the new terminology here: active and reactive components. The branches of a parallel circuit are made up of simple R, X_L or X_C series values, and all calculation work done in this regard remains altered. For a parallel circuit the same voltage is applied to all branches and it is usual to work with V as the reference for the phasor diagram. The current condition may be written as $\overline{I} = \overline{I_1} + \overline{I_2} + \overline{I_3}$ etc, the dash above the I indicating it is a phasor summation and not an arithmetical one. Thus all correct operations for a phasor summation must be performed.

Inductive impedances in parallel

Assume 2 inductive impedances connected in parallel (figure 11.2). Impedance Z_1 is made up of a resistance R_1 and inductive reactance X_1 while Z_2 is made up of resistance R_2 and inductive reactance X_2. The phasor diagram and circuit relationships are shown. As V is common to both branches it is used as the reference phasor. The problem is to find I where $\overline{I} = \overline{I_1} + \overline{I_2}$.

Here $I_1 = \dfrac{V}{Z_1}$ and $I_2 = \dfrac{V}{Z_2}$

Resolving into active and reactive components and using arbitrary signs: $I_a = I_1 \cos \phi_1 + I_2 \cos \phi_2$

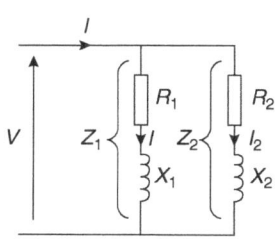
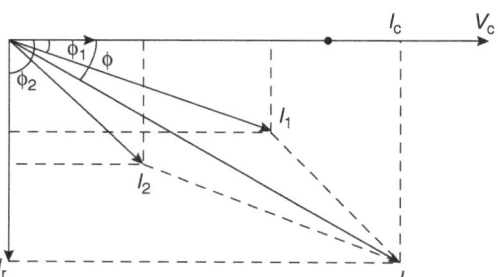

▲ **Figure 11.2** *Inductive impedances in parallel*

And, $I_r = -I_1 \sin \phi_1 - I_2 \cos \phi_2$. It must be remembered that −ve phasors are drawn vertically down.

Then $I = \sqrt{I_a^2 + I_r^2}$ and $\cos \phi = \dfrac{I_a}{I}$ where $\cos \phi$ is the power factor of the whole circuit.

Example 11.2. In the circuit above, let $R_1 = 3\Omega$ and $X_1 = 4\Omega$ while $R_2 = 8\Omega$ and $X_2 = 6\Omega$. If the applied voltage is 20V, find the total current supplied (2 decimal places) and the power factor (2 decimal places) of the complete circuit. Find also the total power expended (2 significant figures).

$$Z_1 = \sqrt{R_2^2 + X_1^2} = \sqrt{3^2 + 4^2} = 5\Omega$$

$$\text{Then } I_1 = \frac{20}{5} = 4A$$

$$Z_2 = \sqrt{R_2^2 + X_2^2} = \sqrt{8^2 + 6^2} = 10\Omega$$

$$\text{and } I_2 = \frac{20}{10} = 2A$$

$$\cos \phi_1 = \frac{3}{5} = 0.6 \text{ (lagging) and, } \sin \phi_1 = \frac{4}{5} = 0.8$$

$$\cos \phi_2 = \frac{8}{10} = 0.8 \text{ (lagging) and, } \sin \phi_2 = \frac{6}{10} = 0.6$$

$$\text{Also } I_a = (4 \times 0.6) + (2 \times 0.8) = 4A$$

$$I_r = -(4 \times 0.8) - (2 \times 0.6) = -4.4A$$

$$\text{Such that } I = \sqrt{4^2 + 4.4^2} = 5.95A$$

$$\text{Circuit power factor } \cos \phi = \frac{4}{5.95} = 0.67 \text{ (lagging)}$$

Power expended = $20 \times 5.95 \times 0.67 = 80W$

The power expended can be checked thus:

$$\text{Power in branch 1} = I_1^2 R_1 = 4^2 \times 3 = 48W$$

Power in branch $2 = I_2^2 R_2 = 2^2 \times 8 = 32W$

Total 80W.

Inductive and capacitive impedances in parallel

The procedure for solving problems associated with this circuit type follows that outlined above, except that allowance is made for the directions and signs when adding reactive components. Thus in figure 11.3, impedance Z_2 is made up of resistance R_2 and capacitive reactance X_2 in *series*. The phasor for the reactive component of current is drawn vertically up and allocated a +ve sign, while the reactive component of current for branch 1 is allocated a −ve sign. The total reactive component is thus a *difference*. Voltage is again used as the reference for the phasor diagram.

As before, $I_a = I_1 \cos \phi_2 + I_2 \cos \phi_2$ and $I_r = -I_1 \sin \phi_1 + I_2 \sin \phi_2$. The sign of I_r is either +ve or −ve, decided by the relative values of $I_1 \sin \phi_1$ and $I_2 \sin \phi_2$. Thus the resulting reactive component will act either up or down and the resultant circuit current may lag or lead. As before $I = \sqrt{(I_a^2 + I_r^2)}$ and $\cos \phi = \dfrac{I_a}{I}$. The term lagging or leading is decided by the sign of I_r.

Example 11.3. A circuit consists of 2 parallel branches. Branch A consists of a 20Ω resistor in series with a 0.07H inductor, while branch B consists of a 60μF capacitor in series with a 50Ω resistor. Calculate the mains current and the circuit power factor if the voltage is 200V at 50Hz (all 1 decimal place).

Branch A. $X_A = 2\pi fL = 2 \times 3.14 \times 50 \times 0.07 = 22\Omega$ $R_A = 20\Omega$

$$\therefore Z_A = \sqrt{20^2 + 22^2} = 29.7\Omega$$

Thus $I_A = \dfrac{200}{29.7} = 6.74A$ and $\phi_A = \dfrac{R_A}{Z_A} = \dfrac{20}{29.7} = 0.674 \text{ (lagging)}$

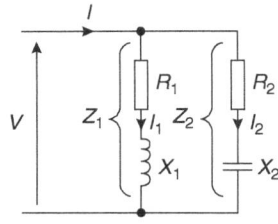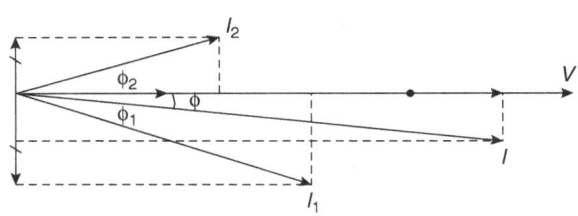

▲ **Figure 11.3** *Inductive and capacitive impedances in parallel*

$$\sin \phi_A = \frac{X_A}{Z_A} = \frac{22}{29.7} = 0.74$$

Branch B. $X_B = \dfrac{1}{2\pi fC} = \dfrac{10^6}{2 \times 3.14 \times 50 \times 60} = 53\Omega$

$R_B = 50\Omega$

$\therefore Z_B = \sqrt{50^2 + 53^2} = 72.8\Omega$

$$\text{Thus } I_B = \frac{200}{72.8} = 2.75\text{A}$$

$$\cos \phi_B = \frac{R_B}{Z_B} = \frac{50}{72.8} = 0.686 \text{ (leading)}$$

$$\sin \phi_B = \frac{X_B}{Z_B} = \frac{53}{72.8} = 0.728$$

Then $I_a = (6.74 \times 0.674) + (2.75 \times 0.68)$

$\quad = 4.55 + 1.885 = 6.43\text{A}$

$I_r = -(6.74 \times 0.74) + (2.75 \times 0.728)$

$\quad = -5 + 2.005 = -2.995\text{A}.$

Note. The mains current will lag as the inductive branch dominates.

$$I = \sqrt{I_a^2 + I_r^2} = \sqrt{6.43^2 + 2.995^2}$$

$$I = 7.1\text{A}$$

$$\cos \phi = \frac{I_a}{I} = \frac{6.43}{7.1} = 0.9 \text{ (lagging)}.$$

The mains current is 7.1A and the circuit operates with a lagging power factor of 0.9.

Parallel resonance

It is worth pointing out that a resonance condition can also occur in a parallel circuit and is termed 'current resonance' to distinguish it from 'voltage resonance' as dealt with for the series circuit. From Example 11.3 a condition arises when $I_A \sin \phi_A = I_B \sin \phi_B$ and as these are the reactive components of currents in inductive and capacitive branches, they oppose each other, producing a total reactive component of zero value. The remaining active components then add to give the line current, as $I = \sqrt{I_a^2 + 0} = I_a$ and the combined circuit will operate at unity power factor. This is illustrated by the phasor diagram (figure 11.4). Since the power factors of both branches are low, the phase angles ϕ_A and ϕ_B are large yet $I_A \cos \phi_A$ and $I_B \cos \phi_B$ are small compared to the reactive components.

Large currents can flow in inductive and capacitor branches, which are much greater than the main supply current and are not supplied from the line. On examining the power waves for an inductor and capacitor, it is seen they are directly opposite in-phase, as are the current waves. It is assumed that as the capacitor discharges

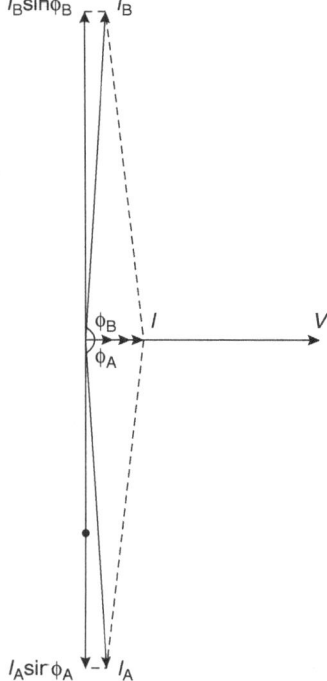

▲ **Figure 11.4** *Parallel resonance*

power, the power given out is absorbed by the inductor building up its field. When the inductive field collapses, power released charges the capacitor and there is a current due to oscillation of power between inductor and capacitor. This resonance creates oscillator circuits and has many applications in radio and electrical filter circuit design. If no supply is available current is not maintained due to energy loss in the circuit resistance, which although small cannot be neglected. To maintain an oscillatory current, the resistance loss must be supplied at the correct frequency from the external supply force.

Power-Factor Improvement

Power-factor improvement is often used in practical electrical engineering work and is best illustrated by the following example.

Example 11.4. (a) Two inductive coils of resistance values 5Ω and 8Ω and inductance values of 0.02H and 0.01H respectively are connected in parallel across a 240V, 50Hz mains supply. Find the coil currents (1 decimal place), the circuit current (1 decimal place) and its power factor (3 decimal places).

The arrangement is shown in figure 11.5. A phasor diagram is also drawn.

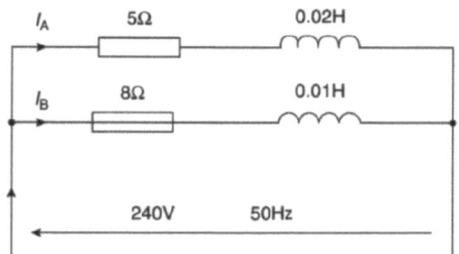
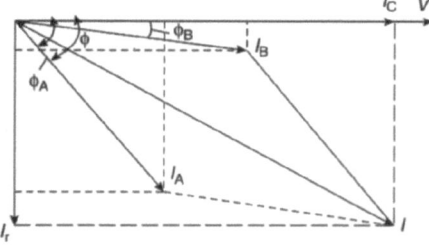

▲ **Figure 11.5** *Power-factor improvement*

Branch A. $X_A = 2\pi fL = 2 \times \pi \times 50 \times 0.02 = 6.28\Omega$

$$Z_A = \sqrt{5^2 + 6.28^2} = 8.02\Omega$$

$$I_A = \frac{240}{8.02} = 29.8A \quad \cos\phi_A = \frac{5}{80.2} = 0.622 \quad \text{lagging}$$

$$\sin \phi_A = \frac{6.28}{8.02} = 0.78$$

Branch B. $X_B = \frac{1}{2}$ that of branch A, since L is halved $= 3.14\Omega$

$$Z_B = \sqrt{8^2 + 3.14^2} = 8.6\Omega$$

$$I_B = \frac{240}{8.6} = 27.9A \cos \phi_B = \frac{8}{8.6} = 0.93 \text{ lagging}$$

$$\sin \phi_B = \frac{3.14}{8.6} = 0.366$$

Then $I_a = (29.8 \times 0.622) + (27.9 \times 0.93) = 44.6A$

$I_r = (29.8 \times 0.78) + (27.9 \times 0.366) = 33.5A$

It is noted that the arbitrary −ve sign is not used as both branches are inductive and there is no doubt as to the resultant current being lagging.

Then $I = \sqrt{(44.6^2 + 33.5^2)}$ $I = 55.6A$

$$\text{P.F.} = \cos \phi = \frac{44.6}{55.6} = 0.801 \text{ (lagging)}$$

Example 11.4. (b) Find the effect on the main circuit current (1 decimal place) and power factor (1 decimal place), f a capacitor of 400μF is connected across the supply in *parallel* with the coils.

The phasor diagram (figure 11.6), shows the new conditions.

Branch C. Reactance of capacito $X_C = \dfrac{1}{2\pi fC}$

Thus $X_C = \dfrac{10^6}{2 \times \pi \times 50 \times 400} = 7.95\Omega$

$$\therefore I_C = \frac{240}{7.95} = 30.25A$$

As there is no resistance in branch C, only capacitive reactance, then $\cos \phi_C = 0$ and $\sin \phi_C = 1$.

Again I_C acts at 90° to the vo tage and is wholly reactive, there being no active component. Then I_a as before $= 44.6A$ and $I_r = -23.3 - 10.2 + 30.25 = -3.25A$.

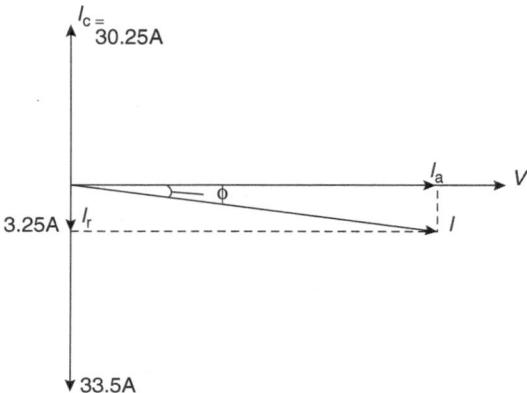

▲ **Figure 11.6** *Phasor diagram showing the new impedance condition*

It is seen that the arbitrary signs have been introduced because the reactive current of branch C acts in the *opposite* direction to that of branches A and B.

The circuit current is now:

$$I = \sqrt{I_a^2 + I_r^2} = \sqrt{44.6^2 + 3.25^2}$$

$$\therefore I = 44.6A$$

and $\cos \phi = \dfrac{44.6}{44.6} = 1.0$ i.e. unity.

From this example it is seen that by connecting a capacitor in parallel with inductive loads, the total line current is reduced from 55.6 to 44.6A and overall circuit power factor is improved from 0.8 (lagging) to unity. The advantages of this arrangement are now considered in detail.

Advantages of power-factor improvement

For most commercial loads, current lags voltage, due to the system's inductance or the operating characteristics of motors and control gear. Typical power-factor values are as follows:

System supplying lighting loads only: power factor (lagging) = 0.95.

System supplying lighting and power loads: power factor (lagging) = 0.75 to 0.85.

System supplying power loads: power factor (lagging) = 0.5 to 0.7.

The lower the power factor, the *greater* the line current must be for a given load kW or output power rating, and following are the disadvantages of this:

(1) The transmission losses in supply cables or power lines increase in accordance with I^2R, where R is the cable or line resistance. For a given transmitted power, the current at 0.7 power factor is $\dfrac{1}{0.7} \times$ current at unity power factor $= 1.43 \times$ current at unity power factor. Also the transmission loss at 0.7 power factor is $(1.43)^2 \times$ loss at unity power factor $= 2 \times$ loss at unity power factor.

(2) Because of the larger currents resulting from a low power factor, there is a greater voltage drop in the supply lines resulting in a lower voltage at the load, thus conductor size must be increased to keep the voltage drop to an acceptable value!

(3) Since the larger current results from a low power factor, the size of the current-carrying conductors in transformers, control gear and alternators must be larger than the minimum possible. This means the physical dimensions of equipment must be larger – an inferior design. Equipment is also more costly.

(4) 'Regulation' – a term used for the stabilising or 'sitting-down' of the generating and transmitting plant voltage – is adversely affected by a low power factor. The lower the power factor, the greater the internal voltage drop in this equipment, i.e. armature reaction and its effects will be worse.

Example 11.5. A 40kW load, operating at 0.707 power factor (lagging), is supplied from 500V, 50Hz mains supply. Calculate (a) the capacitor value required to raise the line power factor to unity (3 significant figures) and (b) the capacitance required to raise the power factor to 0.95 (lagging) (3 significant figures).

(a) Load kVA $= \dfrac{kW}{\cos \phi} = \dfrac{40}{0.707}$

Load current $= \dfrac{40 \times 10^3}{0.707 \times 500} = 113.15A$

Active component of load current $I_1 = I_1 \cos \phi_1$

$$= 113.15 \times 0.707$$

$$= 79.997A = 80A$$

Reactive component of load current $I_1 = I_1 \sin \phi_1$

$$= 113.15 \times 0.707$$

$$= 80A.$$

To cancel the reactive current, a capacitor is fitted to operate in *parallel* with the load, which must pass a similar reactive current value as shown by the phasor diagram (figure 11.7).

Thus I_C must be 80A. Reactance X_C of capacitor must be: $\dfrac{500}{80} = 6.25\Omega$

Since $X_C = \dfrac{1}{2\pi f C}$ then $\dfrac{10^6}{2\pi f C} = 6.25$ (where C is in microfarads)

Thus $C = \dfrac{10^6}{2\times3.14\times50\times6.25}$ $C = 510\mu F$.

(b) *Note.* $I_1 \sin \phi_1$ is not to be cancelled completely as the line phase angle is *reduced* from ϕ_1 to ϕ_2 and the line current to a new value I_2 as illustrated (figure 11.7b). The power or active component remains the same, $VI_2 \cos \phi_2 = 40\ 000$ as before.

$$\therefore I_2 = \frac{40\times10^3}{500\times0.95} = 84.2A$$

Since $\cos \phi_2 = 0.95$ (lagging) $\sin \phi_2 = 0.312$ and $I_2 \sin \phi_2 = 84.2 \times 0.312 = 26.1A$.

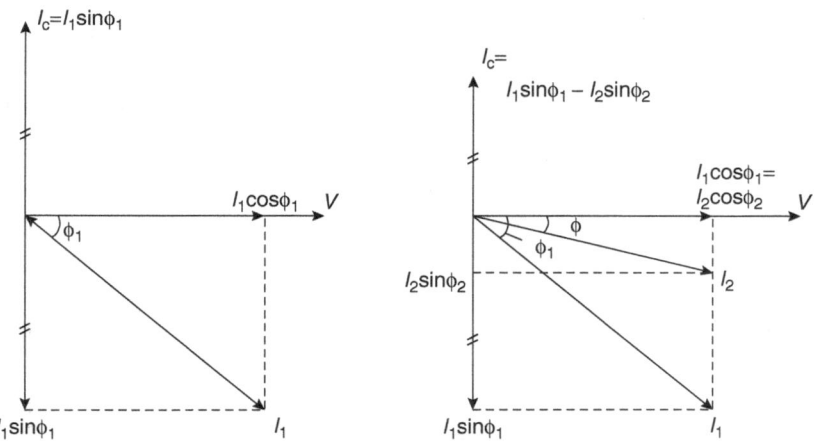

▲ **Figure 11.7** *Required phasor diagram*

The line reactive current compcnent is reduced from 80A to 26.1A $= 53.9$A . This will be the new value of I_c or a capacitor is used that takes a current of 53.9A.

Thus $X_c = \dfrac{500}{53.9} = \dfrac{10^6}{2\pi fC}$ (where C is in microfarads)

or $C = \dfrac{10^6 \times 53.9}{500 \times 2 \times 3.14 \times 50}$ $C = 343\mu F$.

Since 343μF brings the line power factor to 0.95, a further $(510 - 363) = 167\mu F$ would bring the value to unity. As the capacitor cost depends on its capacitance value and little advantage is gained by improving the power factor above 0.95, it is not necessary to achieve unity power factor working. It is vital to note that, although power factor is improved, an increased power output is *not* obtained from the load. Students often have the wrong idea, for example, that if the power factor of a circuit supplying a 5kW motor is improved, a motor will give an output greater than 5kW. This is wrong. All that is achieved is that, by connecting an extra capacitor across the motor, the total line current is reduced, i.e. a condition is attained when the *minimum* supply current required for a specified power output is used. This minimum supply current reduces all the disadvantages already stated but the motor current itself is unaltered.

kW, kVA and kVAr

As mentioned, kW, kVA and kVA~ terminology is still used in electrical engineering and some revision is needed, so the diagram (figures 11.8a and 11.8b) is considered. For a circuit, where current and voltage are out-of-phase, the phasor diagram is as shown. Current I is resolved into an in-phase or active component $I \cos \phi$ and an out-of-phase or reactive component $I \sin \phi$. $I \cos \phi$ is responsible for all the power dissipated in the circuit, since $P = VI \cos \phi$, while $I \sin \phi$ is responsible for no power, being at right angles to the voltage.

From the expression $P = VI \cos \phi$ it is seen that P can be the 'active' component of VI (symbol S – see *Note*). The term volt amperes or kilovolt amperes is used for VI then the kVA (kilovolt amperes) is regarded as resolved into 2 components, one of which is the power component. The term W or kW (kilowatts) describes this component and the other component is termed the 'volt amperes reactive' or 'reactive kilovolt amperes' and designated by VAr or kVAr. If the current phasors of figure 11.8a are multiplied by V, the new condition becomes more apparent and leads to a power diagram. The product VI (S) is shown as the volt amperes (*VA*) or $\dfrac{VI}{1000} = $ kVA and referred to as the 'apparent power'. Since $VI \cos \phi = P$ then $VA \cos \phi = $ W and kVA $\cos \phi = $ kW.

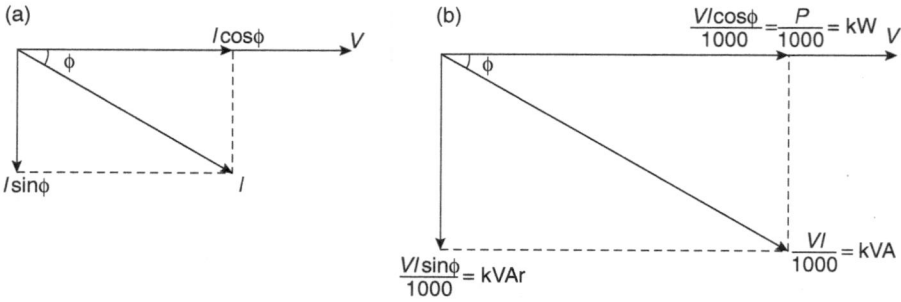

▲ **Figure 11.8** *IV and kVAr curves*

kW is a measure of 'active power', in line with the original definition for power factor, i.e. the ratio of active power to apparent power.

Thus: power factor or $\cos\phi = \dfrac{kW}{kVA}$

Similarly, $VI \sin\phi$ (Q – see *Note*), or kVA $\sin\phi$ is the 'reactive power' or volt amperes reactive designated by kVAr and from the power diagram (figure 11.8b).

$$\text{Apparent power} = \sqrt{\text{Active Power}^2 + \text{Reactive Power}^2}$$

$$\text{Summarising } kW = kVA \cos\phi. \qquad kVAr = kVA \sin\phi$$

$$kVA = \sqrt{kW^2 + kVAr^2}$$

$$\cos\phi = \frac{kW}{kVA} \qquad \sin\phi = \frac{kVAr}{kVA}$$

Note. Symbols *S*, *P* and *Q* are recommended substitutes for *VI*, *VI* cos ϕ and *VI* sin ϕ but it is likely the units – kVA, kW and KVAr – will continue to be shown on phasor diagrams, as this is the older electrical power engineer practice. The appropriate alternative has been introduced and is shown where appropriate.

In summary we have:

$P = VI \cos\phi\ Q = VI \sin\phi$ and $S = VI$

Thus $S = \sqrt{P^2 + Q^2}$ and $P = S\cos\phi \quad Q = S\sin\phi$

$\cos\phi = \dfrac{P}{S}$ and $\sin\phi = \dfrac{Q}{S}$.

It is noted that the kVA values of various loads are *not* in-phase and don't add arithmetically. kW values are all active components, are in-phase and add. kVAr values are reactive components, they can be in-phase or in anti-phase and are added, provided allowance is made for the sign as shown in the following examples.

Example 11.6. Two loads are connected in parallel. Load A is 800kVA at 0.6 (lagging). Load B is 700kVA at 0.8 power factor (lagging). Find the total kW, kVA and overall power factor of the joint loads.

See the diagram (figure 11.9).

For load A. cos ϕ_A = 0.6, sin ϕ_A = 0.8

Active power, $P_A = VI_A \cos \phi_A = 800 \times 0.6 = 480$kW

Reactive power, $Q_A = VI_A \sin \phi_A = 800 \times 0.8 = 640$kVAr

For load B. cos ϕ_B = 0.8, sin ϕ_B = 0.6

Active power, $P_B = VI_B \cos \phi_B = 700 \times 0.8 = 560$kW

Reactive power, $Q_B = VI_B \sin \phi_B = 700 \times 0.6 = 420$kVAr

Total active power, $P = 480 + 560 = 1040$kW

Total reactive power, $Q = 640 + 420 = 1060$kVAr

Total apparent power, $S = \sqrt{(1040^2 + 1060^2)} = 1485$kVA

Overall power factor $= \dfrac{1040}{1485} = 0.7 \,(\text{lagging})$

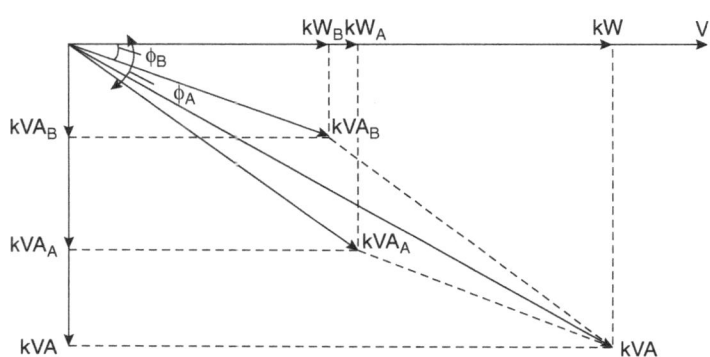

▲ **Figure 11.9** *kVA vs V curves*

Problems involving multiple loads are best treated by setting out the power-diagram components in tabular fashion. An arrow illustrates the phasor direction and reminds students which columns can be added arithmetically.

Example 11.7. A 220V, single-phase alternator supplies the following loads:

(A) 20kW at unity power factor for lighting and heating.
(B) A 75kW induction motor having an efficiency of 90.5% operating at a power factor of 0.8 (lagging).
(C) A synchronous motor taking 50kVA at a power factor of 0.5 (leading).

Find the total kVA, current and the power factor of the combined load.

Load A can be set into the columns directly as shown.

Load B Motor power output = rating as given = 75kW

$$\text{Motor input active power} = \frac{75}{0.905} = 82.9\text{kW}$$

$$\text{Apparent power} = \frac{82.9}{0.8} = 103.6\text{kVA}$$

Load C can also be set into the columns directly.

$$\text{Total apparent power (S)} = \sqrt{127.9^2 + 18.9^2} = 129\text{kV A}$$

$$\text{Total current} = \frac{129 \times 1000}{220} = 588\text{A}$$

Table 11.1 *Table with respect to Example 11.7 (above)*

Load	kVA (s) $\searrow$	kW or $\rightarrow$ kVA cos ϕ (P)	kVAr or $\updownarrow$ kVA sin ϕ (Q)	cos ϕ	sin ϕ
a	20 $\rightarrow$	20 $\rightarrow$	0	1	0
b	103.6 $\searrow$	82.9 $\rightarrow$	−62.16 $\downarrow$	0.8	0.6
c	50 $\nearrow$	25 $\rightarrow$	43.3 $\uparrow$	0.5	0.866
		127.9 $\rightarrow$	−18.86 $\downarrow$		

$$\text{Resultant power factor} = \frac{127.9}{129} = 0.99 \text{ (lagging)}$$

Note. The reactive inductive load component dominates, hence a resultant lagging power-factor condition.

Power-factor improvement (kVA method)

Treatment of power-factor improvement problems is similar to the 'current method'. The diagram for the load condition is made by splitting the original load kVA into its kW (real expended power) and kVAr (imaginary) components. Since the kW remains the same, then for a new power-factor condition for the supply, the final kVAr value is obtained by reducing the original kVAr by an amount equal to the kVAr of the components added. Such components must use no power and a static capacitor is such a device component. The added kVAr leading reduces the lagging supply kVAr. If a synchronous motor is used to obtain a better overall power factor, this contributes output power, which must be taken into account.

Example 11.8. A 400V, 50Hz, 20kW, single-phase induction motor has a full-load efficiency of 91.15% and operates at a power factor of 0.87 (lagging). Find the kVAr value of the capacitor to be connected in parallel to improve the circuit power factor to 0.95 (lagging) (2 decimal places). Find also the capacitance value of this capacitor (3 significant figures). Figure 11.10 illustrates the problem and solution.

Motor output = motor rating as given = 20kW

$$\text{Motor active input power } P_1 = \frac{20}{0.9115} = 21.94\text{kW}$$

$$\text{Motor apparent power } S_1 = \frac{21.94}{0.87} = 25.22\text{kVA}$$

As $\cos \phi_1 = 0.87$ then $\sin \phi_1 = 0.493$

Thus Q_1 or $S_1 \sin \phi_1 = 25.22 \times 0.493 = 12.44\text{kVAr}$

Although the circuit power factor is improved to $\cos \phi_2$ the *power of the circuit* is not altered $\therefore P_1 = P_2$ or $S_1 \cos \phi_1 = S_2 \cos \phi_2$ whence

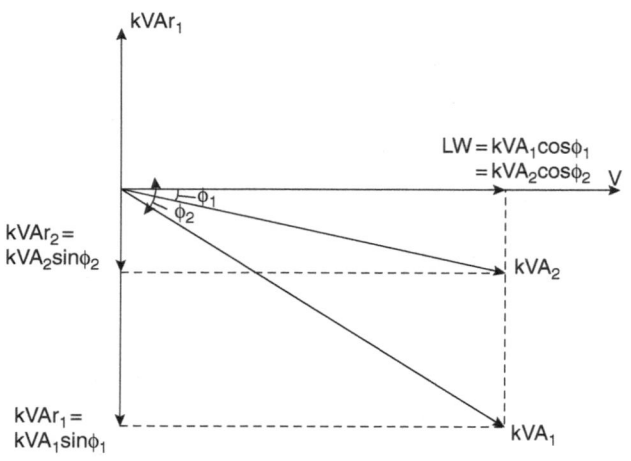

$$S_2 = S_1 \cos \phi_1 / \cos \phi_2$$

$$\text{or } S_2 = 25.22 \times \frac{0.87}{0.95} = 23.1\text{kVA}$$

Again $\cos \phi_2 = 0.95$ therefore, from tables, $\sin \phi_2 = 0.3123$ and

$$Q_2 = 23.1 \times 0.3123 = 7.21\text{kVAr}$$

Required Q value $= 12.44 - 7.21 = 5.23\text{kVAr}$. This should be the capacitor's rating.

$$\text{Capacitor current } I_c = \frac{5230}{400} = 13.75\text{A}$$

$$\text{Capacitor reactance } X_c = \frac{400}{13.75} = 30.59\Omega$$

$$\text{or } X_c = 30.59 = \frac{10^6}{2 \times \pi \times 50 \times C}$$

(where C is in microfarads.)

$$\text{Hence } C = \frac{10^4}{30.59 \times \pi} = 104\mu\text{F}.$$

Polyphase Working

The student intending to have a good practical knowledge of electrotechnology must understand the terms, relationships and theory of polyphase working. The importance of the work now covered cannot be too strongly stressed. Most students consider this part of theory to be 'that small straw that breaks the camel's back' and accordingly give it little attention. The result is that much hasty revision is needed later when A.C. machines are studied. Although Volume 7 is devoted to more advanced A.C. technology, the subject matter must now be considered as *basic and essential* to further studies.

Polyphase or split-phase power systems achieve their high conductor efficiency and improved safety by splitting up the total voltage into smaller parts and powering multiple loads at these reduced voltages, while drawing currents at levels typical of a full-voltage system, achieving a balance between system efficiency (low conductor current) and safety (low load voltage). Incidentally, this approach will work just as well for D.C. power systems as well as A.C.

Three-phase systems

Universal practice has established 3-phase systems to be the most advantageous for polyphase working. A single-phase supply, as used for small installations, can always be obtained from a 3-phase system and in this way the relative advantage of either system is available. 2-phase systems are rarely used and don't warrant study here. More than 3-phase arrangements, such as 6-phase, have fewer and more specialised applications and we will confine our investigation to 3-phase working only.

Consider a 2-pole magnet, as shown (figure 11.11), rotated inside an external stationary armature or stator. Three coils are shown equally displaced, with 'starts' and 'finishes' marked symmetrically in a regular arrangement. Induced e.m.f.s result in each coil, identical in magnitude but displaced in-phase by 120 electrical degrees. A phase sequence Red–Yellow–Blue (R–Y–B) is assumed, i.e. the rotor turns so that the red-phase voltage reaches its maximum 120° before the yellow-phase voltage reaches its maximum and the latter 120° before the blue-phase voltage, as shown by the waveform diagram. A sinusoidal distribution of rotor flux is assumed and sine-wave e.m.f.s induced. The methods by which this is achieved will be discussed for the alternator. The 3 separate coils can be used to supply 3 independent single-phase loads, but

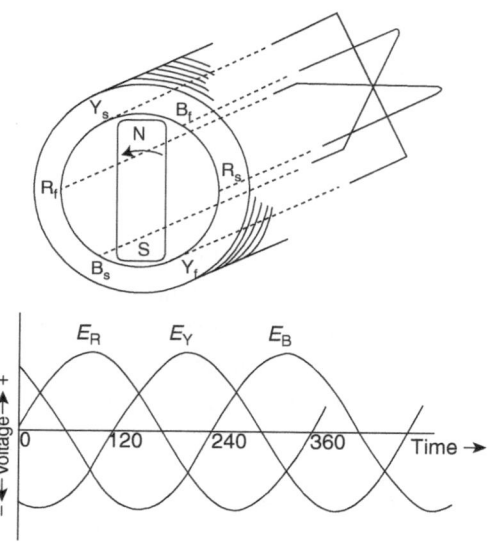

▲ **Figure 11.11** *Three-phase system diagram*

advantages are obtained by connecting the coils or phase windings. Two important methods are described: either the STAR or the DELTA connection.

Star or Y connection

Figure 11.12 shows the star arrangement with the 3 coils or phase windings connected so that either all the *starts* or *finishes* are joined together to form a star-point, i.e. 'corresponding ends' are connected together. Similarly, supply lines are connected to the free ends, remote from the star-point.

The phasor diagram is drawn in terms of voltage with the red-phase voltage (V_R) as the reference. This notation is used from now on. The small letter suffix denotes the phase value, while the capital letter denotes the line value. Lines have been identified with the colours of the phases to whose 'starts' are connected. The double suffix such as V_{R-Y} denotes the voltage *between* lines, the example being the red to yellow line voltage.

Assume the condition when the red-phase voltage is positive and the 'start' of the red-phase winding +ve with respect to the 'finish' or neutral point. Current flows through the lines and load as shown. For this example it is possible because, for the yellow phase at the same instant, its start is −ve with respect to its finish, as the yellow-phase waveform is in its −ve half cycle. Thus for the phasor diagram, the voltage between the red and yellow lines is obtained by the phasor *difference* of V_R and V_Y.

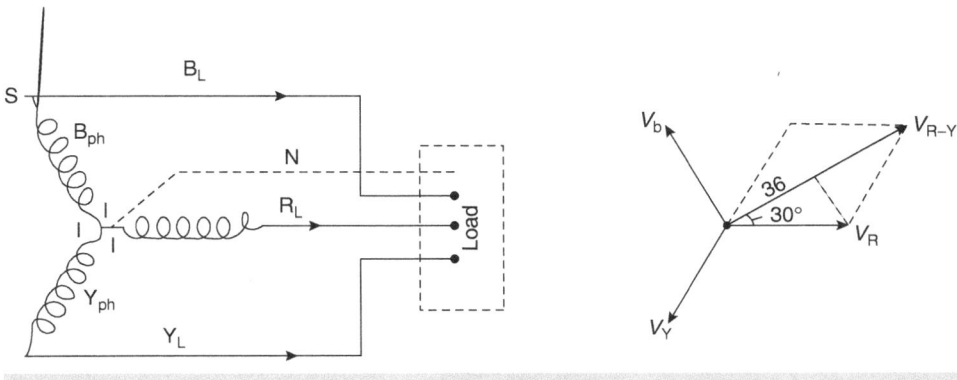

▲ **Figure 11.12** *Star or Y connection*

As a phasor difference is considered, the resultant is obtained by *reversing* one phasor with respect to the other and completing the parallelogram. From the deduction below, it is seen that the line voltage is $\sqrt{3}$ times the phase voltage. This relation holds for the other lines and their associated phases. A further point of importance for the star connection is that the line current equals the phase current or $I_L = I_{ph}$

Consider the phasor diagram. Let the line voltage $V_{R-Y} = 2x$

$$\text{But } \frac{x}{V_r} = \cos 30° \therefore x = \frac{\sqrt{3}}{2}V_r \quad \text{or} \quad 2x = \sqrt{3}V_r$$

Hence $V_{R-Y} = \sqrt{3}V_r$

or the voltage between lines $= \sqrt{3} \times$ the phase voltage.

Thus $V_L = \sqrt{3} \times V_{ph}$

For a star connection the following must be remembered.

$$\text{Line voltage} = \sqrt{3}\text{-phase voltage or } V = \sqrt{3}V_{ph} = 1.732V_{ph}$$

$$\text{Line current} = \text{Phase current or } I = I_{ph}$$

Note. The subscript L, as in V_L and I_L, is omitted when generalising – this is usual and both V and I are assumed to be line values. Again the relations are derived for an alternator or supply source but they also relate to a star-connected load, as the example shows.

Example 11.9. Three 50Ω resistors are connected in a star configuration across 415V, 3-phase mains supply. Calculate the line and phase currents (1 decimal place) and the power taken from the supply (2 significant figures).

As the load is balanced, the voltage across each resistance is the phase voltage, so

$$V_{ph} = \frac{415}{\sqrt{3}} = 240V$$

Phase current = line current or $I_{ph} = I = \frac{240}{50} = 4.8A$

Power dissipated by 1 phase of load $= I_{ph}^2 R_{ph} = 4.8^2 \times 50W$

$$= 1152W \text{ or } 1.152kW$$

and 3-phase power from the supply

$$= 3 \times 1.152 = 3.456kW = 3.5kW.$$

Use of the neutral

An obvious use of the star connection is for distribution, as 2 voltages are available to the consumer, one for lighting and the other for power. Either 1-phase or 3-phase loading is possible and shown (figure 11.13).

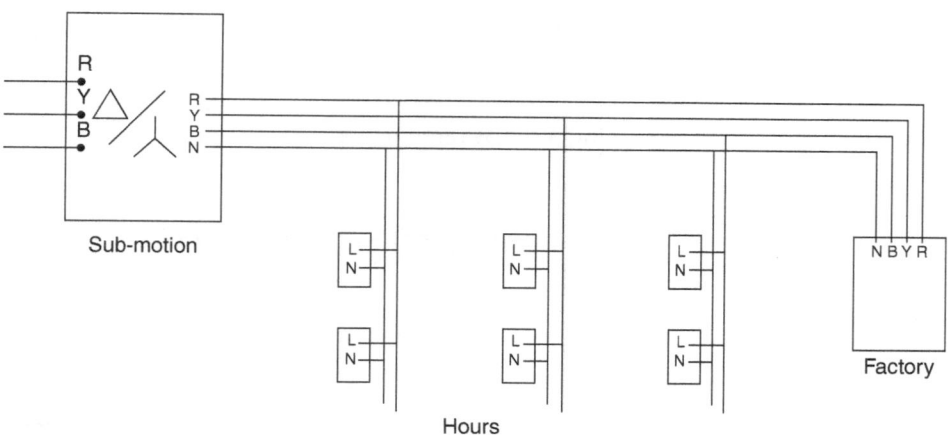

▲ **Figure 11.13** *Use of the neutral line*

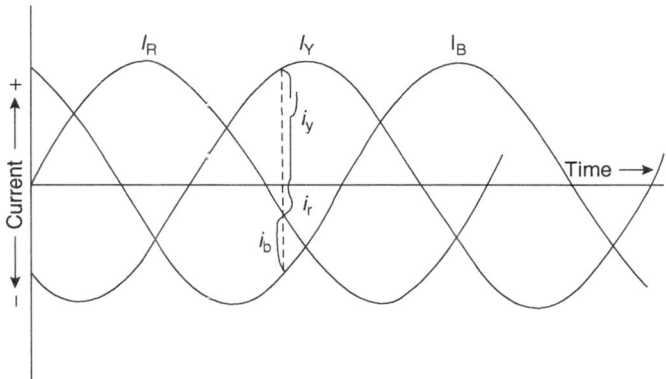

▲ **Figure 11.14** *Current vs time*

Balanced load

A 3-phase load is said to be 'balanced' if the currents in all 3 phases are equal and their phase angles the same. If an instant in time is considered on the diagram (figure 11.14), the sum of *instantaneous* values of the currents $i_r + i_y + i_b = 0$. As these currents meet at the load neutral point and the resultant flows through the neutral line then the neutral carries *no current* and need not be used for balanced loading.

Unbalanced load

A neutral must be used if the load phase currents are unequal or the phase angles different. The neutral line carries the unbalanced current, i.e. the resultant of the 3 line currents. Since this neutral current is a phasor sum, it is obtained graphically or mathematically, as shown.

Example 11.10. The loads of a 4-wire, 3-phase system are: red line to neutral current = 50A, P.F. = 0.707 (lagging).

Yellow line to neutral current = 40A, power factor = 0.866 (lagging)

Blue line to neutral current = 40A, power factor = 0.707 (leading)

Determine the value of the current in the neutral wire. The solution is worked with reference to figure 11.15.

I_r = 50A lagging the voltage by 45° since cos 45 = 0.707

I_y = 40A lagging the voltage by 30° since cos 30 = 0.866

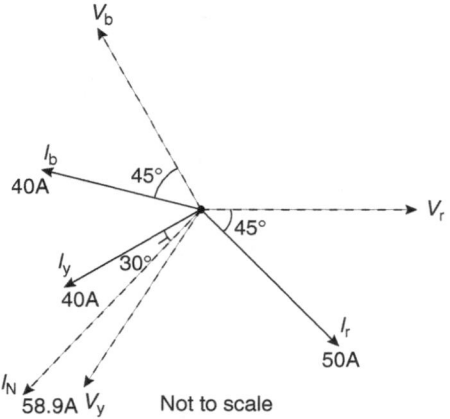

▲ **Figure 11.15** *Full I-V contributions*

$I_b = 40A$ leading the voltage by 45° since cos 45 = 0.707

Resolving into horizontal and vertical components.

$I_H = (50 \times \cos 45) - (40 \times \cos 30) - (40 \times \cos 15)$

$\quad = (50 \times 0.707) - (40 \times 0.866) - (40 \times 0.966)$

$\quad = 35.35 - 34.64 - 38.64 = -37.93A$

$I_v = - (50 \times \sin 45) - (40 \times \sin 30) + (40 \times \sin 15)$

$\quad = - (50 \times 0.707) - (40 \times 0.5) + (40 \times 0.259)$

$\quad = - 35.35 - 20 + 10.36 = -44.99$

Current in the neutral is the resultant.

$$\text{or } I_N = \sqrt{37.93^2 + 44.92^2}$$

$$= 58.9\,A$$

$$\cos \theta = \frac{-37.93}{58.9} = -0.653$$

$$\theta = 49.5°$$

Note. The −ve sign gives the quadrant in which I_N lies, as shown on the diagram.

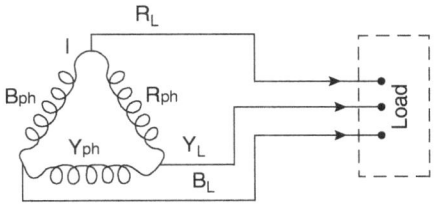
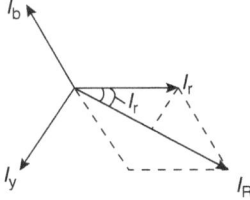

▲ **Figure 11.16** *Delta Δ or mesh connection*

Delta Δ or mesh connection

This arrangement is shown (figure 11.16). For this connection, the 3-phase windings are arranged into a closed circuit by connecting 'uncorresponding ends', i.e. the *start* of one phase to the *finish* of another phase. Thus R start is connected to B finish, Y start to R finish, etc. The same reasoning as introduced for figure 11.17 is applied here, except that voltages are considered. Thus if the diagram represented the 3 equal phase voltages, it is seen that at any instant the sum of the instantaneous voltage values $v_r + v_y + v_b = 0$. For the mesh or closed winding, as the sum of the instantaneous voltages is zero, no circulating current flows round the mesh. Lines are taken from the junction points and for this connection, it is clear that the voltage developed across a phase is the voltage provided for the connected lines.

Thus V_L or $V = V_{ph}$

The lines have been identified with the colours of the phases to whose *starts* they are connected. Assume the condition when the red-phase voltage is positive, i.e. the start of the red-phase winding is +ve with respect to the finish. Current flows through from R phase into R line as shown. At this same instant the voltage in B phase is negative, i.e. its finish is +ve with respect to its start. Thus it is also correctly connected for feeding current into the R line and a line current is obtained by considering the phasor *difference* of 2 phase currents. The resultant line current is obtained by reversing a phase current (I_b) and combining it with I_r as shown (figure 11.16).

As before $x = I_r \cos 30° = I_{ph} \cos 30°$

Or $x = \dfrac{\sqrt{3}}{2} I_{ph}$ and as $I_L = 2x = \dfrac{2\sqrt{3} I_{ph}}{2}$

Hence I_L or $I = \sqrt{3} I_{ph}$

Thus for a delta connection, Line voltage = Phase voltage and Line current = √3 Phase current.

This relationship is deduced for *any* line and the connected phases and will give the same result. As before, V and I are used for line values and V_{ph} and I_{ph} for phase values.

Example 11.11. Three 50Ω resistors are delta-connected across 415V, 3-phase lines as for Example 11.9. Calculate the line and phase currents (1 decimal place) and the power taken from the mains supply in 1 phase of the load (4 significant figures).

Voltage across 1-phase resistor = 415V

Current in 1 phase of load $= \dfrac{415}{50} = 8.3$A

Since the load is balanced, line current = √3 × 8.3 amperes = 14.4A

Power in 1 phase of load $= 8.3^2 \times 50 = 3445$W

In addition the power in all 3 phases of load will be = 3 × 3.445 = 10.3kW.

Three-phase power

For a star-connected load, $V = \sqrt{3}V_{ph}$ and $I = I_{ph}$

The power expended in 1 phase $= V_{ph}\, I_{Ph}\, \cos \phi$ and the power expended in 3 phases $= 3\, V_{ph}\, I_{ph}\, \cos \phi$

Converting to line values, the above becomes:

Three-phase power $= 3\, \dfrac{V}{\sqrt{3}} I \cos \phi$

$$\text{or } P = \sqrt{3}VI \cos \phi$$

For a Delta-connected load $V = V_{ph}$ and $I = \sqrt{3}I_{ph}$

The power expended in 1 phase $= V_{ph}\, I_{ph}\, \cos \phi$ and the power expended in 3 phases $= 3V_{ph}\, I_{ph}\, \cos \phi$

Converting to line values, the above becomes:

Three-phase power $= 3V \dfrac{I}{\sqrt{3}} \cos \phi$

$$\text{or } P = \sqrt{3}VI \cos \phi$$

Thus the general expression holds, irrespective of the type of connection, namely either star or delta connected: 3-phase power is given by $\sqrt{3}\ VI \cos \phi$.

Example 11.12. A 75kW, 400V, 3-phase, delta-connected induction motor has a full-load efficiency of 91% and operates at a power factor of 0.9 (lagging). Calculate the line and phase currents at full load.

$$\text{Output power} = 75 \times 10^3 \text{ watts}$$

$$\text{Input power} = \frac{75 \times 10^3 \times 100}{91} \text{watts}$$

$$\text{also } P = \sqrt{3}\ VI \cos \phi$$

$$\text{So } \sqrt{3} \times 400 \times I \times 0.9 = \frac{75 \times 10^5}{91}$$

$$\text{And } I = \frac{75 \times 10^4}{1.732 \times 4 \times 9 \times 91} \text{ amperes}$$

$$\text{or } I = 132.2A$$

$$\text{Motor phase current } \frac{132.A}{\sqrt{3}} = 76.3A$$

Three-phase kVA, kW and kVAr

As power factor can be defined as the ratio of true power to apparent power, it can be applied to 3-phase working.

Thus:

$$\text{power factor} = \frac{\text{active power}}{\text{apparent power}}$$

Irrespective of star or delta connection $P = \sqrt{3}\ VI \cos \phi$.

So:

$$\cos \phi = \frac{P}{\sqrt{3}VI}$$

It follows that for 3-phase working, in order that the definition for power factor should apply,

apparent power (S) = $\sqrt{3}$ VI

Note introduction of $\sqrt{3}$ – distinguishes this condition from single-phase working.

Again it is known that $\cos \phi = \frac{P}{S}$ or $\frac{KW}{KVA}$ so it follows that: *3-phase* kVA $= \frac{\sqrt{3}VI}{1000}$

Example 11.13. A 3-phase, 400V motor takes a current of 16.5A when the output is 9kW. Calculate (a) the kVA input (2 decimal places) and (b) the power factor, if the efficiency at this load is 89% (2 decimal places).

(a) kVA input $= \frac{\sqrt{3}VI}{1000} = \frac{\sqrt{3} \times 400 \times 16.5}{1000}$

$\qquad\qquad = 11.43$kVA

(b) Output power = 9kW

$\qquad$ True active power $= \frac{9.0}{0.89} = 10.11$kW

$\qquad$ So Power factor $= \frac{\text{active power}}{\text{apparent power}} = \frac{10.11}{11.43} = 0.88$ (lagging) ·

Combining star and delta networks provides an interesting example.

Example 11.14. A 3-phase, star-connected alternator supplies a delta-connected induction motor at 600V. The current taken is 40A. Find (a) the alternator's phase voltage (3 significant figures) and (b) the current in each phase of the motor (1 decimal place). Refer to figure 11.17.

(a) For a star connection $V = \sqrt{3} V_{ph}$

$\qquad \therefore V_{ph} = \frac{V}{\sqrt{3}} = \frac{600}{\sqrt{3}} = 346$V. This is the alternator phase voltage.

(b) For a delta connection $I_{ph} = \frac{I}{\sqrt{3}} = \frac{40}{\sqrt{3}} = 23.1$A

$\qquad$ This is the current in each motor phase.

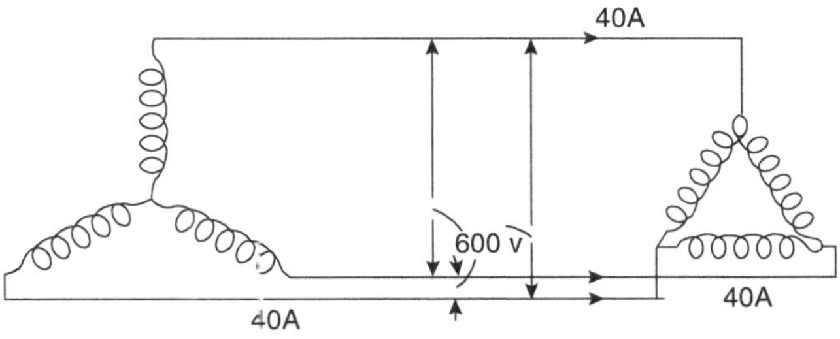

▲ **Figure 11.17** *Star with delta circuit connections*

(c) If the motor operates at a power factor of 0.8 (lagging) and an efficiency of 88%, find the alternator's kVA rating (1 decimal place) and the motor's power output (2 decimal places).

$$\text{Apparent power rating of alternator} = \frac{\sqrt{3}VI}{1000}$$

$$= \frac{\sqrt{3} \times 60 \times 40}{1000} = 41.6\text{KVA}$$

$$\text{Motor apparent input power} = \frac{\sqrt{3}VI}{1000}$$

$$= \frac{1.732 \times 600 \times 40}{1000} = 41.6\text{KVA}$$

$$\text{True active input power} = 41.6 \times 0.8 = 33.28\text{kW}$$

$$\text{Output power} = 33.28 \times \frac{88}{1000} = 29.21\text{kW}$$

Example 11.15. A delta-connected load is as shown (figure 11.18). If the supply voltage is 400V, 50Hz, calculate the red-line current (I_R) (1 decimal place). Assume the currents as shown and maintain the correct phase sequence: R, Y and B. If the red-line current feeds current into the red phase of the load, the blue-phase current will be in the opposite sense so a phasor difference is involved.

Here $I_r = \dfrac{400}{100} 4\text{A}$ in-phase with V_{RY}

Also $I_b = \dfrac{400}{106} = 3.7\text{A}$ leading V_{BR} by 90°

Reversing I_b and using the Cosine Rule

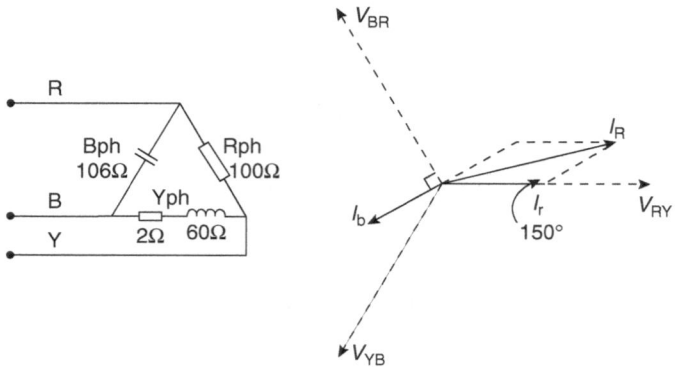

▲ **Figure 11.18** *Delta-connected load I-V contributions*

$$I_R = \sqrt{I_r^2 + I_b^2 + 2I_r I_b \cos 150°}$$

$$= \sqrt{4^2 + 3.7^2 - 2 \times 4 \times 3.7 \cos 150°}$$

or I_R $= 7.4\text{A}.$

Practice Examples

11.1. A coil consumes 300W when connected into a 60V D.C. circuit. It consumes 1200W when connected into an A.C. circuit of 130V. What is the coil's reactance (1 significant figure)?

11.2. A circuit consists of 2 branches A and B in parallel. Branch A has a resistance of 12Ω and a reactance of 3Ω, while the values of branch B are 8Ω and 20Ω respectively. The circuit is supplied at 100V. Calculate the current in each branch (1 decimal place) and the supply current (1 decimal place).

11.3. An inductive circuit of resistance 50Ω and inductance 0.02H is connected in *parallel* with a *25μF* capacitor across a 200V, 50Hz mains supply. Find the total current taken from the supply (1 decimal place) and its phase angle.

11.4. Two coils of resistances 8Ω and 10Ω and inductances 0.02H and 0.05H respectively are connected in parallel across 100V, 50Hz mains supply. A capacitor of capacitance 80μF in *series* with a 20Ω resistor is then connected in *parallel* with the coils. Find the total current taken from the mains supply (2 decimal places) and its phase angle with respect to the applied voltage.

11.5. A single-phase motor has an input of 50.6A at 240V, the power input being 10kW and the output 9kW. Calculate the value of the apparent power (1 decimal place), power factor (2 decimal places) and the efficiency.

11.6. A single-phase motor running from a 230V, 50Hz mains supply takes a current of 11.6A when giving an output of 1.5kW, the efficiency being 80%. Calculate the capacitance required to bring the power factor of the supply current to 0.95 (lagging) (2 significant figures). Calculate the capacitor's kVAr rating (2 decimal places).

11.7. The load taken from a single-phase supply consists of:

(a) Filament lamp load of 10kW at unity power factor.

(b) Motor load of 80kVA at 0.8 power factor (lagging).

(c) Motor load of 40kVA at 0.7 power factor (leading).

Calculate the total load taken from the supply in kW (3 significant figures) and in kVA (2 decimal places) and the power factor of the combined load (2 decimal places). Find the 'mains' current if the supply voltage is 250V (3 significant figures).

11.8. Three equal impedances of 10Ω, each with a phase angle of 30° (lagging), constitute a load on a 3-phase alternator, giving 100V per phase. Find the current per line and the total power when connected as follows: (a) Alternator in star, load in star, (b) Alternator in star, load in delta, (c) Alternator in delta, load in delta, (d) Alternator in delta, load in star (all 1 decimal place).

11.9. A 500V, 3-phase, star-connected alternator supplies a star-connected induction motor rated 45kW. The motor efficiency is 88% and the power factor 0.9 (lagging). The alternator efficiency at this load is 80%. Determine (a) the line current (1 decimal place), (b) the alternator power output (2 decimal places) and (c) the output power of the prime-mover (2 significant figures).

11.10. A 400V, 3-phase system takes 40A at a power factor of 0.8 (lagging). An over-excited synchronous motor is connected to raise the power factor of the combination to unity. If the motor's mechanical output is 12kW with efficiency 91%, find the kVA input to the motor and its power factor. Find also the total power taken from the supply mains (all 2 decimal places).

12

THE D.C. GENERATOR

The origin as well as the progress and improvement of civil society is founded in mechanical and chemical inventions.

Sir Humphry Davy

This chapter will not cover D.C. machine construction, operation and maintenance in detail, as this is better dealt with practically. The basic features of the D.C. generator will be covered and help to consolidate the theory introduced in Chapter 7. Once the D.C. generator construction is outlined, attention can then be given to further theory in Volume 7, and will be considered from a functional viewpoint, under the headings: (1) The Generator and (2) The Motor. Recently there has been a trend to remove detailed study of the D.C. generator from marine electrotechnology syllabi. Nevertheless, study is valued as it helps define the differences between the generator – a machine to convert mechanical into electrical energy – the motor and a machine to convert electrical into mechanical energy.

D.C. Machine Construction

The principal features of D.C. machines are described under (1) the field system or stator and (2) the armature or rotor. Figure 12.1 is a representation for a D.C. compound-wound generator.

Field system

This includes the magnet arrangement of poles and yoke, the field coils and interpoles (if fitted). Interpoles are part of the armature electrical circuit and are mentioned under the heading of POLES AND YOKE. The former are the cores of the machine

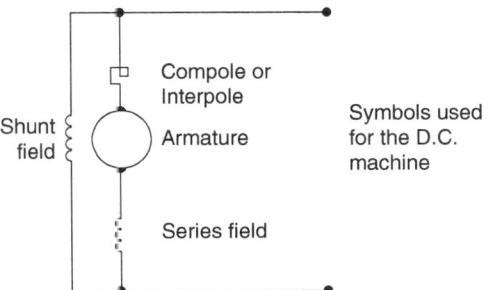

▲ **Figure 12.1** *D.C. machine symbols*

electromagnets and are often fitted with pole-shoes to concentrate the field across the narrow air gaps in which conductors move. The yoke is an extension of a magnet system forming the main frame of the machine carrying flux from and to the poles. Figure 12.2 shows typical field system construction.

Poles and yoke are constructed from cast steel or fabricated from mild steel sheet cut and rolled into shape. Poles may be part of the yoke, but are now usually built up in thin laminate form, riveted together and shaped to include the pole-shoes. Interpoles are similar to the main field poles but located on the yolk between them, having windings in series with the armature to reduce armature reaction effects.

FIELD COILS. Field coils are of 2 types: (1) shunt coils – made of many turns of fine wire and (2) series coils – made from only a few turns of thick cable or conductor. Shunt coils are built on a 'bobbin' or 'former' while series coils may be self-supporting. Figure 12.3 shows a typical construction cross-section with insulation determined by the machine class and its duty.

The armature

This consists of armature core, windings, shaft and commutator. The brushes, although not part of the armature, are considered as they work with the commutator.

ARMATURE CORE. This is built up from iron laminations clamped between 2 end plates. Laminations are insulated from each other to minimise induced circulating or 'eddy currents'. If clamping bolts pass through the core as shown (figure 12.4), they must be insulated.

Modern methods use stamped laminations pressed onto and 'keyed' to a shaft, with end plates screwed onto small machine shafts. For larger designs a 'spider' shape helps ventilation and keeps the iron volume needed to a minimum.

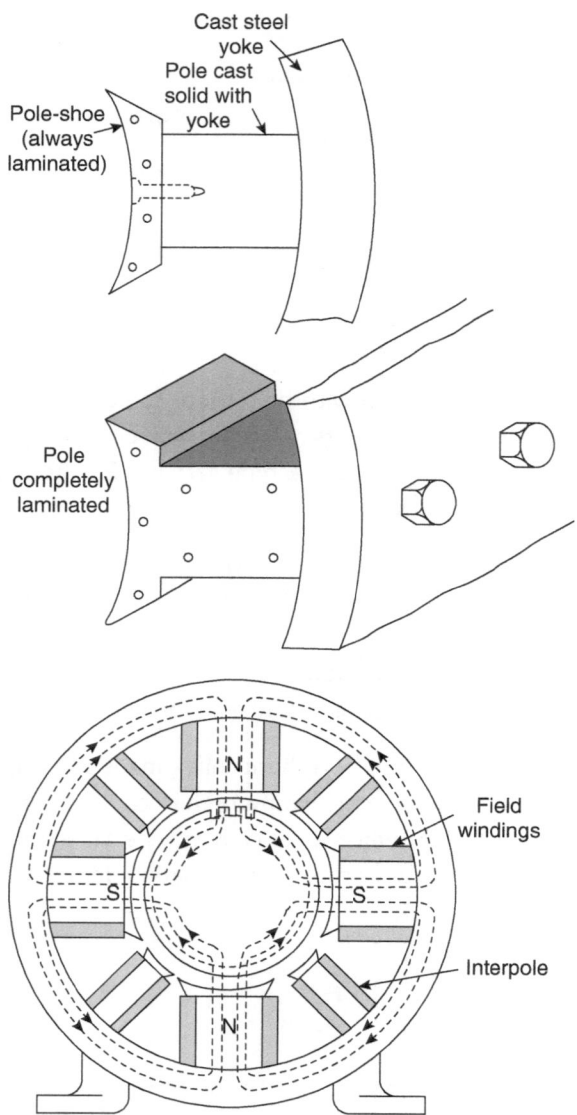

▲ **Figure 12.2** *Typical field system construction*

WINDINGS. The number of conductors, their size, shape, etc. are decided by the machine's design requirements. Figure 12.5 shows a typical method of locating and holding coil sides in place. A wedge of bakelised paper or fabric is shown, but open slots with a piece of fibre and 'binders', made from high-tensile steel wire, can be used. For small machines, for example, motors for vacuum-cleaners, cabin-fans, etc., armature

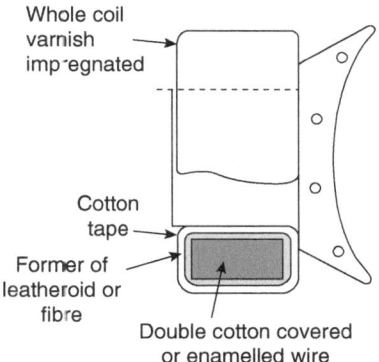

▲ **Figure 12.3** *Typical construction cross-section*

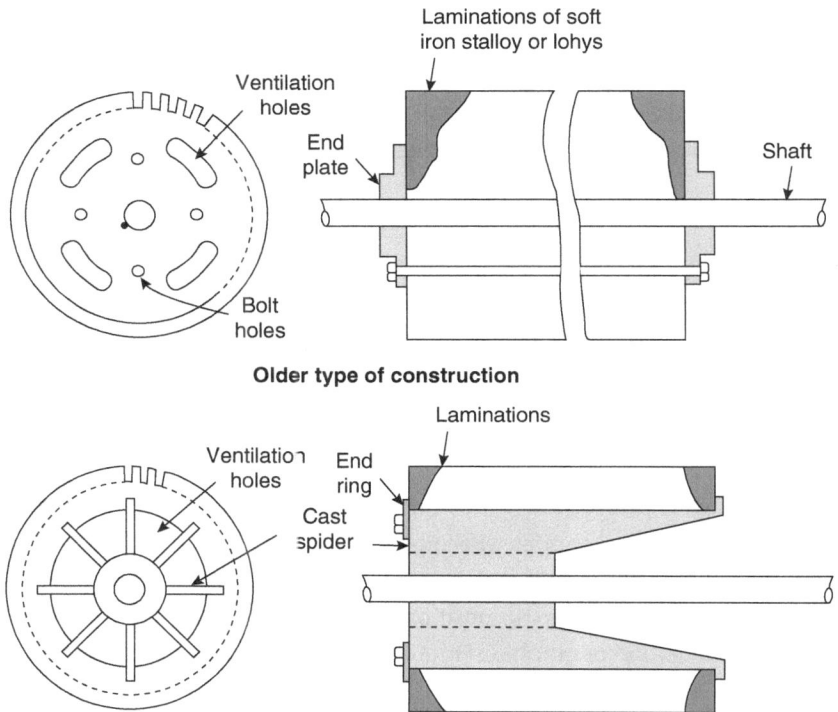

▲ **Figure 12.4** *Core construction*

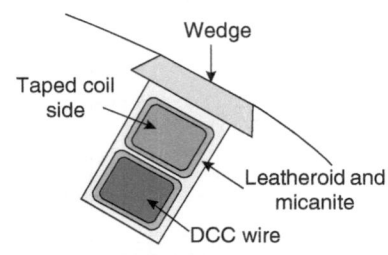

▲ **Figure 12.5** *Typical coil placement*

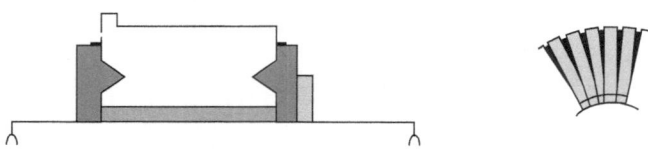

▲ **Figure 12.6** *Commutator placement*

windings are enamelled or cotton-covered wire placed in and wound by hand. Semi-enclosed slots are used with fibre inserts closing the slots. Conductors are arranged into a closed winding and considered in detail after discussing machine construction.

SHAFT. This is made from forged mild steel and designed so it doesn't bend when at maximum speed.

COMMUTATOR. This consists of copper segments insulated from each other by mica. Segments are mounted on but insulated from a sleeve secured to the shaft and clamped by an end-ring, which can be bolted or screwed as shown (figure 12.6). Insulated cone-shaped rings of micanite insulate segments from the steel clamping assembly. Armature windings are soldered to these segments.

BRUSHES. A brush is pressed onto a commutator by a pressure arm and connected to the holder by a braided copper wire 'pigtail' moulded into the brush. One or more brush-holders may be carried on an insulated spindle mounted on the brush rocker-ring, itself clamped once the brush position is set. Modern D.C. machine brushes are made of moulded carbon or graphite. Figure 12.7 is a typical arrangement.

BEARINGS. For most industrial D.C. machines, bearings are a ball or roller type. The advantages are (1) axial length is shorter than the journal type bearings and (2) after packing with grease, long periods of service are possible. For marine work journal bearings give quieter running and are preferred as they resist 'transmitted' vibrations. A steel shaft runs in a brass or cast-iron sleeve lined with metal. For small or medium size machines, bearings are carried in the end shields, but for large machines bearings are separate.

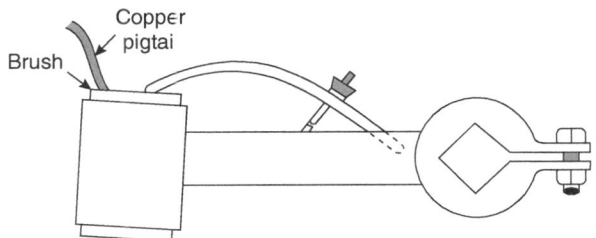

▲ **Figure 12.7** *Typical D.C. machine brush arrangement*

D.C. Armature Winding Arrangements

The simplest winding is built up from single-turn coils of span equal to 1 'pole pitch', i.e. 180° for a 2-pole machine. For a 4-pole machine the coil span is still 1 pole pitch, but is now 90° as shown (figure 12.8). It is unusual to make the span exactly equal to 1 pole pitch and many small machines have an odd number of slots. Each slot carries 2 coil sides, i.e. it contains more than 1 conductor. D.C. windings are usually of a 2-layer type, a coil side lying at the bottom of a slot and another at the top, but 4, 6 or even 8 coil sides may be contained in 1 slot as it may be impracticable to have many slots. There are 2 basic methods of connecting conductors on an armature after forming into either single or multi-turn coils and a complete winding falls into 2 distinct types, namely: (1) a wave or (2) a lap winding.

(1) The WAVE or 2-circuit winding results in 2 parallel paths irrespective of the number of machine poles. Two sets of brushes only are needed but it is usual to fit as many sets of brushes as the machine has poles. Figure 12.9a shows the layout.

(2) The LAP or multi-circuit winding results in as many paths in parallel and as many sets of brushes as the machine has poles. Figure 12.9b shows the layout. In building up a winding it is vital to connect coil elements so the induced e.m.f.s in conductors add, in the same way as cells are connected in series so their e.m.f.s add to give the required battery voltage. Thus conductor X is in series with conductor Y, which occupies relatively the same position as X but is under a pole of *reversed* polarity. A coil element formed by conductors XY is then connected in series with a similarly placed coil element under a pair of poles so the required voltage for a parallel armature path is achieved. For a wave winding, connections are as shown. For a lap winding, the same rule is followed, except that all coil elements under a pair of poles are connected in series *before* the winding connects to conductors.

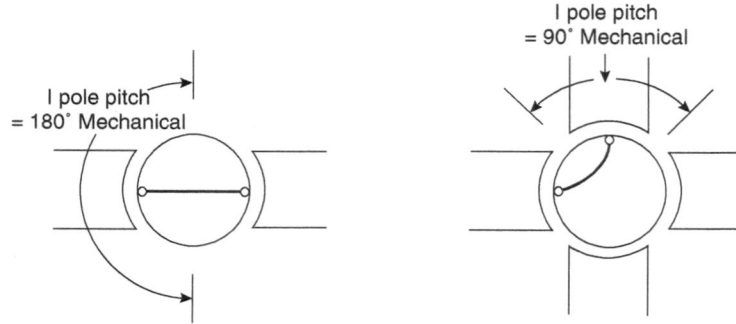

▲ **Figure 12.8** *D.C. winding arrangements*

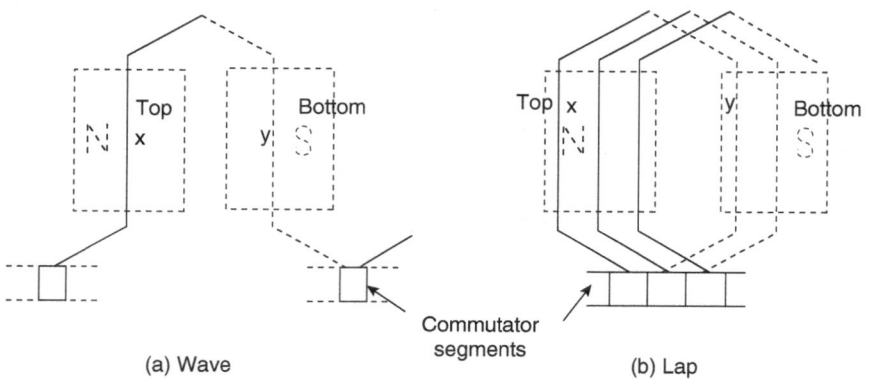

▲ **Figure 12.9** *Lap and wave D.C. windings*

The following example illustrates both lap and wave windings. A small armature is designed to have 8 one-turn coils each having 2 conductors. There is 1 commutator segment to a coil, i.e. 8 commutator segments. If only 2 coil sides are accommodated in a slot there must be 8 armature slots. If a 4-pole system were used there would be 2 slots/pole, with a true pole pitch of 2, the pole pitch being the number of armature slots divided by the number of poles. As the coil sides are under the influence of the correct field poles, winding pitch must be as nearly as possible equal to the pole pitch. So the winding pitch will also be equal to 2 or a coil should embrace 2 teeth.

The LAP winding is considered first. For such a winding, connecting up of conductors is such that the winding progresses round the armature by being pitched alternatively forwards and backwards. If figure 12.10 is considered, it is seen that **conductor No. 1** is connected to No. 6 spaced 2 teeth away (2 commutator segments). No. 6 is then connected to No. 3 and so on. The winding thus progresses by 1 slot until it is closed by all the slots having been occupied and conductor No. 15 is connected to No. 1 through No. 4.

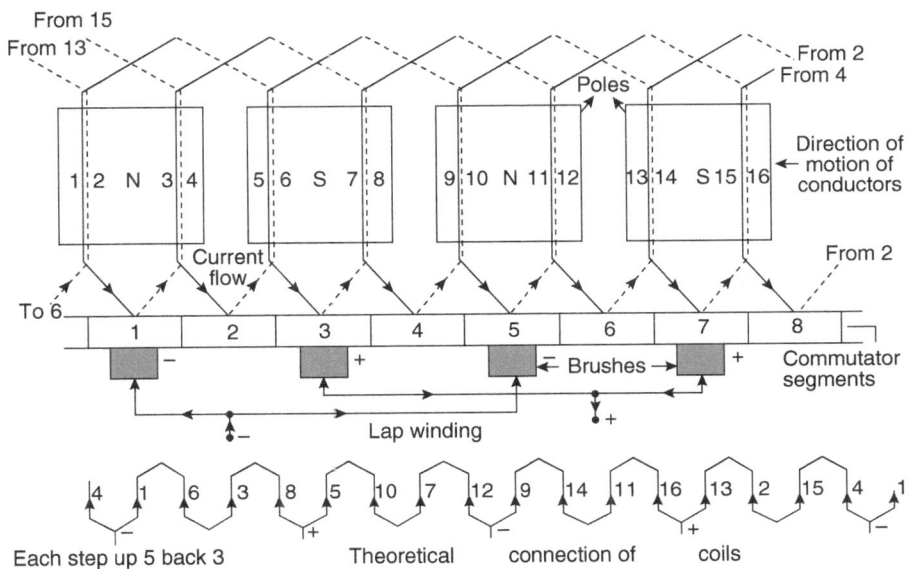

▲ **Figure 12.10** *Armature including lap windings*

If now a WAVE winding is needed a preliminary examination will show that this could not be achieved via the coil connections of figure 12.11. If the winding started at No. 1 proceeded to No. 6 and then on through Nos 9 and 14 it would close back onto conductor No. 1. It is obvious that an armature with 8 slots would not be suitable for such a wave winding and one of 7 or 9 slots must be considered. A 9-slot armature winding will give a winding pitch of length slightly less than the true pole pitch length and is suitable. Consider the diagram (figure 12.11). Conductor No. 1 is connected to No. 6 as before, then connected to Nos 9, 14, 17 and then to No. 4, i.e. the wind ng passes into the slot *next* beyond that which it started. The winding doesn't close immediately and if the connecting-up proceeds as described, the winding will progress 4 times round the armature before the close is made at the starting slot by conductor No. 11 being joined to No. 1 through No. 16. This is a suitable winding but 9 coils will be used with 9 armature slots and 9 commutator segments.

Armature winding details are found elsewhere as machine design and armature winding is specialist work.

The D.C. Generator

D.C. armature and commutator theory shows that commercial D.C. is best obtained using an armature wound with several coils connected so all the coils except those

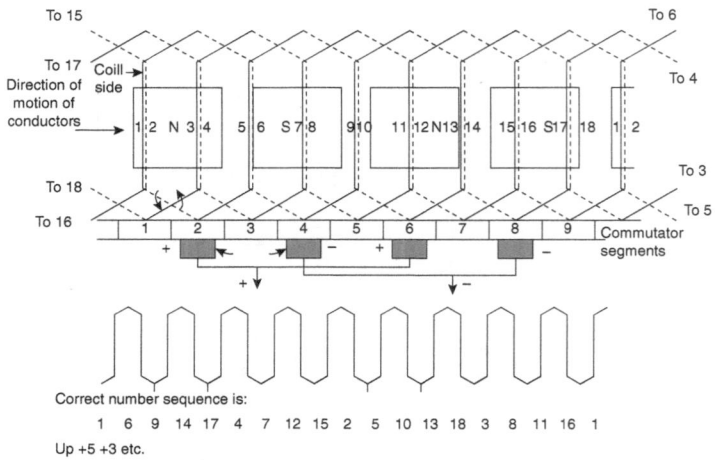

▲ **Figure 12.11** *Armature including wave windings*

short-circuited by brushes are in the circuit. The armature being a continuous closed winding splits itself electrically into several parallel paths. There are 2 key armature winding types: (1) a lap winding or (2) a wave winding. For a lap winding the number of parallel paths in an armature is always equal to the number of poles. For a wave winding the number of parallel paths is always 2, irrespective of the machine's number of poles.

The e.m.f. equation

Consider the diagram (figure 12.12) and the factors for the machine. A simple expression for the composite armature is deduced and it is vital to memorise it. A student must be able to prove it from first principles.

Let N = the machine speed in rev/min, P = the number of poles. Φ = the flux/pole (webers). Z = the number of armature conductors. A = the number of parallel paths of the armature winding.

In 1 second an armature revolves $\dfrac{N}{60}$ times so in 1 revolution, 1 conductor cuts a flux

$= P \times \Phi$ webers

∴ In 1 second, 1 conductor cuts $P\,\Phi\,\dfrac{N}{60}$ webers.

From Faraday's law the magnitude of the e.m.f. generated in volts is given by the flux cut/second

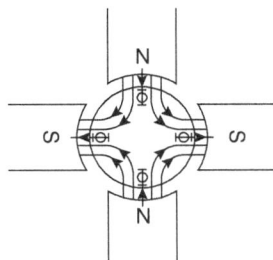

▲ **Figure 12.12** *Flux diagram between magnet poles*

so the e.m.f. generated in 1 conductor $= \dfrac{P\Phi N}{60}$ volts.

If the armature winding divides into A parallel paths, the e.m.f. of a parallel path is also a machine's e.m.f.

In a parallel path there are $\dfrac{Z}{A}$ series conductors, so the e.m.f. of 1 parallel path =

the machine e.m.f. $= \dfrac{P\Phi N}{60} \times \dfrac{Z}{A}$ Thus $E = \dfrac{Z\Phi N}{60} \times \dfrac{P}{A}$ volts, where E = the generated voltage.

Example 12.1. The armature of a 4-pole shunt generator is lap wound and generates 216 volts when run at 600 rev/min. The armature has 144 slots with 6 conductors/slot. If this armature is rewound and *wave connected*, determine the e.m.f. generated at the same speed and flux/pole.

From the e.m.f. equation e.m.f. $= 216 = \dfrac{(6 \times 144) \times \Phi \times 600 \times 4}{60 \times 4}$ or $\Phi = \dfrac{216}{60 \times 144}$

webers

Note. This is a *Lap-wound* armature so $A = P = 4$.

For a *Wave-wound* armature $A = 2$

$\therefore E = \dfrac{6 \times 144}{60} \times \dfrac{216}{60 \times 144} \times \dfrac{600}{2} \times 4 = 216 \times 2 = 432V.$

Characteristics

These are graphs that show the behaviour of any machine type. For example, consider the e.m.f. equation. For a given machine, all the factors except Φ and N are constant. The equation can be written $E = k\,\Phi\,N$ where $k = \dfrac{ZP}{60A}$ Thus $E\,\alpha\,\Phi$ if N is kept constant and $E\,\alpha$ N, if Φ is kept constant.

If Φ and N both vary, E will vary accordingly. Thus the voltage generated is controlled by varying the machine speed or flux as shown by deducing the 'no-load' characteristics.

Associated Magnetic Circuit Effects

In Chapter 6 the iron or steel cored electromagnet was considered. As the magnetic circuit is an essential part of a D.C. machine, it is vital to consider 2 effects influencing generator characteristics.

The first effect is that of residual magnetism. Experiment shows that if an electromagnet's iron core is magnetised by passing a current through the energising coil, when the current is switched off and the magnetising m.m.f. removed, the magnetism or magnetic flux *will not* completely disappear, although in theory it should not exist. It is important for readers to appreciate that this effect occurs.

The second factor is the *saturation effect* of iron when subject to an m.m.f. If a magnetic circuit uses iron as the medium for conveying flux then, as the m.m.f. increases, the flux increases up to a point when the straight-line relation between m.m.f. and flux Φ and consequent flux density B is no longer followed. Thus for an iron sample, if B is plotted against H (the magnetising force/metre), the 'B–H curve' is obtained. The resulting graph is a straight line for *only* a short part of its length. It appears as shown (figure 12.13) indicating that iron *saturates*, i.e. no matter how much the m.m.f. increases, once the curve bends over and flattens out no increased flux Φ or consequent B value is produced, irrespective of the magnetising force's strength.

The flattening out or saturation effect is considered to be due to all the molecular magnets aligning with the magnetic field. The saturating effect is apparent when investigating the relationship of generated voltage E with flux Φ in a D.C. machine's magnetic circuit.

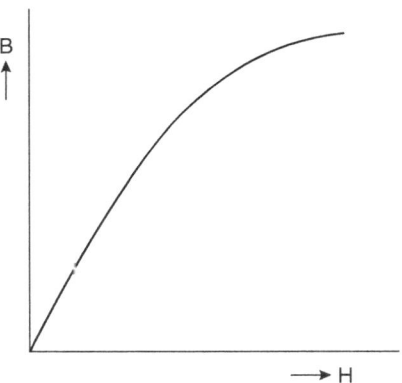

▲ **Figure 12.13** *B-H characteristics*

The no-load characteristic

Theory shows that generated voltage depends on machine flux and speed and the no-load characteristics can be considered under the following headings.

VARIATION OF E WITH N (Flux Φ constant). A permanent-magnet generator is rarely used for practical applications, but the test can be made by controlling the current of separately energised field electromagnets. This current or *field* current I_f flowing through the field coils creates an m.m.f. which results in flux in the air gaps. If this current is a constant value I_{f1}, the flux will be constant. Tests are made by varying the speed at which a machine is driven and noting the voltage generated. As flux Φ is constant and $E \propto N$, a straight-line graph as shown by (1) of the diagram (figure 12.14), results. If the field current is then adjusted to a smaller but constant value I_{f2}, and the test repeated, a straight-line graph such as (2) results. The deduction assumed, namely E varies directly with N, is proven.

VARIATION OF E WITH Φ (Speed N constant). Flux variation is achieved by controlling the energising current I_f in the field coils or 'exciting current'. If no residual magnetism is present then, if I_f increases, m.m.f. increases and flux in the air gaps increases.

The generated e.m.f. increases accordingly and a B–H type of curve (1) is as shown (figure 12.15) if E is plotted to a base of I_f. *Note.* Φ cannot be readily measured but its effects are assessed by knowing the appropriate exciting current values.

Curve (1) at first increases as a straight line, flattening out to horizontal as the magnet saturates. When saturation occurs, if the field current is reduced, curve (2) results. This

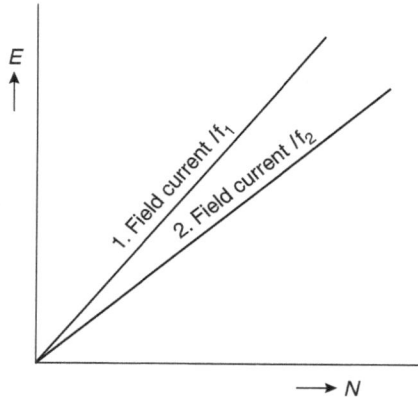

▲ **Figure 12.14** *E vs N characteristics*

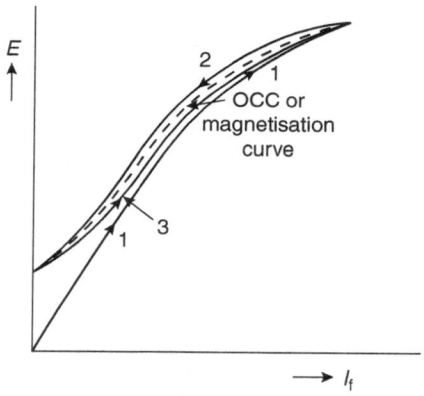

▲ **Figure 12.15** *E vs I characteristics*

curve lies slightly *above* the original curve (1) and for decreasing values of I_f the values of E are above those obtained for the ascending curve (1). The cause of the difference between curves (1) and (2) is magnetic hysteresis (discussed in Chapter 6). When the field current is finally reduced to zero, some generated e.m.f. is present while the machine runs at constant speed N. The e.m.f. is due to residual magnetism, which is essential if a generator is to be self-exciting. The e.m.f. due to residual magnetism can only be removed by demagnetising the field system. If the value of I_f is increased again, curve (3) is followed, which closes up on curve (1). The diagram overemphasises the difference between curves (1) and (2). In a modern machine this difference is not large and if a mean curve is drawn (dotted) this is the *magnetisation* or *open-circuit characteristic* (O.C.C.) curve, which is important, plotted as generated voltage or e.m.f. to a base of field current. In theory relating to A.C. and D.C. generators and motors problems will refer to A.C. or D.C. before they are solved.

Types of D.C. Generator

The machine is classified in different ways, as generator types change by varying the magnet system, i.e. either the magnetic material or connection of the field energising coils differ. In this book generators are described as follows: (1) the permanent-magnet type of generator, (2) the separately excited type of generator, (3) the self-exciting type of generator, further subdivided under the headings of: (3a) shunt-connected, (3b) series-connected and (3c) compound-connected.

(1) The permanent-magnet type of generator

This type is not used much as it is hard to make large permanent magnets or to vary the magnetic field to control generator output. The most common uses are as electrical tachometers (speed indicators), hand-operated insulation testers and primary exciters for large alternators.

THE LOAD CHARACTERISTIC. Φ is constant so the load characteristic is almost identical to the no-load characteristic. A tachometer arrangement is shown (figure 12.16).

If a sensitive voltmeter is used, i.e. one requiring negligible current, generator output current will be small so the armature voltage drop $(I_a R_a)$ will be negligible. R_a is the armature's ohmic resistance and I_a the armature current. The load terminal voltage V is nearly equal to the generated e.m.f. E and the voltmeter calibrated in revolutions per minute (rev/min).

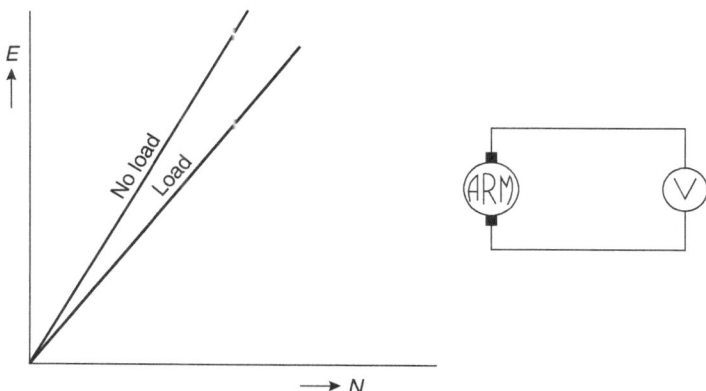

▲ **Figure 12.16** *Generator E vs N characteristics*

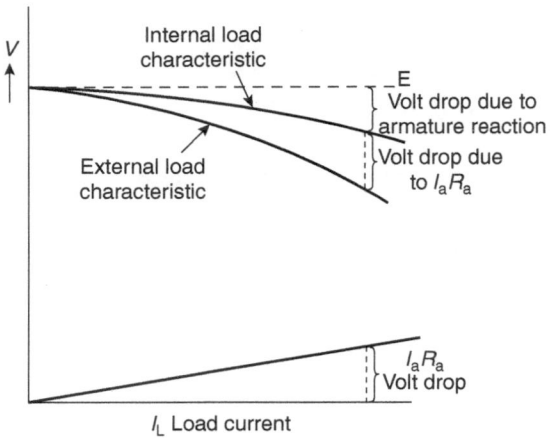

▲ **Figure 12.17** *Generator V vs Load current characteristics*

(2) The separately excited type of generator

Knowing $E \propto N$ if Φ is constant, the no-load characteristic will be a straight line as discussed. However, if N is constant and Φ varied the characteristic will vary as the *B–H* curve and an O.C.C. will result. The 2 characteristic variations are shown (figures 12.14 and 12.15).

THE LOAD CHARACTERISTIC. This is obtained by setting the field current to give the normal rated voltage at the correct speed and by applying load in stages takes current between 0 to 25% overload. For a small generator, loading is applied by switching in banks of similar wattage lamps connected in parallel. If terminal voltage V is plotted versus load current I_L, the external load characteristic is obtained (figure 12.17).

If a machine is stopped and armature resistance R_a measured by the ammeter/voltmeter method with a separate low-voltage supply, the $I_a R_a$ voltage-drop line can be plotted. If various $I_a R_a$ voltage-drop values are added to the external characteristic, the *internal load characteristic* is obtained by construction. The difference between this line and the horizontal line of the theoretical generated e.m.f. *E* shows the voltage drop due to any *armature reaction* effects, which were explained in Volume 7, but are described here briefly. It is the passage of current through an armature that sets up a magnetic field, which interacts with the main field, weakening and distorting the latter. The magnitude of generated e.m.f. reduces and commutation is effected negatively.

Load characteristics are introduced to illustrate the effects responsible for a voltage drop inside the generator, when the machine is on load. In problems armature reaction is seldom mentioned but the armature is usually credited with a resistance value greater than its ohmic value to allow for a total internal voltage drop. The voltage equation will be:

$$E = V + I_a R_a$$

The separately excited D.C. generator is used for applications such as machines supplying current to electroplating vats, with some 6–10kA needed at voltages of 6–12V with output controlled by varying a separately excited field.

(3a) The shunt-connected generator

Figure 12.18 shows the typical connections for this machine. Armature current is fed to both the load circuit and the parallel field circuit, which although taking a small current in comparison with the load must be considered.

$$\text{Thus}: I_a = I_f + I_{l.}$$

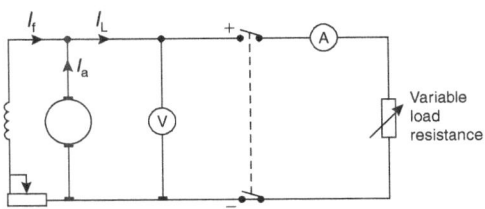

▲ **Figure 12.18** *Shunt-connected generator load circuit*

A machine's shunt field is connected across the generator's terminals. Field coils form a high-resistance circuit wound with many turns of fine wire, i.e. the ampere-turns are produced by a small current value and many turns.

$$\text{As before}: E = V + I_a R_a$$

Example 12.2. A 4-pole, wave-wound generator delivers 40kW at 200V. Its armature has 181 turns and a resistance of 0.01Ω. The air-gap flux/pole is 0.02Wb. Calculate the machine's speed, neglecting the voltage drop at the brushes and taking the shunt-field resistance as 50Ω.

$$\text{Load current} = \frac{\text{kilowatts}}{\text{voltage}} \quad \text{or} \quad I_L = \frac{40\,000}{200} = 200\text{A}$$

$$\text{Shunt} - \text{field current} \; I_f \quad \frac{V}{R_f} = \frac{200}{50} = 4\text{A}$$

So armature current $I_a = 200 + 4 = 204\text{A}$

Armature voltage drop $= 204 \times 0.01 = 2.04\text{V}$

Generated voltage = terminal voltage + voltage drop in armature

$E = V + I_a R_a = 200 + 2.04 = 202.04\text{V}$ Thus $N = 840$ rev/min.

Note. The armature is assumed to be wound with single-turn coils, i.e. 2 conductors/turn.

THEORY OF SELF-EXCITATION. As shunt-connected generators use the principle of self-excitation, it is key to explaining the theory involved. If a field system has residual magnetism, armature rotation generates a small e.m.f. that causes a field current to produce more flux, which in turn causes more e.m.f. to give a continuous positive feedback condition. Voltage rises and only flattens off when the voltage drop across the field equals the terminal voltage.

Note. Field current must be in the correct direction through coils to aid the original residual flux build-up.

Summarising, the following are the conditions necessary for self-excitation:

1. There must be residual magnetism sufficient to generate a small e.m.f. when an armature rotates at the correct speed.
2. The shunt-field circuit must be continuous and connected so current flow causes flux to build up, assisting the original residual flux.
3. The shunt-field circuit resistance must be less than the *critical resistance* determined from the O.C.C. when the machine runs at a particular speed.

Critical resistance is helpful to understand the conditions for self-excitation to occur, which is very important and also the subject of examination questions.

The magnetisation curve or O.C.C. applied to self-excitation, critical resistance

Figure 12.19 illustrates a circuit and its characteristic obtained with a separately excited generator. The initial part of the O.C.C. graph is complex, due to the effects of

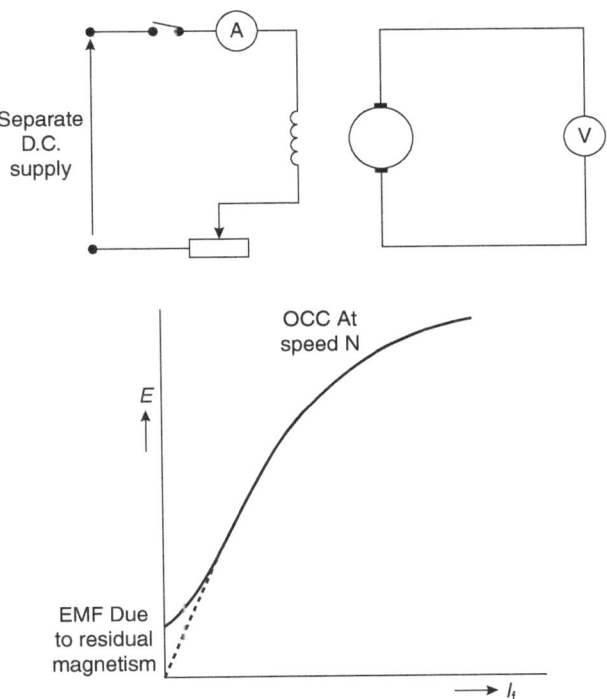

▲ **Figure 12.19** *E vs I characteristics*

residual magnetism. If the graph starts from zero it is usually straight, as the machine's magnetic circuit involves air gaps and saturation conditions that are reached gradually. Full saturation conditions are seldom attained. It is stressed that the O.C.C. depends on speed. When the field is *shunt-connected*, provided the conditions set out above are fulfilled, a generator will self-excite and an O.C. voltage value is attained where the voltage drop in the shunt field is equal to the generated terminal voltage.

This condition is illustrated (figure 12.20) and best understood by considering the O.C.C. and field voltage-drop line. The magnetisation curve for any speed N is drawn from results obtained by separate excitation. Imagine a shunt field and regulator with a resistance of R_f ohms, assuming a current of value I_f amperes to flow, the field voltage drop will be $I_f R_f$ volts.

Plot this value (e.g. point R) and extend the straight line through R from zero to cut the O.C.C. at point P. At this intersection point (P), the voltage drop across the field equals the applied terminal voltage and conditions balance. Consider the I_f condition shown, where generated voltage SQ is greater than the field voltage drop RQ by SR volts. More current flows in the field circuit because of this voltage difference and both graphs rise until a point of intersection is reached.

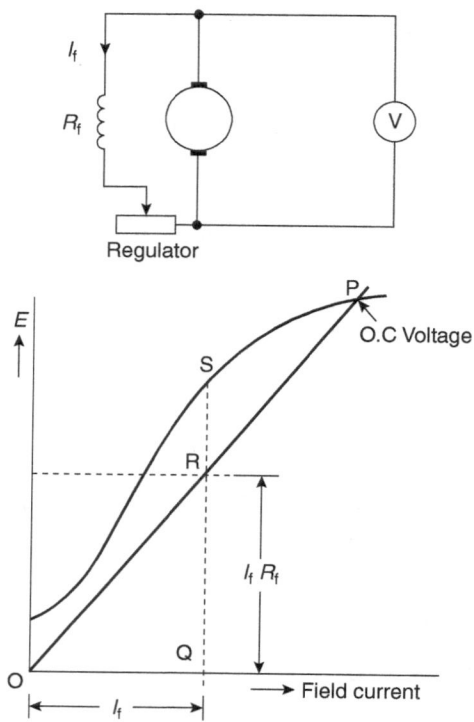

▲ **Figure 12.20** *E vs field current*

Example 12.3. A D.C. generator when separately excited and driven at 1000 rev/min gave the following test values on O.C.

Field current (A)	0	0.16	0.48	0.66	0.8	1.0	1.29
O.C. e.m.f. (V)	6.25	50	150	200	225	250	275

Field windings are then shunt-connected. Find (a) the voltage to which the machine will self-excite on O.C. when driven at 1000 rev/min and the shunt-field circuit resistance is 240Ω, (b) the value of the regulator resistance to be added or subtracted from the field circuit to allow the generator to self-excite to 237.5V and (c) the value of the critical resistance at this speed.

(a) Plot the O.C.C. as shown (figure 12.21). Using the graph, take any value of field current, viz. 1A. The voltage across the field circuit with 1A flowing will be 1 × 240 = 240V. Plot point (R) and draw the field voltage-drop line through the origin as shown. The O.C. voltage to which the machine self-excites is 257V.

(b) For the machine to excite to 237.5V, note this value on the O.C.C. and join it to zero to obtain the new field resistance voltage-drop line. Note the field current for 237.5V, this is 0.88A. From Ohm's law, the field-circuit resistance is $\dfrac{237.5}{0.88} = 269.9 = 270\Omega$.

∴ Resistance to be added $= 270 - 240 = 30\Omega$.

(c) Neglecting the start of the graph (due to residual magnetism), draw a tangent through the origin. Read the voltage value on this tangent and the corresponding field current. For example: 155V and 0.5A. The critical resistance will be $\dfrac{155}{0.5}$ ohms $= 310\Omega$.

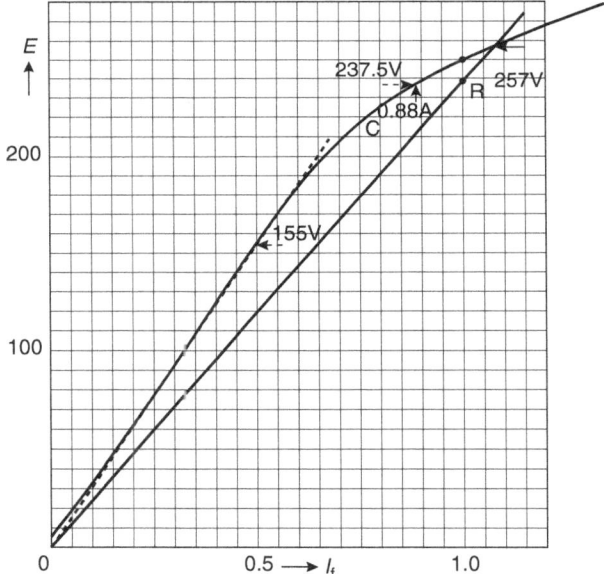

▲ **Figure 12.21** *E vs I graph*

CRITICAL RESISTANCE. The effect of altering the shunt-field resistance or regulator is seen by reference to the example and graph (figure 12.21). Reduce R_f and the slope of the field resistance voltage-drop line becomes less and the intersection point with the O.C.C. moves higher up, i.e. the generator O.C. voltage is raised. The opposite occurs with R_f increased. If R_f is increased until the field voltage-drop line lies outside the magnetisation curve, there will be no point of intersection and the generator will not self-excite. If R_f is reduced, the slope of the field voltage-drop line decreases until the line lies along or becomes tangential to the O.C.C. The resistance value deduced from the field voltage-drop line falls until it reaches the value given by the line tangential to the O.C.C. The resistance value for this condition is the *critical resistance*, and depends on speed. For any speed, if field resistance is less than the critical resistance, a machine will self-excite if the other conditions are satisfied.

THE LOAD CHARACTERISTIC. The test circuit graph is shown (figure 12.22). Armature resistance R_a is measured. The external load characteristic is plotted from test results and the internal characteristic drawn as described for a separately excited machine. Features of the load characteristics are (1) rapid fall-off of terminal voltage and (2) a 'bend-back' of the characteristic on itself.

(1) When the external circuit is connected to a load, there is a voltage drop in the armature. Terminal voltage falls, resulting in a decrease of field exciting current. This in turn causes the external characteristic to 'slump down' more than for the separately excited machine. The armature reaction effect is as for the separately excited machine, i.e. it is responsible for a voltage decrease, equivalent to an increased armature voltage drop, accounted for by giving the armature an R_a value *greater* than its ohmic resistance.

(2) As load resistance decreases, load current increases at first with a resulting terminal voltage fall, tending to slow an increase of load current. At first external load resistance decreases and consequent load current rise dominating and a rising current with

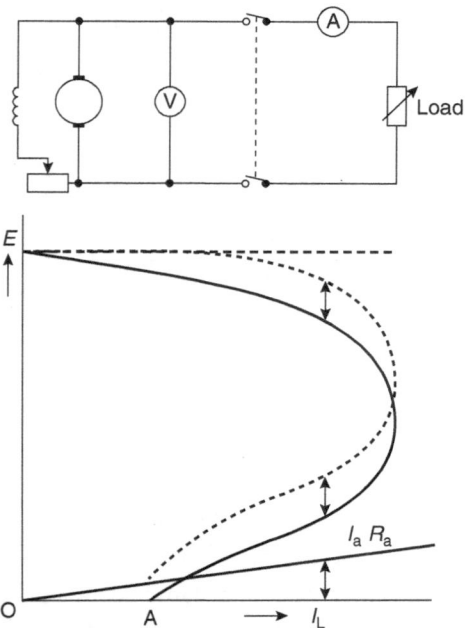

▲ **Figure 12.22** *Load characteristics*

falling terminal voltage is shown. At a certain current value the demagnetising effect of armature reaction, the armature-resistance voltage drops and the loss of field current due to reduced voltage combine to produce a terminal voltage resulting in *less* load current even though load resistance is decreased and the curve bends back on itself! The armature may be short-circuited – a self-protecting effect. OA in figure 12.22 is caused by residual magnetism, but a sudden short-circuit may cause an excess armature reaction effect tending to cancel residual magnetism, thus demagnetising the machine, which may then fail to self-excite when short-circuit is removed. The machine will need to be remagnetised before it can operate again. A shunt-connected machine can be used for most purposes where a simple generator is needed. Examples are battery chargers and small lighting-sets, the motor-car dynamo and some electrical systems.

(3b) The series-connected generator

Figure 12.23 shows connections for this machine used in specialist work and in the compound generator.

The series field of this generator is designed to be connected in the main armature circuit to the load. Field coils are wound with a few turns of thick cable, i.e. the field ampere-turns are produced by a large current and a small number of turns. Thus $I_a = I_f = I_L$. Terminal voltage on load is V *and* the generated e.m.f. E is greater than V by the internal voltage drops in the armature and series field. Thus:

$$E = V + I_f R_f + I_a R_a = V + I_a (R_a + R_f)$$

SELF-EXCITATION. The same theory and conditions apply as for the shunt-connected machine. It is noted that load resistance constitutes the field regulating resistance and so for any chosen speed there is a critical resistance value. If load resistance, i.e. the field-circuit resistance, is less than the critical value, the machine will self-excite. If the circuit resistance

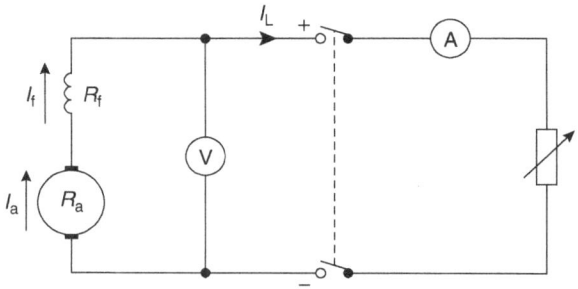

▲ **Figure 12.23** *Series-connected generator self-excitation circuit*

is above the critical value for that machine speed, self-excitation and voltage build-up will not occur.

THE LOAD CHARACTERISTIC. Consider the circuit shown (figure 12.24). A circuit switch is closed with load resistance at maximum, which is reduced until the machine self-excites. Load current and terminal voltage settle at a fixed value but if the load is altered, new voltage and current values occur. This is repeated for decreasing and increasing load resistance, until a full external load characteristic is obtained (figure 12.24).

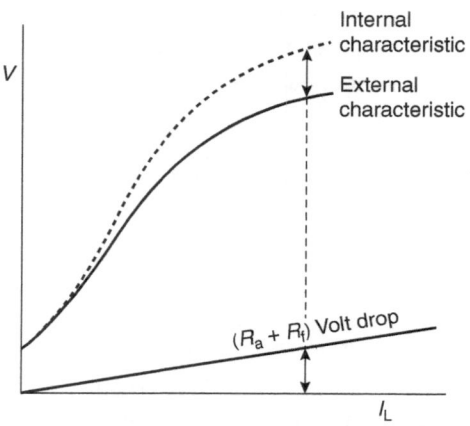

▲ **Figure 12.24** *V vs I load characteristics*

The machine is then shut down and R_a and R_f measured separately. The armature and series-field resistance voltage-drop lines are drawn and the internal load characteristic deduced. The effects of armature reaction may be investigated if an O.C.C. (obtained by separate excitation at the correct speed) was superimposed on the characteristics. The machine has the following disadvantages:

(1) It cannot self-excite until the load circuit is completed and its resistance value is less than the critical resistance.

(2) The voltage to which it self-excites depends on the load current and very little voltage control is possible.

(3) The load characteristic is a rising one and is unsuitable, in fact dangerous, and may result in load 'burn-out'.

A series generator is never used for normal generating purposes, but only certain applications. Machines of this type are specialist marine electrical systems only, for example, specific electric propulsion and winch controls.

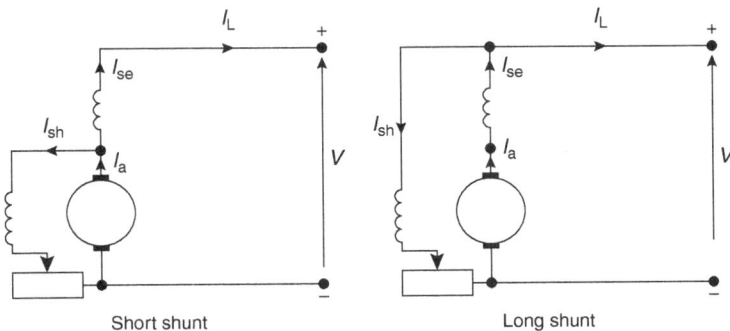

▲ **Figure 12.25** *Short shunt and long shunt circuit diagrams*

(3c) The compound-connected generator

This generator uses series and shunt fields; its characteristic is treated as made up of shunt and series-machine characteristics. The final characteristic depends on the relative strengths of each field. It is noted that the shunt field is the basic requirement; it is the shunt generator's performance that is improved on.

TYPES OF ELECTRICAL CONNECTION. Figure 12.25 shows how a machine is connected in either 'short' or 'long' shunt. There is no appreciable difference in the resulting generated voltage as is seen from the example.

Example 12.4. A 110V compound generator has armature, shunt and series-field resistances of 0.06Ω, 25Ω and 0.04Ω respectively. A load consists of 200 lamps each rated 55W, 110V. Find the generated e.m.f. and armature current if the generator is connected (a) long shunt and (b) short shunt.

(a) LONG SHUNT. Load current $I_L = \dfrac{200 \times 55}{110} = 100$A

Shunt-field current $I_{sh} = \dfrac{110}{25} = 4.4$A

Series-field current I_{se} and armature current $I_a = 100 + 4.4 = 104.4$A

Generated voltage $E = V + I_a(0.06 + 0.04) = 110 + 104.4 \times (0.1)$ or $E = 120.44$V.

There are 2 fields with different current values; symbol I_f is not used for field current, but instead symbols I_{sh} and I_{se} are used in both figure 12.25 and this example.

(b) SHORT SHUNT. Load current I_L (as before) = 100A = I_{se} Voltage drop in the series field = $I_{se} R_{se}$

= 100 × 0.04 = 4V

Voltage applied to the shunt field = terminal voltage + voltage drop in series field = 110 + 4 = 114V

Shunt-field current $I_{sh} = \dfrac{114}{25} = 4.56A$

Armature current $I_a = 100 + 4.56 = 104.56A$

$E = V +$ voltage drop in series field + voltage drop in armature

$E = 110 + 4 + (104.56 \times 0.06) = 120.27V.$

TYPES OF FIELD ARRANGEMENT. The series field is often connected so the flux produced adds to the shunt-field flux. For this common arrangement the machine is said to be *cumulatively* connected. All generators, used for supplying lighting and power for electrically driven auxiliary machinery aboard ship, have this connection. If the series field is connected to weaken the shunt field, the generator is *differentially* connected. This arrangement is used for specialist work such as certain types of welding generator.

THE LOAD CHARACTERISTIC. Figure 12.26a shows graphs due to directly loading a machine in the manner described. Curve (a) shows the characteristic if a shunt field only is used. Curve (b) is obtained with the series field only. Curve (c) results from use of both fields. Any point on this characteristic is obtained by adding voltages obtained from graphs (a) and (b), for any one value of load current.

Figure 12.26b shows how the load characteristics of a compound generator vary by altering the relative strength of the series field, so-called flat-compounding as required by most regulations. The curve is not quite flat and a rise in voltage between no load and full is called 'the hump'. It may be 6–7% for small generators, but in most cases is about 2–3%. Over-compounding compensates for the supply line voltage drop. This is shown by the example.

Example 12.5. A factory is sited some way from a generating-station and takes 100A at 200V. The supply cable resistance is 0.02Ω/core. Find the percentage compounding required for the generator.

Voltage drop in the line on full load = 100 × 2 × 0.02 = 4V

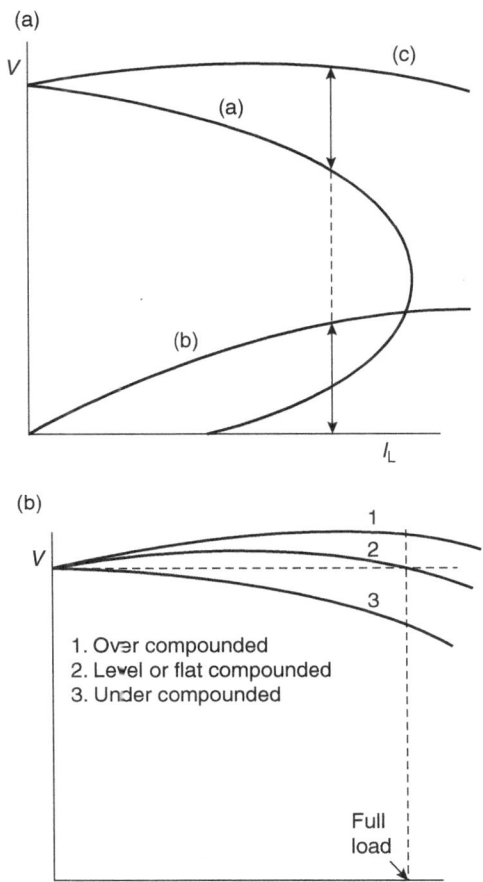

▲ **Figure 12.26** *V vs I load characteristics*

To supply 200V to a factory, generated voltage must be 204V. A suitable over-compounded characteristic is shown (figure 12.27) and the compounding can be expressed as the rise on full load to the O.C. voltage.

Thus percentage compounding

$$= \frac{DC}{CB} = \frac{DC}{AO} = \frac{204 - 200}{200} = 4/200 = 0.02 \text{ or } = 0.02 \times 100 = 2\%$$

Thus a generator is required to be 2% over-compounded. Further work on the D.C. generator is covered in more advanced studies, but sufficient knowledge allows a study of the motor to be made. The following examples form a useful conclusion to this chapter.

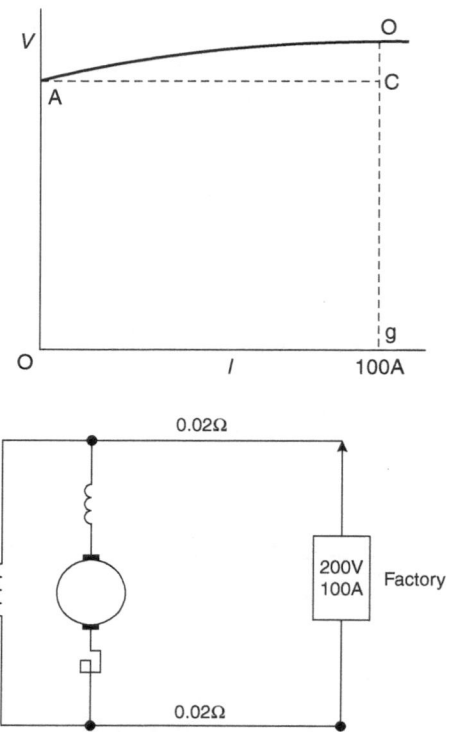

▲ **Figure 12.27** *Over-compounded V vs I characteristics*

Example 12.6. A shunt generator is to be converted into a compound generator by addition of a series-field winding. From a test on the machine with shunt excitation a field current of 3A gives 440V on no load and 4A gives 440V at the full-load current of 200A. A shunt winding has 1600 turns/pole. Find the number of series turns required/pole.

Ampere-turns/pole required to give 440V on O.C.

$= 3 \times 1600 = 4800$At

Ampere-turns/pole required to give 200A at 440V on load

$= 4 \times 1600 = 6400$At

Full-load ampere-turns must be increased by $6400 - 4800 = 1600$At

But these 1600At/pole are obtained from the series field which passes 200A

Thus the required number of series turns/pole $= \dfrac{1600}{200} = 8.$

Example 12.7. A 4-pole compound generator has a lap-wound armature and is connected in short shunt. The resistances of the armature and fields are 0.1Ω and 50Ω (shunt), 0.08Ω (series). The machine supplies a load consisting of sixty 100V, 40W lamps in parallel. Calculate the total armature current, the current/armature path and the generated e.m.f.

Since this is a lap-connected armature $A = P$

For 1 lamp, since $P = VI$ ∴ $I = \dfrac{40}{100} = 0.4A$

The load current $I_L = 60 \times 0.4 = 24A$

Voltage drop in series field $= 24 \times 0.08 = 1.92V$

Voltage across shunt field $= 101.92V$

Shunt-field current $= \dfrac{101.92}{50} = 2.04A$

Armature current $= 24 + 2.04 = 26.04A$

Current per armature path $= \dfrac{26.04}{4} = 6.51A$

Generated voltage = terminal voltage + voltage drop in series field + voltage drop in armature

$= 100 + 1.92 + (26.04 \times 0.1) = 104.52V.$

Practice Examples

12.1. The armature of a 4-pole shunt generator is lap wound and generates 216V when running at 600 rev/min. The armature has 144 slots with 6 conductors/slot. If the armature is rewound to be wave connected, find the e.m.f. generated at the same speed and flux/pole (3 significant figures).

12.2. A compound-wound, long-shunt D.C. generator has an output of 250A at 220V. The equivalent resistances of the armature, series and shunt windings are 0.025, 0.015 and 176Ω respectively. There is a 2V voltage drop across the brushes. Find the induced voltage (2 decimal places).

12.3. The curve of induced e.m.f. against excitation current for a separately excited generator when run on no load at 1200 rev/min is given by:

E.m.f. (V)	15	88	146	196	226	244	254
Excitation current (A)	0	0.4	0.8	1.2	1.6	2.0	2.4

Deduce the voltage to which a machine will self-excite if the shunt-field resistance is set at 90Ω and the machine run at 900 rev/min (2 decimal places).

12.4. A 220V, 4-pole, wave-wound shunt generator has an armature resistance of 0.1Ω and a field resistance of 50Ω. Calculate the flux/pole, if the machine has 700 armature conductors, runs at 800 rev/min and is supplying a 38kW load (4 decimal places).

12.5. In a 250kW, 440/480V, over-compounded generator, the flux/pole required to generate 440V on no load is 0.055Wb at 620 rev/min. The resistances of the armature, interpoles and series field are 0.01, 0.005 and 0.005Ω respectively. Find the flux/pole required at full load, the speed now being 600 rev/min. Neglect the current taken by the shunt field (4 decimal places).

12.6. Estimate the series turns/pole needed for a 50kW compound generator to develop 500V on no load and 550V on full load. Assume a long-shunt connection and that the ampere-turns required per pole on no load are 7900 whereas the ampere-turns required per pole on full load are 11 200 (1 significant figure).

12.7. A 4-pole machine has a lap-wound armature with 90 slots each containing 6 conductors. If the machine runs at 1500 rev/min and the flux/pole is 0.03Wb, calculate from first principles the e.m.f. generated. If the machine is run as a shunt generator with the same field flux, the armature and field resistances being 1.0Ω and 200Ω respectively, calculate output current when the armature current is 25A. Through a fall in speed the e.m.f. becomes 380V. Calculate load current in a 40Ω load (2 decimal places).

12.8. A D.C. generator gave the following O.C.C. when driven at 1000 rev/min.

Field current (A)	0.2	0.4	0.6	0.8	1.0	1.2	1.4	1.6
Armature voltage (V)	32	58	78	93	104	113	120	125

If the machine is run as a shunt generator at 1000 rev/min, the shunt-field resistance being 100Ω, find (a) the O.C. voltage, (b) the critical value of the shunt-field resistance and (c) the O.C. voltage if the speed was raised to 1100 rev/min, the field resistance being kept constant at 100Ω (all 3 significant figures).

12.9. Calculate the input power to drive a shunt generator having an output of 50kW at 230V, if under these conditions the bearing, friction, windage and core loss is 1.6kW and the total voltage drop at the brushes is 2V. The resistance of the armature is 0.034Ω and that of the field circuit 55Ω (4 significant figures).

12.10. A D.C. generator when separately excited and run at 200 rev/min gave the following test results:

Field current (A)	0	1	2	3	4	5	6	7	8	9
O.C. voltage (V)	10	38	61	78	93	106	115	123	130	135

The field is then shunt-connected and run at 400 rev/min. Determine (a) the e.m.f. to which a machine will excite when field-circuit resistance is 36Ω (3 significant figures), (b) the critical field-circuit resistance value (2 significant figures), (c) the extra resistance needed in a shunt-field circuit to reduce e.m.f. to 220V (1 decimal place) and (d) the critical speed when field-circuit resistance is 36Ω (3 significant figures).

13

THE D.C. MOTOR

Every discovery opens a new field for investigation of facts, shows us the imperfection of our theories. It has justly been said, that the greater the circle of light, the greater the boundary of darkness by which it is surrounded.

Sir Humphry Davy

A D.C. machine will run as a motor if its field and armature are connected to a suitable supply. The 'motoring' action is based on the fundamental law described in Chapter 5, which stated that a force acts on a conductor if it lies in a magnetic field and carries current. Figure 13.1 shows one such arrangement.

Direction of Force

The 4 small diagrams (figure 13.2) show that to reverse the force direction and thus the direction the armature will rotate in, the conductor current must be reversed with respect to the magnetic flux.

Practically it must be remembered that if a motor runs in the 'wrong' direction when first connected, rotation reversal is obtained by reversing the supply leads to the armature circuit. A hand rule exists to help memorise motor action and is comparable with that shown in Chapter 12 for the generator.

LEFT-HAND RULE (Fleming's). Figure 13.3 shows this rule. The first and second fingers represent flux and current respectively, as for the right-hand rule. The direction of force on the conductor is represented by the thumb. *Note*. As for the right-hand rule, thumb, index finger and second finger *must* be placed at right angles to each other.

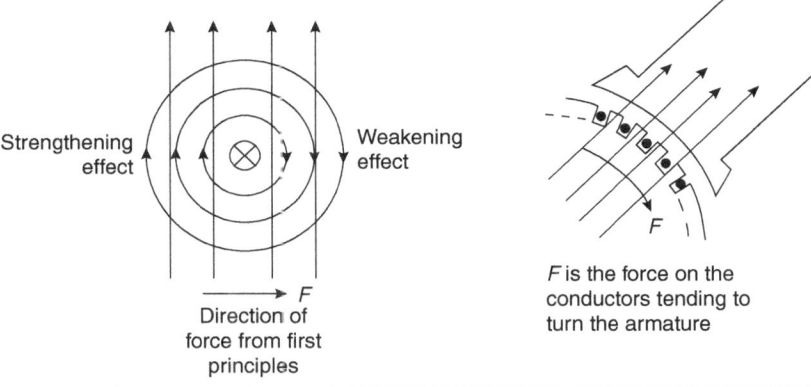

▲ **Figure 13.1** *Force acting on a conductor lying in a magnetic field and carrying current*

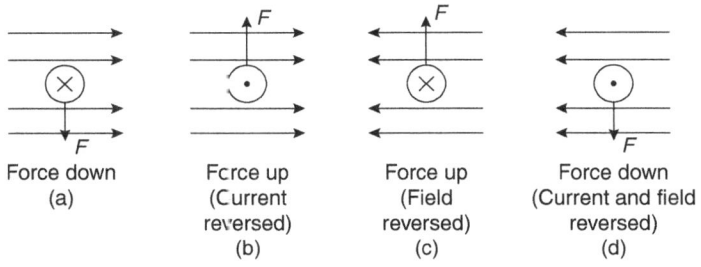

| Force down (a) | Force up (Current reversed) (b) | Force up (Field reversed) (c) | Force down (Current and field reversed) (d) |

▲ **Figure 13.2** *Changing force direction*

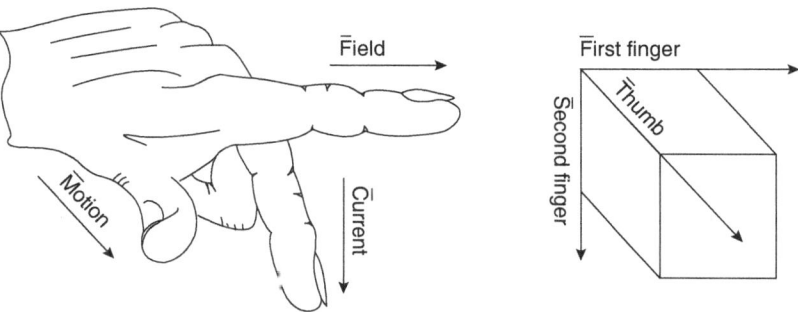

▲ **Figure 13.3** *Fleming's Left-Hand Rule*

Magnitude of Force

From the principles laid out in Chapter 5, it was shown that the force acting on a conductor in a magnetic field is proportional to flux density, current and the active conductor length in the field. The law is summarised by the formula: $F = BI\ell$ newtons but the force *magnitude* also depends on the angle of the conductor to the field direction, which is *maximum* when they are at right angles.

Example 13.1. Calculate the force in newtons, on a conductor, 0.5m long, carrying a current of 500A at right angles to a magnetic field of uniform density 0.8T (3 significant figures).

$$\text{Since } F = BI\ell$$
$$\text{Then } F = 0.8 \times 500 \times 0.5 = 200 \text{ newtons.}$$

If the conductor is situated on an armature at a radius of r metres, the torque produced on the shaft is expressed as $F \times r$ newton metres.

Back e.m.f. of a motor

If a motor rotates due to the torque produced by the armature conductors, then these same conductors will cut the magnetic field. From Faraday's law an e.m.f. is induced, the magnitude of which is given by the generator expression developed in Chapter 12, namely:

$$E = \frac{Z\Phi N}{60} \times \frac{P}{A} \text{volts}$$

From first principles the induced e.m.f. direction is such as to *oppose* the applied voltage and a condition of balance results. Since the rotation direction will be opposite to that for a generator operated under the same directions of flux and current in the armature conductors, the induced e.m.f. *opposes* current flow and is termed a 'back e.m.f.' This e.m.f. must be less than the terminal voltage V to permit motoring. Thus the armature starts as a passive load, but as it rotates, it accelerates until a balance condition is reached when supply voltage equals the armature voltage drop plus the generated back e.m.f. This condition is expressed by the voltage equation and the motor armature operates as an active load.

Voltage equation

$V = E_b + I_a R_a$. This equation explains the voltage conditions for an armature circuit. Here V is the voltage applied to the armature, E_b is the back e.m.f. generated and $I_a R_a$ is the armature voltage drop caused by the armature current I_a passing through the armature resistance R_a. If a problem is encountered where a brush voltage drop is given, this must be added. The equation is comparable with the generator terminal voltage equation $V = E - I_a R_a$. A thought about the difference in the equations summarises the basics of generator and motor action.

Current equation

$$\text{Since } V = E_b + I_a R_a \text{ then } I_a R_a = V - E_b$$

$$\text{and } I_a = \frac{V - E_b}{R_a}$$

This equation shows how motor current depends on the back e.m.f. generated. Starting conditions are illustrated.

$$\text{At start } E_b = 0 \therefore I_{as} = \frac{V}{R_a}$$

Usually R_a is small to minimise the armature-resistance voltage drop for working conditions, so I_a will be very large. For example, a 220V motor with a 0.4Ω armature resistance may take a full-load current of 52A, but if started without taking special precautions, the starting current I_{as} will be $I_{as} = \frac{220}{0.4} = 550\text{A}$. This large starting current may give rise to undesirable starting conditions. It could easily 'blow' a fuse, or accelerate too rapidly, resulting in mechanical or electrical damage through excess sparking at the commutator. It is for this reason that, the starting current I_{as} is limited by use of a 'starter'. The basic feature of a starter is a variable resistance inserted into the armature circuit at starting which is gradually reduced or cut out as the motor accelerates up to normal working speed. At the 'instant of starting':

$$I_{as} = \frac{V}{R_{as} + R_s} \quad \text{where } R_s \text{ is the full value of the starting resistance.}$$

Speed equation

The equation is vital for understanding motor action. It is obtained by rearranging the terminal voltage equation and using the generator expression:

Since $V = E_b + I_a R_a$ Then $E_b = V - I_a R_a$.

But $E_b = \dfrac{Z\Phi N}{60} \times \dfrac{P}{A}$, E_b is the generated e.m.f. and magnitude determined from the generator formula.

Hence $\dfrac{Z\Phi N}{60} \times \dfrac{P}{A} = V - I_a R_a$

or $N = \dfrac{(V - I_a R_a)}{Z\Phi} \times \dfrac{60A}{P}$ rev/min

Example 13.2. Calculate the full-load speed of a motor operating from a 440V supply, given $R_a = 0.75\Omega$, full-load armature current = 55A, the flux/pole = 0.02Wb and it is a 4-pole machine with a simple wave-wound armature of 43 slots and 12 conductors per slot (4 significant figures).

Number of armature conductors $Z = 43 \times 12 = 516$ For a wave-wound armature $A = 2$.

$$P = 4 \text{ and } \Phi = 0.02Wb \text{ Then } N = \frac{440 - (55 \times 0.75)}{516 \times 0.02} \times \frac{60 \times 2}{4} = 1160 \text{ rev/min}$$

Speed controlling factors

The deductions below are derived from the speed equation and are of vital importance. Students should ensure they understand the implication of each deduction in detail.

Since $N = \dfrac{(V - I_a R_a)}{Z\Phi} \times \dfrac{60A}{P}$, it is clear that for a particular machine only certain variables affect the expression. Thus 60, A, Z and P are all constants and written as k.

Then we have $N = \dfrac{k(V - I_a R_a)}{\Phi}$ or $N = k\dfrac{E_b}{\Phi}$ since $E_b = V - I_a R_a$.

If the expression $N = \dfrac{k(V - I_a R_a)}{\Phi}$ is considered, for the purposes of approximation, the

small voltage drop $I_a R_a$ can be neglected and we have $N = \dfrac{kV}{\Phi}$ or $N \propto \dfrac{V}{\Phi}$ (approx.).

Thus speed can be controlled by varying V or Φ.

Variation of V gives *direct* speed control, while varying Φ gives *inverse* speed control. It should be remembered that the real relation is $N \propto \dfrac{E_b}{\Phi}$ but under working conditions E_b is not very different from that of V, the $I_a R_a$ voltage drop being small. The practical application of the deduction leads to the basic systems of motor speed control in that:

> Variation of voltage across the armature terminals produces a direct variation of speed, i.e. raising the armature voltage increases speed and vice versa.

In contrast:

> Variation of field flux produces an inverse variation of speed, i.e. lowering flux increases speed and vice versa. Thus the relationship $N \propto \dfrac{V}{\Phi}$ (approx.) determines the motor speed characteristic shape.

Example 13.3. The armature resistance of a 200V shunt motor is 0.4Ω. The no-load and armature current = 2A (a term for when a motor runs unloaded). When loaded and taking an armature current = 50A, motor speed = 1200 rev/min. Find the approximate no-load speed (4 significant figures).

On no-load. Back e.m.f. $E_{b0} = V - I_{a0} R_a$

$$= 200 - (2 \times 0.4) = 199.2V$$

On load. Back e.m.f. $= E_{b1} = V - I_{a1} R_a$

$$= 200 - (50 \times 0.4) = 180V$$

Also since $N = k\dfrac{E_b}{\Phi}$, then $N_0 = \dfrac{kE_{b0}}{\Phi_0}$ and $N_1 = \dfrac{kE_{b1}}{\Phi_1}$

Since this is a shunt motor, the field is unaffected by loading the armature so $\Phi_1 = \Phi_0$.

$$\therefore \frac{N_0}{N_1} = \frac{kE_{b0}}{\Phi_0} \bigg/ \frac{kE_{b1}}{\Phi_1} = \frac{E_{b0}}{E_{b1}} \text{ since } k, \Phi_1 \text{ and } \Phi_0 \text{ cancel}$$

$$N_0 = N_1 \times \frac{E_{b0}}{E_{b1}} = 1200 \times \frac{199.2}{180} \text{ or } N_0 = 1328 \text{ rev/min}$$

Types of D.C. Motor

As for the generator, motor-field windings can be connected in shunt, series or a combination to give a compound arrangement. The motor is a machine that takes current from the supply and the fields are considered a load *added* to the armature circuit.

The shunt motor

The arrangement is shown (figure 13.4). It is seen that $I_L = I_a + I_{sh}$. A supply voltage V is applied to both the armature and the field circuits 'equivalent resistance' treatment as a parallel circuit *cannot* be supplied to find I_L because, although R_{sh} is a passive load, the armature is an active load when the machine runs. The shunt motor is essentially a constant-speed machine used for most applications.

$$\text{Here } I_{sh} = \frac{V}{R_{sh}} \text{ and } I_a = I_L - I_{sh}$$

The series motor

The series motor arrangement is as shown (figure 13.4). Here $I_L = I_{se} = I_a$. The voltage equation is modified slightly as, if V is taken as the supply voltage, allowance is made for the voltage drop in the series field. The equation must be written as:

$$V = I_{se}R_{se} + I_aR_a + E_b = E_b + I_aR_a + I_aR_{se} \quad \text{or} \quad V = E_b + I_a(R_a + R_{se}).$$

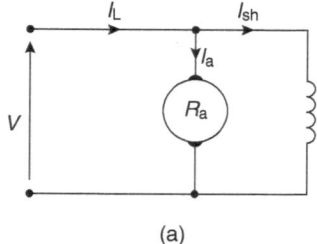

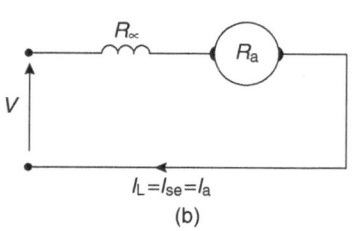

(a)   (b)

▲ **Figure 13.4** *The series motor*

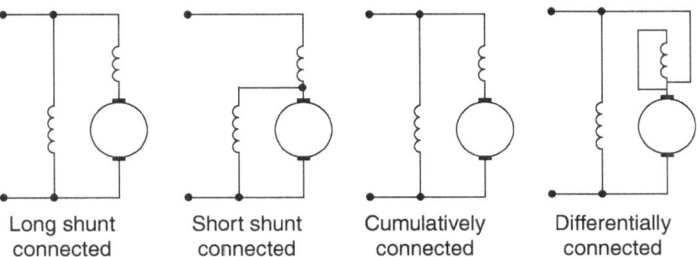

Long shunt   Short shunt   Cumulatively   Differentially
connected   connected   connected   connected

▲ **Figure 13.5** *Different compouna connections*

The voltage applied to the armature is equal to *V minus* the voltage drop in the series field and a voltage drop at the brushes (if present). As machine current rises with increasing load, the voltage across the armature falls and speed changes accordingly. The ohmic value R_{se} is kept as small as possible for this machine type. This is a variable speed motor used mainly for traction, hoist, crane and winch work.

The compound motor

As shown by the diagram (figure 13.5), the motor, like the generator, can be connected as a long-shunt or short-shunt machine. The 2 fields can be connected to assist or oppose each other. If the resultant flux is strengthened by the arrangement, fields are said to be 'cumulatively' connected. If fields are, however, connected to *weaken* each other, the motor is 'differentially' connected, which is rarely used.

Most marine motors are cumulatively compounded machines. The relative strengths of the shunt and series fields are chosen by the performance required and is considered when characteristics are studied in detail.

Equations

The power equation

This equation shows the conversion from electrical to mechanical power and the causes of electrical loss. It is used for deducing the torque equation, and gives the student little difficulty provided he properly understands the voltage equation.

Since $V = E_b + I_aR_a$ and the armature is the cause by which electrical energy supplied is converted into mechanical energy, the following deduction is possible. Multiply the expression by I_a and study the result.

$$\text{Thus } V = E_b + I_aR_a \text{ becomes } VI_a = E_bI_a + I_a^2R_a$$

VI_a is a measure of the power input to the armature circuit, $I_a^2R_a$ indicates a resistance loss and is the power lost, being converted into heat in the armature. It is known as a *copper loss* and is due to armature resistance. E_bI_a is a measure of the armature power developed, as seen if the expression is rearranged:

$$\underset{\text{Input power}}{VI_a} \quad - \quad \underset{\text{Copper loss}}{I_a^2R_a} \quad = \quad \underset{\text{Output power}}{E_bI_a}$$

Note. Output power E_bI_a is in watts and is the mechanical power developed by the armature conductors and is *not* a true measure of the shaft output until the machine's mechanical losses, for example, those due to friction and windage, have been subtracted. When data concerning mechanical losses is not given, an estimate of shaft output power can still be obtained.

Example 13.4. A 4-pole motor has a wave-wound armature of 594 conductors. Armature current = 30A and the flux per pole = 0.009Wb. Calculate the total power developed when running at 1400 rev/min. Estimate the shaft output power if mechanical losses absorb 10% of developed power.

For this machine $P = 4$, $A = 2$, $Z = 594$ and $\Phi = 0.009$Wb

$$E_b = \frac{Z\Phi N}{60} \times \frac{P}{A} = \frac{594 \times 0.009 \times 1400}{60} \times \frac{4}{2} = 249.48\text{V}$$

The power developed $= E_bI_a = 249.48 \times 30 = 7484.4$W

$$= 7.5\text{kW (approx.)}$$

As mechanical power loss $= 10\%$ of 7.5kW

$$= 0.75\text{kW then shaft output power} = 7.5 - 0.75 = 6.75\text{kW}.$$

The torque equation

This is an important expression. often developed from first principles for examination purposes. The method here involves the power and voltage equations and is considered to be the simplest. As the armature's electrical power output $= E_b I_a$ watts and the mechanical power developed is given by:

$$\frac{2\pi \times \text{speed (rev/min)} \times \text{torque (newton metres)}}{60}$$

we can write: $E_b I_a = \dfrac{2\pi NT}{60}$

or $T = \dfrac{60}{2\pi N} \times E_b I_a = \dfrac{60}{2 \times 3.14} \times \dfrac{E_b I_a}{N}$

Substituting for E_b in terms of machine data, we have:

$$T = \frac{60}{2 \times 3.14} \times \frac{Z\Phi N}{60} \times \frac{P}{A} \times I_a$$

$$\frac{60}{2 \times 3.14 \times 60} \times Z\Phi I_a \frac{P}{A} = 0.159 Z\Phi I_a \frac{P}{A} \quad \text{or} \quad T = 0.159 Z\Phi I_a \frac{P}{A} \text{Nm.}$$

Torque controlling factors

For the torque equation, the factors that affect torque are fixed. Thus for any machine 0.159, Z, P and A are all constants and can be written together as k, giving the expression $T = k\Phi I_a$ or $T \propto \Phi I_a$. The torque developed varies directly with either flux and/or the armature current and this fact is used for problems and determining machine characteristics. With a shunt motor, for different loading conditions, Φ is basically constant with: $T \propto I_a$. For a series motor, however, Φ is *not* constant and is often taken as proportional to I_a. Therefore, if $\Phi \propto I_a$ and $T \propto \Phi I_a$, we can write for a series motor $T \propto \Phi I_a^2$. This deduction is used in Example 13.5.

Example 13.5. A series motor running at a speed $= 600$ rev/min develops 3kW and takes a current of 40A. If the starting current is limited by means of a starter to 60A, find the starting torque. Neglect armature reaction effects and assume the magnetic circuit is unsaturated (1 decimal place).

As the magnetic circuit is unsaturated, it is assumed that $\Phi \propto I_{se} \propto I_a$, so $T \propto \Phi I_a$ or $T = kI_a^2$. There are 2 possible torque conditions.

$$\text{Thus: when running } T_1 = kI_{a1}^2.$$

$$\text{At starting } T_2 = kI_{as}^2 \text{ or } kI_{a2}^2.$$

When running at 600 rev/min, output = 3kW and T_1 is given by:

$$3000 = \frac{2\pi NT_1}{60} \quad \text{or} \quad T_1 = \frac{3000 \times 60}{2 \times \pi \times 600} = 47.8\text{Nm}.$$

$$\text{Also } \frac{T_2}{T_1} = \frac{kI_{a2}^2}{kI_{a1}^2} \quad \text{or} \quad T_2 = T_1 \left(\frac{I_{a2}}{I_{a1}}\right)^2 = 47.8 \left(\frac{60}{40}\right)^2$$

So the starting torque $T_2 = 107.6$Nm.

Motor Characteristics

The behaviour of shunt, series and compound motors is illustrated by their characteristics which are considered under (1) electrical load characteristics and (2) mechanical characteristic. Electrical characteristics show speed and torque in terms of armature current while the mechanical characteristic shows speed related to torque, assuming a constant applied terminal voltage. Electrical characteristics are important as they show machine performance when loaded. The mechanical characteristic shows the motor's suitability for a particular application. Characteristics are checked by making a load test on a motor, but its theoretical performance is reasoned from the 2 expressions: $N \propto \dfrac{V}{\Phi}$ (approx.) and $T \propto \Phi I_a$.

The shunt motor

Electrical characteristics

SPEED. If flux Φ is constant, assuming a constant applied voltage V, N is considered constant over the load range, as $N \propto V$ and V is constant. Speed is unaffected by

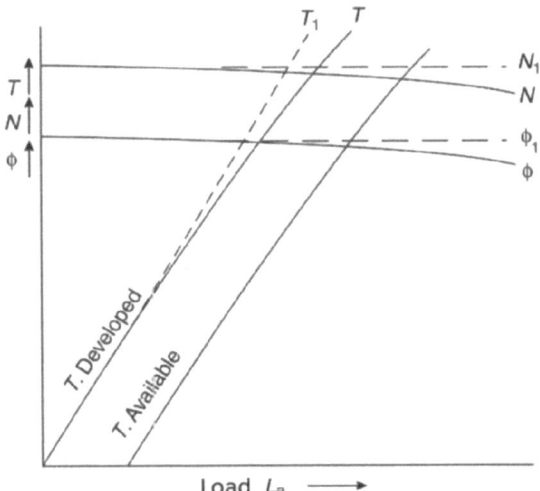

▲ **Figure 13.6** *Various shunt motor electrical characteristics*

I_a and the theoretical graph shown dotted in figure 13.6, as N_1 against I_a. This motor is considered to be a constant-speed machine, although in practice speed falls slightly with load, as shown. This is because the back e.m.f. reduces slightly (the speed drop from no load to full load is about 2% for large machines and 6% for small machines), as the armature voltage drop $I_a R_a$ increases. Although field current I_{sh} and flux Φ are constant, the armature reaction effect causes the overall resulting flux Φ to drop slightly. As $N \propto \dfrac{E_b}{\Phi}$ it should be constant if both E_b and Φ variations are proportional. Weakening flux however, means a rise in armature current due to the corresponding drop in E_b. The $I_a R_a$ drop increases as a result, and so the speed lowering effect of a reduced E_b is greater than the speed raising effect of a falling Φ! The net result is speed falling slightly over the load range of I_a.

TORQUE. T varies as I_a giving a straight line through the origin, since Φ is assumed constant. In practice Φ is weakened by the armature reaction and T drops as a result, departing from a theoretical straight line T_1. Torque available at the shaft is everywhere lower due to rotational losses. Thus 2 torque characteristics are shown in figure 13.6.

Mechanical characteristic

As illustrated (figure 13.7), this is obtained by plotting N against T and is seen to drop slightly. Shunt motors are considered to be constant-speed machines and have about a 4% drop in speed from no load to full load. They are used for all constant-speed drives, for example, machine tools, centrifugal pumps, purifiers, etc.

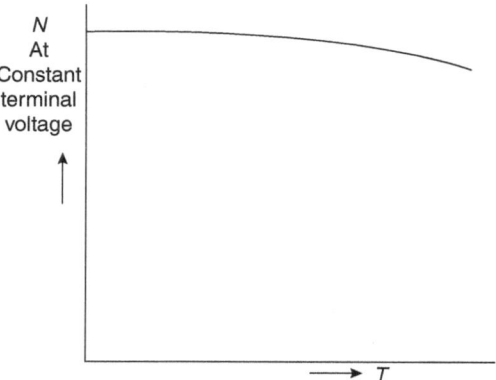

▲ **Figure 13.7** *N and At characteristics vs time*

The series motor

Electrical characteristics

SPEED. For this machine, the load current value is also that of the field current and, allowing for the armature reaction effects, it is seen (figure 13.8) that the useful flux Φ is slightly less than that given by the magnetisation curve Φ_1. As Φ increases with load and N varies as $\frac{1}{\Phi}$, the speed must drop and the curve conforms to an inverse variation, flattening out as saturation of Φ occurs. The no-load flux is small but speed can be excessive. For this reason a series motor should never run 'light' as it is liable to 'race' and may be damaged by a virtual centrifugal force. Like a shunt motor, N is lower than N_1 for the reasons described.

TORQUE. Field saturation is not usually achieved over the working load range and Φ is assumed proportional to I_a. As torque is proportional to $\Phi \times I_a$ we have from earlier that: $T_1 \propto I_a^2$ so the curve follows a parabola. On heavy loads, as Φ starts to saturate, $T \propto I_a$ so the graph tends to a straight line passing through the origin for low values of I_a, as shown (figure 13.8). As for the shunt motor, due to machine losses, torque available at the shaft is less than the developed torque. At start $T \propto I_a^2$ and so the starting torque is very high, being one advantage of this type of motor.

Mechanical characteristic

This is shown (figure 13.9) and given by plotting the N and T values, for the same armature-current values as for the electrical characteristics. The result is a curve similar

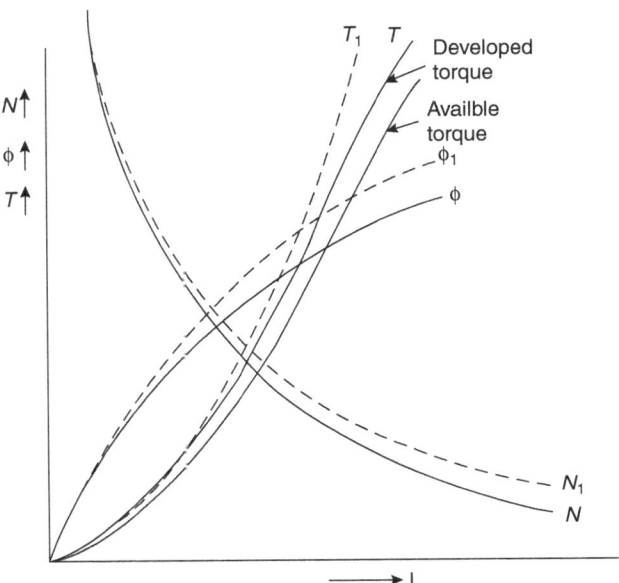

▲ **Figure 13.8** *Various series motor characteristics*

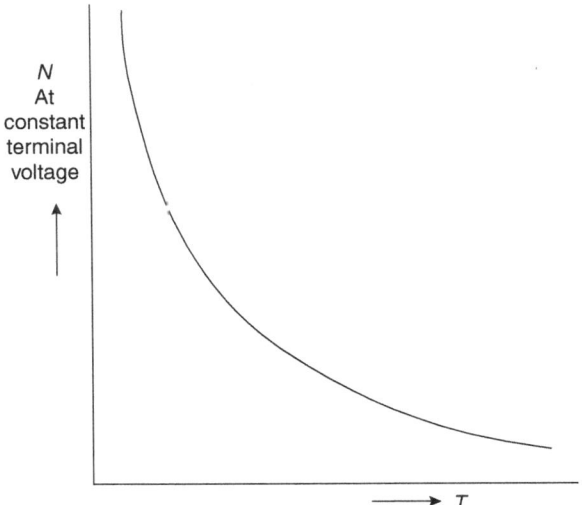

▲ **Figure 13.9** *N and At characteristics vs time*

in shape to the speed-current curve (figure 13.8). Series motors are variable speed machines, giving low speed on heavy loads, which are ideal for traction, winch, hoist and fan work. Their excellent starting torque is used to advantage when heavy masses must be accelerated quickly, for example, for lifting or traction.

The compound motor

Field connections give either cumulative or differential flux result. The former is usual and the latter used only for exceptional motor duties. The shunt and series motor have such good characteristics that compounding is only used as a means to minimise disadvantages that occur in simple machines. For example, the series motor tends to race on no load, which can be limited by providing a stabilising shunt field, and so the compounding for a machine may be arranged to give either a *strong shunt, weak series* effect or a *strong series, weak shunt* field combination. The characteristics are considered for these arrangements.

CUMULATIVE CONNECTION OF FIELDS. Here the 2 fields assist each other to give a resultant strengthening flux. Machine characteristics depend on the relative field strengths.

(1) *Strong shunt – weak series*. The characteristics of the shunt motor are so good that in practice it is suitable for most motor drive duties. The provision of a weak series field doesn't materially alter the load characteristics, but the field gives an improved starting torque – as explained below.

Electrical characteristics

SPEED. Figure 13.10 shows the characteristic. As flux rises due to the series field, this will have a speed-lowering effect since $N \propto \dfrac{V}{\Phi}$ (approx.).

Speed tends to 'sit down' more than it would for the same machine *without* a series field. If the series field is weak, its effect is unobserved in the speed characteristic, differing little from the shunt motor. However, when a machine is coupled to a flywheel, a stronger series field can be used, so sudden load application momentarily slows down with a rise of I_a. Motor speed tends to 'sit down' and the required driving power is obtained from the flywheel. This arrangement enables a motor and the electrical system to be protected from undue shock and is used for motors driving specialised loads, such as the rolls in steelworks, as well as presses and hammers.

TORQUE. During starting, when voltage is applied to a shunt field, due to its self-inductance (a winding of thin wire and many turns), a back e.m.f. is induced, which opposes the shunt field current. The shunt field current builds up slowly and the torque ($T \propto \Phi I_a$) is small in spite of a large armature current. A series field that passes the starting current I_{as} will produce a flux to strengthen the shunt flux, the net flux at starting will be much larger and an improved starting torque will be obtained, which may be

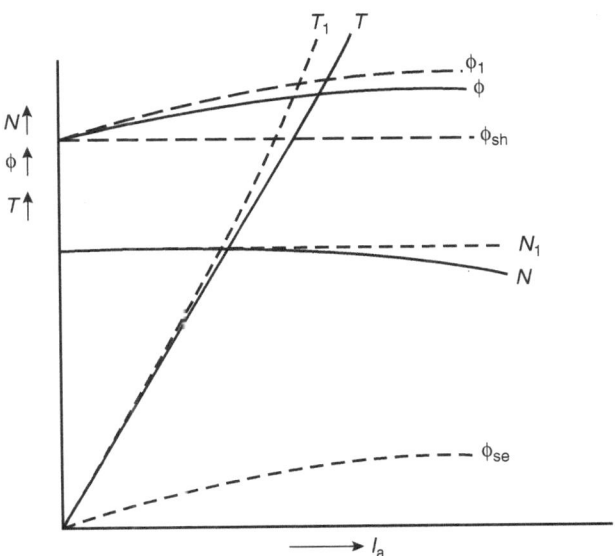

used for starting with heavy loads, for example, compressors, centrifugal pumps, certain machine tools. Once the machine accelerates, the characteristic follows that of a shunt motor, and the armature reaction effect alters the theoretical characteristic from T_1 to T as shown (figure 13.10).

Mechanical characteristic

This characteristic is generally similar to that of a shunt motor.

(2) *Strong series – weak shunt.* Again the characteristic of the series motor is suitable for appropriate applications. Its main disadvantage, for example, a tendency to race on light load, must be removed and this is the purpose of the shunt field.

Electrical characteristics

SPEED. It is seen from the diagram (figure 13.11) that although net flux varies, it follows the magnetisation curve, i.e. it never falls to zero as is the case for the series motor. In effect the shunt field dominates on light loads and the machine runs as a shunt motor at a fixed speed. Once load is applied, the series field asserts itself and the speed characteristic passes from that of a shunt machine to that of a series machine. The tendency for racing on no load is removed and this is the typical characteristic of a

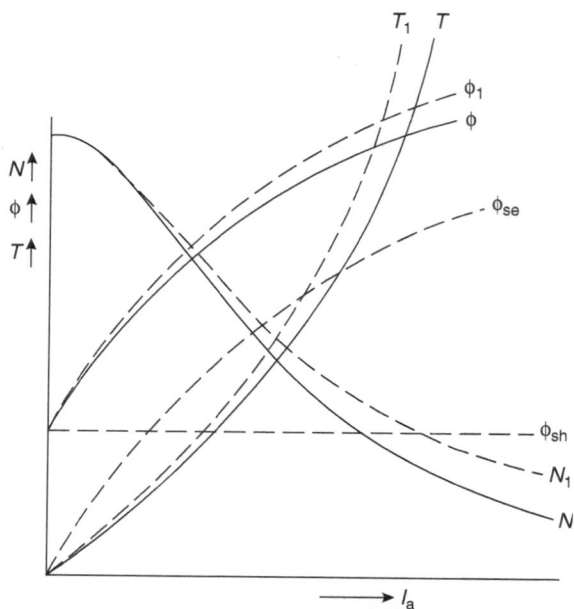

▲ **Figure 13.11** *Various electrical characteristics vs I*

ship's D.C. winch. *Note.* The effect of armature reaction and voltage drop on E_b is seen at higher current values and speed N is lower than the theoretical value N_1.

TORQUE. Since a motor behaves like a shunt machine on light loads, the torque characteristic commences as a straight line through the origin but becomes parabolic as the series field increases. The armature reaction effect gives a slight reduction of net flux with a consequent falling off of torque T from the theoretical graph T_1 (shown dotted, figure 13.11).

Mechanical characteristic

The characteristic for this motor is shown (figure 13.12). It is similar to the electrical speed characteristic and can be deduced, as described for the series and shunt machines. The exact graph shape and position depends on the shunt and series fields' relative strengths.

DIFFERENTIAL CONNECTION OF FIELDS. This can help maintain a constant speed, for example, for an alternator drive. Increase of load results in an increase in the series flux and, as fields are opposed, resultant flux *decreases*. Speed, being inversely proportional to flux, increases to compensate for the fall due to application of a load.

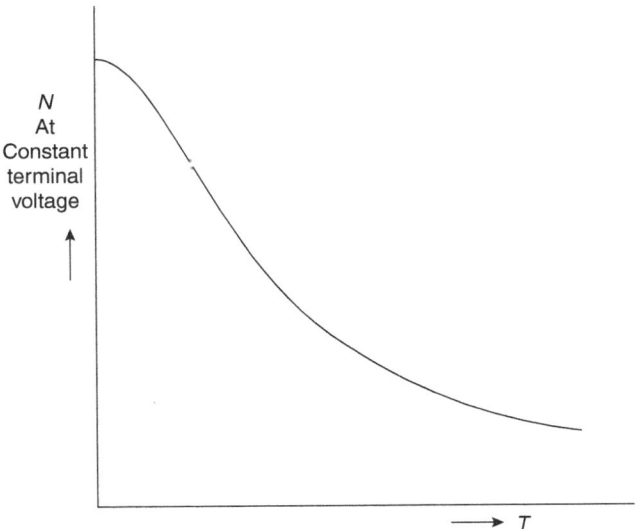

▲ Figure 13.12 *N and At characteristics vs time*

Machine speed and alternator frequency tend to remain constant but armature current increases appreciably to provide the required torque with reduced flux. A failing of the arrangement is the field cancelling effect at starting and that, due to the series field establishing itself quicker, a machine may start to run in reverse. Special arrangements must be made when starting a motor with a differentially connected field system.

Example 13.6. A 220V shunt motor runs on light load at a speed of 1250 rev/min and takes a current of 2.8A. On full load the current taken from the mains is 40A. Owing to armature reaction, the flux per pole is 4% less than the no-load value. Calculate the speed on full load if armature resistance = 0.29Ω and field resistance = 165Ω (4 significant figures.)

No load. Voltage across shunt field = 220V

Current through shunt field $= \dfrac{220}{165} = 1.332A$

Armature current = 2.8 − 1.33 = 1.47A

Voltage drop across armature $= I_{a0}R_a = 1.47 \times 0.29 = 0.426V$

and $E_{b0} = 220 - 0.426 = 219.574V$

Full load. Current through shunt field as before = 1.332A

Armature current = 40 − 1.332 = 38.67A Voltage drop across armature = $I_a R_a$ = 38.67 × 0.29 = 11.23V and E_{b1} = 220 − 11.23 = 208.77V

Now since $E_b \propto \Phi N$ $\therefore E_b = k\Phi N$ and $\dfrac{E_{b0}}{E_{b1}} = \dfrac{\Phi_0 N_0}{\Phi_1 N_1}$ But $\Phi_1 = 0.96\Phi_0$

$$\therefore \frac{E_{b0}}{E_{b1}} = \frac{\Phi_0 N_0}{0.96\Phi_1 N_1} \text{ or } N_1 \frac{N_0 \times E_{b1}}{0.96 \times E_{b0}} \frac{1250 \times 208.77}{0.96 \times 219.57}$$

Thus N_1 = 1238 i.e. speed on full load = 1238 rev/min.

Example 13.7. A 220V series motor works with an unsaturated field taking a current of 100A and run at 800 rev/min. Calculate the speed the motor will run at when developing half the torque. Motor total resistance = 0.1Ω (4 significant figures).

$$\text{Here } T = kI_a^2 \quad \therefore \frac{T_1}{T_2} = \frac{I_{a1}^2}{I_{a2}^2} \text{ But } T_2 = 0.5T_1$$

$$\text{So } \frac{T_1}{0.5T_1} = \frac{100^2}{I_{a2}^2} \text{ or } I_{a2}^2 = 100^2 \times 0.5 = 5000$$

$$\therefore I_{a2} = \sqrt{5000} = 70.7\text{A}$$

Also under the first condition $E_{b1} = V − I_{a1}(R_a + R_{se})$

$$= 220 − (100 \times 0.1) = 210\text{V}$$

Under the second condition $E_{b2} = V − I_{a2}(R_a + R_{se})$

$$= 220 − (70.7 \times 0.1) = 212.93\text{V}$$

But $E_b = k\Phi N$ or $\dfrac{E_{b1}}{E_{b2}} = \dfrac{\Phi_1 N_1}{\Phi_2 N_2}$ also $\Phi \propto I_a$

$$\text{so } \frac{E_{b1}}{E_{b2}} = \frac{I_{a1} N_1}{I_{a2} N_2} \text{ or } N_2 = \frac{I_{a1} N_1 E_{b2}}{I_{a2} E_{b1}} = \frac{100 \times 800 \times 212.93}{70.7 \times 210}$$

$$\text{so } N_2 = 1147 \text{ rev/min}$$

Motor Starters

The need for a starter to work with a motor was mentioned earlier, as at the instant of starting as the machine is *not* rotating, there is no back e.m.f. The current is thus limited by armature resistance alone, unless an arrangement is made to add further/extra resistance in the armature circuit. Thus for all but 'fractional output power' motors, with appreciable resistance, a resistor is inserted into the armature circuit and then removed in steps, as the motor accelerates up to its correct running speed. The arrangement is incorporated in a unit called a 'motor starter' or more simply a 'starter' and consists of a tapped resistor and a switching device, which enables resistance to be gradually reduced and finally eliminated altogether. A starter may incorporate other attachments as needed for safe motor operation, which may include protective safeguards against the effects of a reduced working voltage or an overcurrent.

Although motor starters are studied elsewhere, it is appropriate to mention that the form of starter needed for a particular machine is mostly decided by the duty for which the motor is used. It may be a manually operated or automatic type; it may be designed for starting and stopping a motor and may require to be done only once a day. In contrast, the duty may require the motor to be started and stopped continually for long periods, with a winch or hoist. Such a starter is often referred to as a 'controller'. These observations show that a starter is an important item requiring attention, one that needs careful and routine maintenance and a thorough knowledge of its function from a theoretical and practical viewpoint.

Speed Control

As for the starter, so for the full treatment of speed control, further study has yet to be made. The basic methods whereby the speed of a D.C. motor can be controlled are discussed here, the reader is reminded of the basic deduction $N \propto \dfrac{E_b}{\Phi}$ or $N \propto \dfrac{V}{\Phi}$ (approx.). Varying the voltage applied to a motor armature and keeping flux constant varies the speed in direct proportion, and is termed 'voltage control'. Varying a machine's flux and keeping voltage constant will vary the speed in inverse proportions and is termed 'field control'.

FIELD CONTROL. This is the most common control type. When a motor is loaded, its speed varies with the load. It may be desired to adjust the speed for any load condition, i.e. keep it constant over the working range or raise it above normal running speed. Field control is used because its addition into the field circuit is easily achieved, control is smooth and little energy is wasted as heat.

It must be remembered that this type of control gives speed variation in an *upward* direction only. It is used for raising speed above normal and as flux is weakened, for the same driving torque, armature current rises. *Note.* $T \propto \Phi I_a$. Thus a motor may be of larger dimensions if speed variation is required and interpoles fitted to ensure good commutation across the working range.

VOLTAGE CONTROL. This is achieved in various ways for different kinds of D.C. motor but the key requirement is to reduce the voltage applied to the machine armature. Thus a large variable rheostat is connected in series with the armature or the latter supplied from a variable voltage supply. The method is used to lower speed and control is in a *downward* direction only.

A wide range in motor speed is obtained by combining field and voltage control and the methods of applying these are important enough to require further study. To meet the requirements of the duty for which a motor is required, the starter and speed controller may be incorporated into one unit. Since correct application and use of a motor is of vital importance to practical engineers, it is hoped the treatment given to the D.C. machine in Volume 7, will be seen as a valued addition.

Example 13.8. The armature of a motor has 660 conductors whose effective length is 410mm; of these, only 70% are simultaneously in the magnetic field. Flux density = 0.65T, the effective diameter of the armature = 300mm and each conductor carries a current = 80A. If armature speed = 800 rev/min, calculate the output power developed (3 significant figures).

Force on one conductor is given by $F = B I \ell$ newtons

$\therefore F = 0.65 \times 80 \times 410 \times 10^{-3}$ $F = 21.32N$

Number of conductors in the field at any given instant = 0.7×660

$\therefore$ Total force = $21.32 \times 0.7 \times 660 = 9.85kN$

Torque = force × radius or $T = 9850 \times \dfrac{0.3}{2}$ newton metres

Thus $T = 9850 \times 0.15 = 1477.5Nm$

And power developed $= \dfrac{2 \times 3.14 \times 800 \times 9850 \times 0.15}{60} = 124kW$

Example 13.9. A shunt motor takes 180A. The supply voltage = 400V, the shunt-field resistance = 200Ω and that of the armature = 0.02Ω. If there is a 2V voltage drop at the brushes, calculate (a) the motor back e.m.f. (2 decimal figures), (b) the output power developed (3 significant figures) and (c) the efficiency, neglecting all losses for which information is not given (1 decimal place).

Shunt-field current $= \dfrac{400}{200} = 2A$

Armature current $= 180 - 2 = 178A$

Armature voltage drop $= 178 \times 0.02 = 3.56V$

(a) Back e.m.f. $= 400 - 3.56 - 2$ (voltage drop at brushes) $= 394.44V$

(b) Output power developed $= \dfrac{394.4 \times 178}{1000} = 70.2kW$

Efficiency $= \dfrac{\text{output}}{\text{input}} = \dfrac{394.4 \times 178}{400 \times 180}$

Efficiency or $\eta = 0.975$ or 97.5%

Example 13.10. A 4-pole D.C. motor with a lap winding is connected to a 200V supply mains. The armature carries 600 conductors and has a resistance $= 0.3\Omega$. The shunt-field circuit resistance $= 100\Omega$ and the flux per pole $= 0.02Wb$. On no load armature current $= 3A$. If the normal full-load current in the armature is 50A, determine the drop in the motor speed from no load to full load. Neglect the effect of armature reaction (1 decimal place).

Back e.m.f. on no load $E_{b0} = 200 - I_{a0}R_a$

Shunt-field current $= \dfrac{200}{100} = 2A$ $I_{a0} = 3 - 2 = 1A$

$\therefore E_{b0} = 200 - (1 \times 0.3) = 200 - 0.3 = 199.7V$

No-load speed is given by N_0 where:

$E_{b0} = \dfrac{Z\Phi_0 N_0}{60} \times \dfrac{P}{A}$

or $199.7 = \dfrac{600 \times 0.02 \times N_0}{60} \times \dfrac{4}{4}$

Thus $199.7 = 0.2 \times N_0$

or $N_0 = \dfrac{199.7}{0.2} = 998.5 \text{rev/min}.$

Back e.m.f. E_{b1} on full load is given by:

$E_{b1} = 200 - I_{a1} = 200 - (50 - 2)\,0.3$

$E_{b1} = 200 - (48 \times 0.3)$

or $E_{b1} = 185.6V$

Since $E_{b1} = k\Phi_1 N_1$ and assuming a constant flux, then $\Phi_0 = \Phi_1$

or $\dfrac{E_{b0}}{E_{b1}} = \dfrac{k\Phi_1 N_1}{k\Phi_0 N_0}$ whence $N_1 = \dfrac{N_0 E_{b1}}{E_{b0}}$

Thus $N_1 = \dfrac{998.5 \times 185.6}{199.7}$

Full-load speed = 928 rev/min.

Example 13.11. Calculate the first resistance step of a starter for a 240V shunt motor with armature resistance = 0.5Ω, if the maximum current limit = 60A and the lower limit about 45A.

Let R_s = the total resistance of the series resistor put into the armature circuit. If the armature current at start = I_{as} 60A and also $I_{as} = \dfrac{240}{R_a + R_s} = \dfrac{240}{0.5 + R_s}$

or $R_s + 0.5 = \dfrac{240}{60}$ giving $R_s = 4 - 0.5 = 3.5Ω$

As the motor starts and accelerates up to speed, the starter handle is kept in position until the current falls to 45A. Thus the starting resistance is still in circuit, but a back e.m.f. is building to a final value given by E_{b1}.

Here $E_{b1} = 240 - 45(3.5 + 0.5) = 240 - 180 = 60V$

At this stage the handle is moved and a section of the starting resistor cut out. Let R_1 be the new value of the total starter resistance. Current rises to 60A but the back e.m.f. does not change until the motor speed changes. Thus at the instant of moving the handle:

$$240 = E_{b1} + I_a (R_a + R_1)$$

or 240 = 60 + 60 (0.5 + R_1) so:

$$0.5 + R_1 = \dfrac{240 - 60}{60} = 3 \text{ or } R_1 = 3 - 0.5 = 2.5Ω.$$

Thus the resistance removed during the first movement of the handle after switching on, is:

$$3.5 - 2.5 = 1Ω \quad \text{The first resistance step is thus 1Ω.}$$

Estimation of D.C. machine efficiency

The efficiency of a D.C. machine can be assessed by measuring its output power and comparing this with the input power. Thus in general:

$$\text{Efficiency} = \frac{\text{Power output}}{\text{Power input}}$$

For small motors, output is measured with a calibrated brake or a dynamometer and input power measured by electrical instrumentation. However, on large motors it is difficult to measure output power with any accuracy. Similarly, mechanical input power to a generator is difficult to measure, and with large generators, the electrical output involves dissipation of much energy, usually as heat.

The Swinburne test was a technique devised to assess losses and estimate efficiency at any load from:

$$\eta = \frac{\text{Power input} - \text{losses}}{\text{Power input}} \ (\text{for a motor})$$

$$\eta = \frac{\text{Power input}}{\text{Power output} + \text{losses}} \ (\text{for a generator})$$

Although this test is rarely now considered in electrotechnology examinations, brief mention of the loss issues are discussed here. Possible losses are divided into 2 groups, those that vary with load and those that remain largely constant at all loads. These were introduced in Chapter 6 and summarised here:

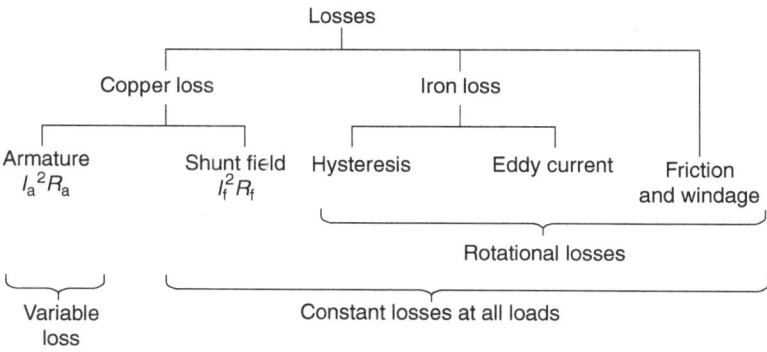

▲ **Figure 13.13** *Various D.C. machine losses*

Rotational losses due to friction and windage are constant if the speed is constant. *Iron loss* is due to the core material (hysteresis) magnetic properties and eddy currents, which are minimised by laminating the cores. Iron loss varies with load, but such variation has little effect on overall constant losses. Losses due to the windings resistance, or *Copper loss*, occurs in the armature and in field windings. Field copper loss is constant providing the supply voltage remains unaltered and the winding is shunt-connected. The only big losses that vary with load are due to the armature circuit resistance, the armature copper losses $I_a^2 R_a$.

Hence for a motor:

$$\eta = \frac{\text{Power input} - \text{variable loss} - \text{constant loss}}{\text{Power input}}$$

$$\eta = \frac{VI - I_a^2 R_a - I_f^2 R_f - P_{WIF}}{VI} \times 100\%$$

For a generator:

$$\eta = \frac{\text{Power output}}{\text{Power output} + \text{variable loss} + \text{constant loss}}$$

$$\eta = \frac{VI}{VI + I_a^2 R_a + I_f^2 R_f + P_{WIF}} \times 100\%$$

To assess a machine's constant losses it is run as a motor on no load. As no output power is developed, all the input power overcomes the machine's constant losses. On a shunt-connected machine, this no-load input power supplies the shunt-field loss (VI_f or $I_f^2 R_f$) and armature loss. This no-load armature loss comprises iron, windage and friction losses (P_{WIF}), and a small armature copper loss, which is usually ignored. To make an accurate estimate of a machine's efficiency one must account for a change in copper loss due to resistance change as temperature rises. This involves calculating the new resistance value at the higher temperature using the temperature coefficient of resistance equation $R_T = R_0 (1 + \alpha T)$, with rotational losses unaltered.

Example 13.12. A 250V D.C. shunt motor takes a current of 7A when running on no load. The armature and shunt field-circuit resistances are 0.15Ω and 125Ω respectively.

Find the machine efficiency when it (a) runs as a shunt motor taking a total current of 52A and (b) is driven as a shunt generator delivering 15kW at its output terminals (2 decimal places each).

On no load:

Shunt-field current $I_f = \dfrac{V}{R_f} = \dfrac{250}{125} = 2A$

Armature current $I_{a_0} = I_L - I_f = 7 - 2 = 5A$

No load armature power input = windage and friction losses + small $I_a^2 R_a$ (often negligible losses).

$$\therefore P_0 = VI_{a_0} = 250 \times 5 = 1250W$$

$$\text{Field Copper loss } I_f^2 R_f = VI_f = 250 \times 2$$

(a) At 52A load $I_a = 52 - 2 = 50A$ (I_f constant).

$$\text{Armature Copper loss} = I_a^2 R_a = 50^2 \times 0.15 = 375W$$

$$\eta = \frac{VI - I_a^2 R_a - I_f^2 R_f - P_{WIF}}{VI} \times 100 \text{ \%}$$

$$\eta = \frac{(250 \times 52) - 375 - 500 - 1250}{(250 \times 52)} \times 100\% \quad \eta = 83.65\%$$

(b) Load current $I_L = \dfrac{15 \times 10^3}{250} = 60A$

$$\text{Shunt} - \text{field current } I_f = 2A$$

$$\text{Armature current } I_a = 60 + 2 = 62A$$

$$\text{Armature Copper Loss } I_a^2 R_a = 622 \times 0.15 = 576.6W$$

$$\eta = \frac{VI}{VI + I_a^2 R_a + I_f^2 R_f + P_{WIF}} \times 100\%$$

$$\eta = \frac{(15 \times 10^3)}{(15 \times 10^3) + 576.6 + 500 + 1250} \times 100\% \quad \eta = 86.57\%$$

The efficiency at any load is estimated and a graph of efficiency against load current plotted, indicating the efficiency trend to give an accurate assessment of the load current where maximum efficiency occurs.

Practice Examples

13.1 A 110V series motor has a resistance of 0.120Ω. Determine its back e.m.f. when developing a shaft output of 7.5kW when efficiency is 85% (1 decimal place).

13.2 A 500V D.C. shunt motor has an input of 90kW when loaded. The armature and field resistances are 0.1Ω and 100Ω respectively. Calculate the back e.m.f. value (1 decimal place).

13.3 A 460V D.C. motor takes an armature current of 10A at no load. At full load the armature current = 300A. If the resistance of the armature is 0.025Ω, what is the back e.m.f. value at no load and full load (1 decimal place)?

13.4 An armature winding of a D.C. motor has 240 conductors arranged in 4 parallel paths on an armature whose effective length and diameter are 400mm and 300mm respectively. Assuming average flux density in the air gap = 1.2T and the input to the armature = 40A, calculate (a) the force in newtons and torque in newton metres developed by one conductor, (b) the total torque developed by the complete winding, assuming all the conductors are effective and (c) the armature power output in watts, if the speed = 800 rev/min (all 1 decimal place).

13.5 A marine shunt motor is used for driving a 'fresh water' pump and found to take an armature current of 25A at 220V, when running on full load. Speed is measured to be 725 rev/min and the armature resistance = 0.2Ω. If field strength is reduced by 10% using a speed regulator and the torque is unchanged, determine the steady speed ultimately attained and the armature current, in rev/min.

13.6 A shunt generator delivers 50kW at 250V and runs at 400 rev/min. The armature and field resistances are 0.02Ω and 50Ω respectively. Calculate the machine's speed in rev/min when running as a shunt motor taking 50kW input at 250V. Allow 2V for brush-contact drop.

13.7 A 105V, 3kW D.C. shunt motor has a full-load efficiency of 82%. The armature and field resistances are 0.25Ω and 90Ω respectively. The full-load speed of the motor is 1000 rev/min. Neglecting armature reaction and brush drop, calculate the speed at which the motor will run at no load if the line current at no load is 3.5A. Calculate the resistance added to the armature circuit to reduce the speed to 800 rev/min, the torque remaining constant at full-load value (2 decimal places).

13.8 A shunt motor runs at 1C00 rev/min when cold, taking 50A from a 230V supply. If the armature and field windings both increase in average temperature from 15°C to 60°C, as the motor warms up, determine the speed in rev/min when the motor is warm, given armature and field resistances are 0.2Ω and 200Ω at 15°C respectively. The total current drawn from the supply is constant. Neglect brush drop and armature react on, and assume an unsaturated magnetic circuit, with resistivity temperature coefficient 0.40% from and at 15°C.

13.9 A 4-pole shunt motor has a wave-wound armature of 294 conductors. The flux per pole = 0.025Wb and the armature resistance = 0.35Ω. Calculate (a) the speed of the armature in rev/min and (b) the torque developed when the armature takes a current of 200A from a 230V supply (1 decimal place).

13.10 A shunt motor runs at 600 rev/min from a 230V supply when taking a line current of 50A. Its armature and field resistances are 0.4Ω and 104.5Ω respectively. Neglecting armature reaction and with a 2V brush drop, calculate (a) the no-load speed in rev/min if the no-load line current is 5A, (b) the resistance placed in the armature circuit to reduce the speed to 500 rev/min when taking a line current = 50A (2 decimal places) and (c) the percentage reduction in the flux per pole so the speed = 750 rev/min, when taking an armature current of 30A with no added resistance in the armature circuit (1 decimal place).

SOLUTIONS TO PRACTICE EXAMPLES

Chapter 1

1.1. Let R be the equivalent resistance of the parallel arrangement.

Then $\dfrac{1}{R} = \dfrac{1}{2} + \dfrac{1}{4} + \dfrac{1}{5} + \dfrac{1}{10} = \dfrac{10 + 5 + 4 + 2}{20}$

$= \dfrac{21}{20} = \dfrac{2.1}{2}$ and $R = \dfrac{2}{2.1} = 0.952\Omega$

Voltage drop across the arrangement $= 8.6 \times 0.952 = 8.19V$

Current I_1 in 2Ω resistor $= \dfrac{8.19}{2} = 4.1A$

Current I_2 in 4Ω resistor $= \dfrac{8.19}{4} = 2.1A$

Current I_3 in 5Ω resistor $= \dfrac{8.19}{5} = 1.6A$

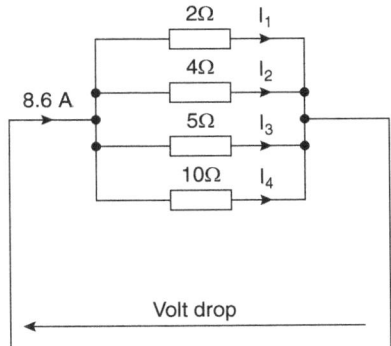

▲ **Figure 1** *Circuit diagram*

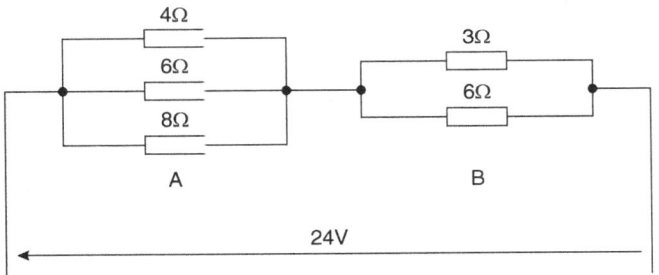

▲ **Figure 2** *Circuit diagram*

Current I_4 in 10Ω resistor $= \dfrac{8.19}{10} = 0.8A$

Check. Total current = 8.6A

1.2. For group A. Let $R_A =$ the equivalent resistance, then:

$$\frac{1}{R_A} = \frac{1}{4} + \frac{1}{6} + \frac{1}{8} = \frac{6+4+3}{24} \text{ or } R_A = \frac{24}{13} = 1.85Ω$$

For group B. Let $R_B =$ the equivalent resistance, then:

$$\frac{1}{R_B} = \frac{1}{3} + \frac{1}{6} = \frac{2+1}{6} = \frac{3}{6} \text{ or } R_B = 2Ω$$

Total circuit resistance $R = R_A + R_B = 1.85 + 2 = 3.85Ω$

Circuit current $= \dfrac{24}{3.85} = 6.23A$

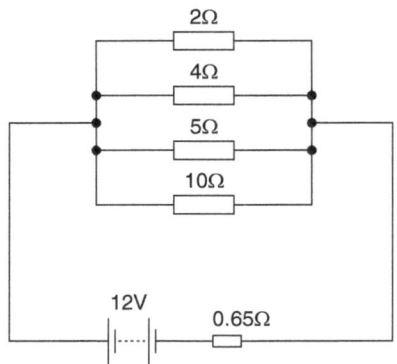

Voltage drop across group A = 1.85 × 6.23 = 11.53V

Voltage drop across group B = 2 × 6.23 = 12.46V

Check. Total voltage drop = 23.99 = 24.00V

Current in resistors – group A

$$\frac{11.53}{4} = 2.88A$$

$$\frac{11.53}{6} = 1.91A$$

$$\frac{11.53}{8} = 1.44A$$

Check $I_{TOTAL} = 6.23A$

Current in resistors – group B

$$\frac{12.46}{3} = 4.153A$$

$$\frac{12.46}{6} = 2.076A$$

Check $I_{TOTAL} = 6.23A$

1.3. From Q1.1 the equivalent resistance *R* of the load = 0.95Ω

The total resistance of the circuit = 0.95 + 0.65 = 1.6Ω

$I = V/R$

The circuit current $= \dfrac{12}{1.6} = 7.5A$

The terminal voltage $= 7.5 \times 0.95 = 7.1V$

Current in 5Ω resistor $= \dfrac{7.125}{5} = 1.4A$

1.4. (a) Ammeter with shunt

Voltage drop across parallel arrangement for f.s.d. $= 10 \times 15 \times 10^{-3} = 0.15V$

Current to be carried by shunt

$$= 25 - (15 \times 10^{-3})$$

$$= 24.985A$$

Resistance of shunt $= \dfrac{0.15}{24.985}$

$$= 0.006Ω$$

(b) Voltmeter with series resistance

Resistance of instrument circuit to drop 500V

$$= \dfrac{500}{15 \times 10^{-3}}$$

$$= 33\ 333Ω$$

∴ Series resistance to be added $= 33\ 333 - 100$

$$= 33\ 323Ω$$

1.5. Parallel section BC has a resistance, given by:

$$\dfrac{1}{R} = \dfrac{1}{40} + \dfrac{1}{40} = \dfrac{2}{40} \text{ or } R = 20kΩ$$

Total resistance of network $= 60 + 20 = 80kΩ$

Current taken by network $= \dfrac{240}{80000}$

$$= 3 \times 10^{-3} A \text{ or } 3mA$$

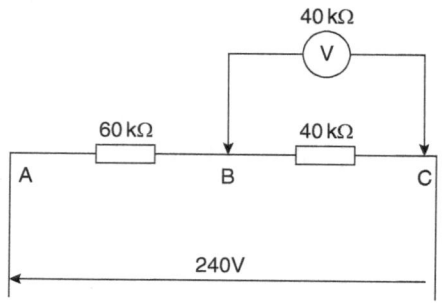

▲ **Figure 4** *Circuit diagram*

Voltage drop across section BC

$$= 3 \times 10^{-3} \times 20000 \text{ volts} = 60V = \text{reading on voltmeter}$$

1.6. Let E = e.m.f. of the battery and R_i its internal resistance.

then $E = 0.18 (10 + R_i)$ ... (i).

and $E = 0.08 (25 + R_i)$ ... (ii).

Equating (i) and (ii) $0.18 (10 + R_i) = 0.08 (25 + R_i)$

or $1.8 + 0.18 R_i = 2 + 0.08 R_i$

∴ $(0.18 - 0.08) R_i = 2 - 1.8$

or $0.1 R_i = 0.2$

or $R_i = 2\Omega$

Substituting in $E = 0.18 (10 + 2)$

$$= 0.18 \times 12$$

$$= 2.16V$$

1.7. P.D. across 4Ω resistor $= 4 \times 1.5 = 6V$

This is also the voltage drop across the other resistors in group A

Current in 2Ω resistor $= \dfrac{6}{2} = 3.0A$

Current in 6Ω resistor $= \dfrac{6}{6} = 1.0A$

Current in 8Ω resistor $= \dfrac{6}{8} = 0.8A$

Current in 4Ω resistor $= \dfrac{6}{4} = 1.5A$

Total current $= 6.3A$

The equivalent resistance P_B of parallel group B is obtained from:

$$\frac{1}{R_B} = \frac{1}{10} + \frac{1}{15} = \frac{3+2}{30} = \frac{5}{30} \text{ or } R_B = \frac{30}{5} = 6\Omega$$

So voltage drop across group B = 6 × 6.25 = 37.5V

$$\text{Current in 10}\Omega\text{ resistor} = \frac{37.5}{10} = 3.75A$$

$$\text{Current in 15}\Omega\text{ resistor} = \frac{37.5}{15} = 2.5A$$

Check. Total current = 6.25A

Voltage drop across group A = 6V

Supply voltage = 6 + 37.5 = 43.5V

1.8. Generator O.C. voltage = 110V

Voltage drop in generator for 75A = 110 − 108.8 = 1.2V

$$\text{Internal resistance of generator} = \frac{1.2}{75} = 0.016\Omega$$

Voltage drop in cables = 108.8 − 105 = 3.8V

$$\text{Resistance of cables} = \frac{3.8}{75} = 0.0507\Omega$$

On 'short-circuit', the only limitation to current is the resistance of the generator and the cables.

$$= 0.016 + 0.057 = 0.0667\Omega$$

$$\text{So short-circuit current} = \frac{110}{0.0667} = 1650A$$

1.9. Meter voltage drop for f.s.d = 1 × 0.12 volts = 120mV.

Since shunt voltage drop for 300A is 150mV, then the meter resistance must be increased by an external resistor of value 0.03Ω. Obtained from 1A × (0.03 + 0.12) Ω = 1 × 0.15 = 0.15V or 150mV.

Under this condition the actual current metered would be 301A. 300A will pass through the shunt and 1A through the ammeter.

Note. The resistor must be rated for this current, i.e. 1A. Thus $1^2 \times 0.03 = 0.03W$ (see Chapter 2.)

1.10.

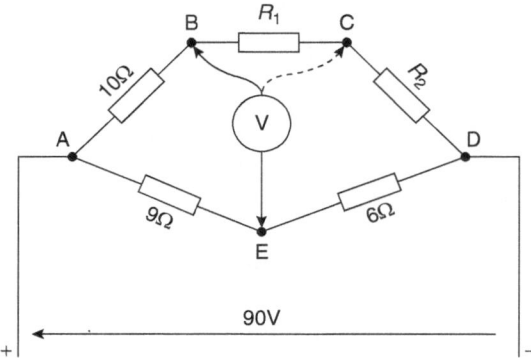

▲ **Figure 5** *Circuit diagram*

P.D. across BC = 34 + 6 = 40V, since E is 6V above C and B in turn is 34V above E.

The P.D. across AE and ED is proportional to their resistance values.

Thus P.D. across AE $= \dfrac{9}{15} \times 90 = 54V$ with A +ve to E.

And P.D. across ED $= \dfrac{6}{15} \times 90 = 36V$ with E +ve to D.

Since A is positive to E by 54V and B is positive to E by 34V (voltmeter reading), then A is +ve to B by (54 – 34) = 20V.

Similarly B is above E by 34V and E is +ve to C by 6V then the P.D. across BC = 34 + 6 = 40V. E is above D by 36V and above C by 6V so C must be +ve to D by (36 – 6) = 30V.

$$\text{Thus the P.D.s across the resistors are } \left.\begin{array}{l} AB = 20V \\ BC = 40V \\ CD = 30V \end{array}\right\} \text{Total 90V.}$$

Also as the branches are *series* circuits, the ohmic values are proportional to the P.D.s.

$$\therefore \frac{\text{P.D. across AB}}{\text{P.D. across BC}} = \frac{10}{R_1} \text{ or } R_1 = \frac{10 \times 40}{20} = 20\Omega$$

And, $\dfrac{\text{P.D. across CD}}{\text{P.D. across AB}} = \dfrac{R_2}{10}$ or $R_2 = \dfrac{10 \times 30}{20} = 15\Omega$

$$\text{Current in branch ABCD} = \frac{90}{10 + 20 + 15} = \frac{90}{45} = 2A$$

$$\text{Current in branch AED} = \frac{90}{9 + 6} = \frac{90}{45} = 6A$$

Supply current = 2 + 6 = 8A

Chapter 2

2.1. Total mass to be lifted = $(2 + 0.25) \times 10^3$kg

Force exerted = $2.25 \times 10^3 \times 9.81$ newtons

Work done = $2.25 \times 10^3 \times 9.81 \times 30$ Nm

$= 66.2175 \times ˙04$Nm or 662.175kJ

$$\text{Hoist output power} = \frac{\text{work done (joules)}}{\text{time (seconds)}}$$

$$= \frac{666175}{90} = 7.36\text{kW}$$

Power input = $200 \times 50 = 11$kW

$$\text{Efficiency} = \frac{7.36}{11} = 0.6689 \text{ or } 66.9\%$$

2.2. Since battery voltage of about 20V is needed, cells must be connected in a series–parallel arrangement. Ten cells in series will give 22V and this will be the e.m.f. of the battery irrespective of the number of identical parallel banks.

An arrangement of 10 cells in series with 3 such banks in parallel is a practical combination.

Battery e.m.f. = e.m.f. of 1 bank = $2.2 \times 10 = 22$V

Internal resistance of 1 bank = $0.3 \times 10 = 30\Omega$

$$\text{Internal resistance of battery} = \frac{3}{3} = 1\Omega$$

$$\text{Resistance of 1 lamp} = \frac{V^2}{P} = \frac{20^2}{10} = 40\Omega$$

or lamp current $= \dfrac{P}{V} = \dfrac{10}{20} = 0.5A$

and lamp resistance $= \dfrac{20}{0.5} = 40\Omega$

Resistance of 3 lamps in parallel $= \dfrac{40}{3} = 13.33\Omega$

Total resistance of the complete circuit $= 13.33 + 1 = 14.33\Omega$

Circuit current $= \dfrac{22}{14.33} = 1.54A$

(a) Voltage drop in battery $= 1.54 \times 1 = 1.54V$

Battery terminal voltage $= 22 - 1.54 = 20.46V$

(b) Current taken by 1 lamp $= \dfrac{1.54}{3} = 0.513A$

(c) Power loss per cell $=$ (current in 1 bank)$^2 \times$ resistance of a cell

$$= \left(\dfrac{1.54}{3}\right)^2 \times 0.3$$

$$= 0.5132 \times 0.3 = 0.079W$$

2.3. The equivalent head of water can be obtained thus: A pressure of 15 bars $=$ $15 \times 10^5 Nm^{-2}$. Specific weight of water is $10^3 \times 9.81 Nm^{-3}$

Then the head of water $= \dfrac{15 \times 10^5}{10^3 \times 9.81}$

$$= 152.85m$$

Force required to lift 12 700 litres or 12.7×10^3 kg is $12.7 \times 103 \times 9.81$ newtons

$$= 124\ 587N$$

Work done per hour $= 124\ 587 \times 152.85$ Nm $= 19.064MJ$

Pump power output $= \dfrac{19.064 \times 10^6}{3600} = 5.296kW$

Input to pump or output of motor

$$= \dfrac{5.296}{0.82} = 6.465kW$$

$$\text{Input to motor } = \frac{6.465}{0.89} = 7.275\text{kW}$$

$$\text{Motor current } = \frac{7275}{220} = 33.1\text{A}$$

2.4. Battery e.m.f. $= 4 \times 2.2 = 8.8\text{V}$

Terminal voltage of battery = voltage drop across resistor

$$= 5 \times 1.4\text{V} = 7\text{V}$$

Voltage drop in battery $= 8.8 - 7 = 1.8\text{V}$

$$\text{Internal resistance of battery} = \frac{1.8}{1.4} = 1.29\Omega$$

$$\text{Internal resistance of 1 cell} = \frac{1.29}{4} = 0.32\Omega$$

For parallel working:

$$\text{Internal resistance of battery} = \frac{0.32}{4} = 0.08\Omega$$

E.m.f. of battery = e.m.f. of ˙ cell $= 2.2\text{V}$

Total circuit resistance $= 5 + 0.08 = 5.08\Omega$

$$\text{Circuit current } = \frac{2.2}{5.08} = 0.43\text{A}$$

2.5. Winch output

$$= 5 \times 1\text{C}^3 \times 9.81 \times 36.5 \text{ Nm per minute}$$

$$= 4.905 \times 36.5 \times 104 \text{ joules per minute}$$

$$= \frac{4.905 \times 36.5 \times 10^4}{60} \text{ joules per second}$$

$$= \frac{179.033}{6} \times 10^3 \text{watts} = 29.84\text{kW}$$

Since the winch is 75% efficient, the input must be $29.84 \times \dfrac{100}{75}$ kilowatts

$$= 29.84 \times \frac{4}{3} = 39.78\text{kW}$$

Input to winch = output of motor

∴ power rating of motor = 39.78kW

If the motor efficiency is taken as 85%,

The electrical input would be $\dfrac{39.78}{0.85}$ = 46.8kW

Current taken from the mains = $\dfrac{46800}{220}$ = $\dfrac{2340}{11}$ = 212.7A

2.6. Lighting load = 100 × 100 = 10 000W and

200 × 60 = 12 000W

= 10 + 12 = 22kW

Heating load = 25kW

Miscellaneous loads = 30 × 220 = 6.6kW

Total load = 22 + 25 + 6.6 = 53.6kW

Generator output = 53.6kW

Generator input = $\dfrac{53.6}{0.85}$ = 63.1kW

Now generator input = engine output.

So engine must develop 63.1kW

2.7. O.C. e.m.f. of battery = 4.3V

O.C. e.m.f. cell = $\dfrac{4.3}{3}$ = 1.43V

Value of load resistor = $\dfrac{4.23}{0.4}$ = 10.575Ω

Voltage drop in battery = 4.3 − 4.23 = 0.07V

Internal resistance of battery = $\dfrac{0.07}{0.4}$ = 0.175Ω

Internal resistance of 1 cell = $\dfrac{0.175}{3}$ = 0.058Ω

With a cell *reversed*, the e.m.f. of 2 cells cancel each other and the effective e.m.f. = that of 1 cell = 1.43V.

Let I be the current under this condition.

Then $1.43 = I(10.57 + 0.175)$ $\therefore I = \dfrac{1.43}{10.75}$ 0.134A

Note. For the solution, the nternal resistance of a cell is assumed to be the same in both the forward and reverse direction in the absence of any further detailed information.

2.8.

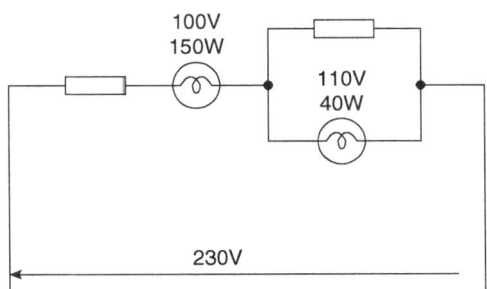

100V
150W

110V
40W

230V

▲ **Figure 6** *Circuit diagram*

Current for 40W lamp $= \dfrac{40}{110} = 0.363A$

Resistance of 40W lamp $= \dfrac{110}{0.363} = 303\Omega$

Current for 150W lamp $= \dfrac{100}{1.5} = 1.5A$

Resistance of 150W lamp $= \dfrac{100}{1.5} = 66.66\Omega$

Parallel circuit has to carry 1.5A

$\therefore$ Current in shunt resistor = 1.5 – 0.363 = 1.137A

Voltage drop across shunt = 110V

Resistance of shunt $= \dfrac{11\mathrm{\Im}}{1.137} = 96.8\Omega$

110V is dropped across the parallel circuit

100V is dropped across the series lamp

$\therefore$ 230 – 210 = 20V must be dropped across the series resistor, which carries 1.5A

$$\therefore \text{ Resistance value of series resistor } = \frac{20}{1.5} = 13.3\Omega$$

2.9.

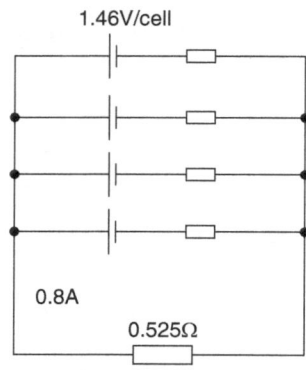

1.46V/cell

0.8A

0.525Ω

▲ **Figure 7** *Circuit diagram*

E.m.f. of battery = e.m.f. of 1 cell for parallel working = 1.46V

Let R_i = the internal resistance of the battery

and R_c = the internal resistance of 1 cell.

The total resistance of the circuit = 0.525 + R_i

Circuit voltage drop = 0.8 (0.525 + R_i)

$$= \text{applied e.m.f.} = 1.46V$$

$\therefore 1.46 = (0.8 \times 0.525) + 0.8\,R_i$

$$= 0.42 + 0.8\,R_i \text{ or } 0.8\,R_i = 1.46 - 0.42 = 1.04$$

So $R_i = \dfrac{1.04}{0.8} = 1.3\Omega$

Now since the cells are in parallel, then

$$\frac{1}{R_i} = \frac{1}{R_c} + \frac{1}{R_c} + \frac{1}{R_c} + \frac{1}{R_c} = \frac{4}{R_c}$$

or $R_c = 4 \times R_i = 4 \times 1.3$

and internal resistance of 1 cell = 5.2Ω

2.10.

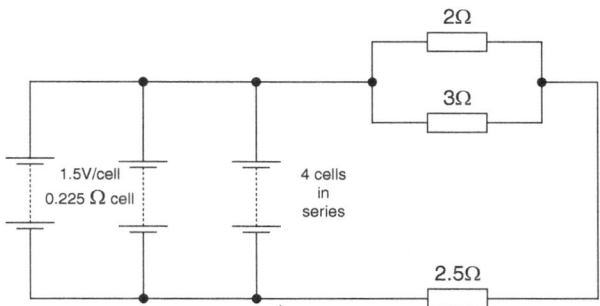

▲ **Figure 8** *Circuit diagram*

Battery e.m.f. = e.m.f. of 1 bank = 4 × 1.5 = 6V

Battery resistance $= \dfrac{\text{resistance of 1 bank}}{3} = \dfrac{4 \times 0.225}{3} = 0.3\Omega$

Load resistance = 2.5 + R (resistance of parallel section)

Here $\dfrac{1}{R} = \dfrac{1}{3} + \dfrac{1}{2} = \dfrac{5}{6}$ or $R = \dfrac{6}{5} = 1.2\Omega$

Load resistance = 2.5 + 1.2 = 3.7Ω

Resistance of complete circuit = 3.7 + 0.3 = 4Ω

Circuit current $= \dfrac{6}{4} = 1.5A$

Voltage drop in battery = 0.3 × 1.5 = 0.45V

Battery terminal voltage = 6 − 0.45 = 5.55V

Power rating of 2.5Ω resistor = I^2R = 1.5^2 × 2.5 = 5.6W

Current in 2Ω resistor $= \dfrac{\text{voltage drop}}{\text{resistance}} = \dfrac{\text{total current} \times R}{2}$

$= \dfrac{1.5 \times 1.2}{2} = 0.9A$

Power rating of 2Ω resistor = 0.9^2 × 2 = 1.6W

Current in 3Ω resistor = 1.5 − 0.9 = 0.6A

Power rating of 3Ω resistor = 0.62 × 3 = 1.1W

Energy conversion = energy in external resistors + energy in battery

= time (total wattage of external resistors + battery resistance power wastage)

$$= t (5.625 + 1.62 + 1.08 + 1.5^2 \times 0.3)$$

$$= 3600 (8.325 + 0.675) = 32400J$$

This could also be obtained thus:

Energy = e.m.f. × current × time

$$= 6 \times 1.5 \times 3600$$

$$= 32\,400J \text{ or } 32.4kJ$$

Chapter 3

3.1. (a) Volume = area × length or $A = \dfrac{V}{l} = \dfrac{10 \times 10^3}{100 + 10^3}$

$$= 0.1mm^2$$

Then $R = \dfrac{el}{A} = \dfrac{17 \times 10^{-6} \times 100 \times 10^3}{10^{-1}} = 17\Omega$

(b) Area of plate = 100 × 100 = 104mm²

Thickness of plate $= \dfrac{10 \times 10^3}{10^4} = 1mm$

This is the length in the expression $R = \dfrac{\rho \ell}{A}$

$$\therefore R = \dfrac{17 \times 10^{-6} \times 1}{10^4} \text{ ohms} = 1.7 \times 10^{-3} \mu\Omega$$

Alternatively using = 1.7 × 10⁻⁸ ohm-metres for (a) – as an example.

$$R = \dfrac{1.7 \times 10^{-8} \times 100}{0.1 \times 10^{-6}} = 17\Omega$$

3.2. Since $R_{20} = R_0(1 + \alpha 20)$ and $R_{60} = R_0(1 + \alpha 60)$

Then $\dfrac{R_{60}}{R_{20}} = \dfrac{R_0(1 + \alpha 60)}{R_0(1 + \alpha 20)}$

and $R_{60} = \dfrac{R_{20}[1 + (60 \times 0.004\,28)]}{[1 + (20 \times 0\,004\,28)]}$

or $R_{60} = \dfrac{90(1 + 0.2568)}{1 + 0.856} = \dfrac{90 \times 1.2568}{1.0856}$ ohms $= 104.4\Omega$

Current taken by coil at 20°C $= \dfrac{230}{90} = 2.56\text{A}$

At 60°C to keep current constant, voltage must be $2.56 \times 104.4 = 267.26\text{V}$. So the voltage is raised by $267.26 - 230 = 37.26\text{V}$

3.3. Assuming 1 litre of water to have a mass of 1 kg so the mass of 0.75 litre of water

$$= 0.75 \times 1 = 0.75\text{kg}$$

Heat required $= 0.75 \times 4.2 \times (100 - 6) = 296\text{kJ}$

The current taken by the heater is $\dfrac{220}{120} = 1.83\text{A}$

and the power rating of the heater $= 220 \times 1.83 = 403.3\text{W}$

Since the heater is 84% efficient, only 403.3×0.84 watts are available to heat the water.

$$\therefore \text{ time of heating } = \dfrac{295 \times 10^3}{403.3 \times 0.84} \text{ seconds } = 873\text{s}$$

$$= \dfrac{873}{60} \text{minutes } = 14\text{min } 33\text{s}$$

3.4. Since $R = \dfrac{\rho\ell}{A}$ then $\ell = \dfrac{RA}{\rho}$

$$\ell = \dfrac{15.7 \times \pi \times 0.315^2}{407 \times 10^{-6} \times 4} \text{mm} = 3\text{cm}$$

3.5. Since $\dfrac{R_2}{R_1} = \dfrac{R_0(1+\alpha T_2)}{R_0(1+\alpha T_1)}$

$$\therefore R_2 = \frac{R_1(1+\alpha T_2)}{1+\alpha T_1}$$

Then $\dfrac{R_2}{R_1} \times (1 + \alpha T_1) = 1 + \alpha T_2$

So $1 + \alpha T_2 = \dfrac{240}{200}\big[1 + (0.0042 \times 15)\big]$

$$= 1.2\,(1 + 0.063)$$

and $\alpha T_2 = 1.2 + 0.0756 - 1 = 0.2756$

thus $T_2 = \dfrac{0.2756}{0.0042} = 65.6°C$

Temperature rise $= 65.6 - 15 = 50.6°C$

3.6.

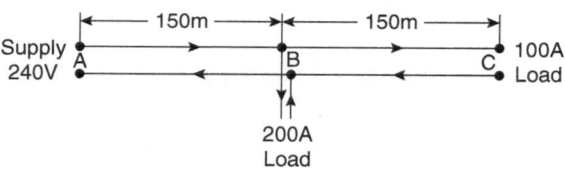

▲ **Figure 9** *Supply-load diagram*

The resistance of a cable core 880m long and area 50mm² = 0.219Ω

then the resistance of a cable core 880m long and area 150 mm² = $\dfrac{0.219}{3} = 0.073\Omega$ and the resistance of a cable core 150m long and area

150mm² $= \dfrac{0.073 \times 150}{880} = 0.0124\Omega$

Current in length AB = 300A

Resistance of length AB = 2 × 0.0124 = 0.0248Ω

Voltage drop in length AB = 300 × 0.0248 = 7.44V

Voltage at 200A load = 240 − 7.44 = 232.56V

Voltage drop in section $BC = \dfrac{1}{3}$ that in AB as the current is $\dfrac{1}{3}$, all else being the same

$\therefore$ Voltage drop in BC $= \dfrac{7.44}{3} = 2.48V$

Voltage at 100A load, i.e. at $C = 232.56 - 2.48 = 230.08V$

3.7. Resistance of 1mm diameter cable $R_1 = \dfrac{el_1}{A_1}$

or $2.47 = \dfrac{e \times 100 \times 10^3 \times 4}{\pi \times 1^2}$

Let $R_2 =$ resistance of $800 \times \dfrac{105}{100} = 840m$ of 1.5mm diameter cable

Then $R_2 = \dfrac{e \times 840 \times 10^3 \times 4}{\pi \times 1.5^2}$

So $\dfrac{R_2}{R_1} = \dfrac{e \times 840 \times 10^3 \times 4}{\pi \times 1.5^2} \bigg/ \dfrac{e \times 100 \times 10^3 \times 4}{\pi \times 1^2} = \dfrac{840 \times 1^2}{100 \times 1.5^2}$

or $R_2 = 2.47 \times 8.4 \times \left(\dfrac{1}{1.5}\right)^2 = 9.22\Omega$

Since there are 19 strands in parallel, the resistance of the complete cable, being inversely proportional to area, will be reduced by 19

$\therefore$ Resistance of cable $= \dfrac{9.22}{19} = 0.49\Omega$

3.8. Since $R = R_0 (1 + \alpha T)$ or $15 = 10[1 + (\alpha \times 100)]$

$$= 10 + (\alpha \times 1000) \text{ or } 5 = 1000\alpha \text{ and } \alpha = 0.005$$

or using the definition

$$\alpha = \dfrac{\text{increase of resistance per } 0°C \text{ rise in temperature}}{\text{resistance at } 0°C}$$

$$= \dfrac{15 - 10}{100} \bigg/ 10 = \dfrac{5}{1000} = 0.005$$

Also since

$R = R_0 (1 + \alpha T)$ then $30 = 10 (1 + 0.005T)$

and $30 = 10 + 0.05T$

or $20 = 0.05T$ and $T = \dfrac{20}{0.05} °C = 400°C$

3.9. Heat required by brass = $500 \times 0.39 \times (910 - 15)$ kilojoules

$$= 5 \times 39 \times 895$$

or energy required by brass = 174.525kJ

Energy taken from supply $= \dfrac{174.525 \times 10^3 \times 100}{80}$

$$= 21.816 \times 10^4 KJ$$

Time taken to expend this energy at the rate of 200kW

$$= \dfrac{21.816 \times 10^4}{200} \text{ seconds}$$

$$= 1091s$$

$$= 18.18min \text{ or } 18 \text{ min } 11s$$

3.10. Electrical energy used $= \dfrac{744}{2} = 372 \text{ units} = 372kW\,h$

$$= 372 \times 3600 KJ$$

Energy passed to heat water = $372 \times 360 \times 0.8$ kilojoules

$$= 1.071 \times 360kJ$$

Heat energy received by water = 1.07×10^6 kilojoules

Temperature rise of water = $82 - 16 = 66$

So quantity of water $= \dfrac{1.07 \times 10^6}{66 \times 4.2}$ kilogrammes $= 3860kg$

Assuming 1 litre has a mass of 1kg then the quantity of water used = 3860 litres

Chapter 4

4.1. Input to accumulator = 6×18 ampere hours

Output from accumulator = 3.5×28 ampere hours

$$\text{Ampere hour efficiency} = \frac{3.5 \times 28}{6 \times 18} = 0.907$$

$$= 90.7\%$$

4.2. Mass of deposit, $m = zIt$

or $(19.34 - 14.52) \times 10^{-3} = 330 \times 10^{-9} \times I \times 50 \times 60$

or $I = \dfrac{4.82 \times 10^{-3}}{50 \times 60 \times 330 \times 10^{-9}}$ amperes

$= 4.869$A

Error in reading $= 5.1 - 4.869$

$= 0.231$A (high)

This is better expressed as a percentage thus:

$$= \frac{\text{Difference between false and true reading}}{\text{true reading}} \times 100$$

$$= \frac{5.1 - 4.869}{4.869} \times 100$$

$$= 4.75\% \text{ (high)}$$

4.3. E.m.f. of battery $= 40 \times 1.9 = 76$V

or $E_b = 76$V

Internal battery resistance $= 40 \times 0.0025$

$$= 0.1\Omega$$

Total resistance of circuit $= 1 + 0.1 = 1.1\Omega = R$

For charging $V = E_b + IR$

Thus $90 = 76 + (I \times 1.1)$

or $I = \dfrac{90 - 76}{1.1} = \dfrac{14}{1.1}$ amperes

$= 12.72$A

4.4. Area of nickel deposited $= \pi \times 100 \times 10^{-3} \times 150 \times 10^{-3} = 0.0471 m^2$

Volume of nickel deposited $= 47.1 \times 10^{-3} \times 0.5 \times 10^{-3}$

$$= 23.55 \times 10^{-6} m^3$$

Mass of nickel deposited, $m = 23.55 \times 10^{-6} \times 8.6 \times 10^3$

$$= 0.202 \ 53 kg$$

Now $m = zIt$ so $0.202 \ 53 = 302 \times 10^{-9} \times I \times 8 \times 3600$

or $202.53 \times 10^{-3} = 30.2 \times 8 \times 36 \times 10^{-6} I$

giving $I = \dfrac{202.53 \times 10^3}{30.2 \times 8 \times 36} = 23.3A$

4.5. Discharge ampere hours $= 6 \times 12 = 72$

Charge ampere hours $= 6 \times 22 = 88$

Ampere hour efficiency $= \dfrac{72}{88}$

$$= 0.82 \text{ or } 82\%$$

Discharge watt hours per cell $= 6 \times 12 \times 1.2 = 86.4$

Charge watt hours per cell $= 4 \times 22 \times 1.5 = 132$

Watt hour efficiency $= \dfrac{86.4}{132} = 0.65 \text{ or } 65\%$

4.6. Battery voltage at start of charge $= 80 \times 1.8 = 144V$

$$= E_{b1}$$

Battery voltage at end of charge $= 80 \times 2.4 = 192V$

$$= E_{b2}$$

No battery resistance is given and is thus neglected.

Let R_1 = control resistance at start of charge

Then $V = E_{b1} + IR_1$ or $230 = 144 + 5 R_1$

Thus $5R_1 = 230 - 144 = 86$

$$R_1 = \dfrac{86}{5} \ 17.2\Omega \text{ (maximum value)}$$

Let R_2 = control resistance at end of charge

Then $230 = 192 + 5 R_2$ or $5 R_2 = 230 - 192 = 38$

$$R_2 = \frac{38}{5} = 7.6\Omega \text{(minimum value)}$$

$$\text{Charging time} = \frac{60}{5} = 12h \text{ (approx.)}$$

(Leave charging say 13h to allow for losses.)

4.7. Area of deposit $= 2 \times 50 \times 150 = 15\,000mm^2$

Volume of deposit $= 15 \times 10^3 \times 10^{-6} \times 0.05 \times 10^{-3}$

$= 0.75 \times 10^{-6}m^3$

Mass of deposit, $m = 0.75 \times 10^{-6} \times 8800 = 0.0066kg$

Now $m = zIt$ so $I = \dfrac{m}{zt}$

or $I = \dfrac{6.6 \times 10^{-3}}{330 \times 10^{-9} \times 30 \times 60}$ Thus $I = 11.1A$

4.8. Discharge ampere hours $= 4 \times 40 = 160$

Charge ampere hours $= 8 \times 24 = 192$

Ampere hour efficiency of battery $= \dfrac{160}{192} = 0.833$ or 83.3%

Discharge watt hours $= 4 \times 40 \times 1.93 \times 40$

Charge watt hours $= 8 \times 24 \times 2.2 \times 40$

Watt hour efficiency of battery $= \dfrac{4 \times 40 \times 1.93 \times 40}{8 \times 24 \times 2.2 \times 40}$

$= 0.731$ or 73.1%

4.9. Voltage drop in leads $= 10 \times 1 = 10V$

Voltage drop due to battery internal resistance $= 10 \times 0.01 \times 30 = 3V$

At start of charge, if R_1 is the external resistance

Then $200 = (30 \times 1.85) + 10 + 3 + 10R_1$,

or $200 = 55.5 + 13 + 10R_1$,

and $200 - 68.5 = 10\,R_1$, thus $R_1 = \dfrac{131.5}{10} = 13.15\Omega$

At end of charge, if R_2 is the external resistance

Then $200 = (30 \times 2.2) + 10 + 3 + 10R_2$

or $200 = 66 + 13 + 10R_2$

and $200 - 79 = 10R_2$ thus $R_2 = \dfrac{121}{10} = 12.1\Omega$

At start 13.15Ω are needed

At end 12.1Ω are needed

4.10. Here $m = zIt$ or $t = \dfrac{m}{zI} = \dfrac{4.2 \times 10^{-3}}{330 \times 10^{-9} \times 3.5}$ seconds

Thus $t = 3636s = 60.6$ min, i.e. 60 min 36s

From the second law of electrolysis or by proportion:

$$\frac{\text{Mass of hydrogen liberated}}{\text{Mass of copper liberated}} = \frac{\text{Chemical equivalent of hydrogen}}{\text{Chemical equivalent of copper}}$$

Thus mass of hydrogen $= \dfrac{1 \times 4.2}{31.8} = 0.1321g$

Chapter 5

5.1. $F = BI\ell$ newtons

$= 0.25 \times 100 \times 1 = 25$ newtons per metre length

5.2. M.m.f. $F = 4 \times 250 = 1000At$

(a) Magnetising force $H = \dfrac{F}{\ell} = $ m.m.f. per metre length

$$= \frac{1000}{500 \times 10^{-3}} = 2000At/m$$

(b) Flux density $B = \mu_0 \times H = 4 \times \pi \times 10^{-7} \times 2000$

$$= 8 \times \pi \times 10^{-4} \text{ teslas}$$

Cross-sectional area of ring $= 400 \times 10^{-6}$ m^2

∴ Flux $\Phi = B \times A = 8 \times \pi \times 10^{-4} \times 400 \times 10^{-6}$

$= 1.0048 \times 10^{-6}$ webers or $1.005\mu\text{Wb}$

5.3. M.m.f. F produced $= 3200 \times 1$

$= 3200\text{At}$

The magnetising force H or m.m.f. $/ m = \dfrac{F}{\ell} = \dfrac{3200}{800 \times 10^{-3}}$

$= 4000\text{At/m}$

Also since $B = \mu_0 \times H$

$3 = 4 \times \pi \times 10^{-7} \times 4000$

$= 16 \times \pi \times 10^{-4}$ teslas

Area of solenoid $= \dfrac{\pi \times c'^2}{4} = \dfrac{\pi \times 20^2 \times 10^{-6}}{4} = \pi \times 10^{-4}$

So $\Phi = B \times A = 16 \times \pi^2 \times 10^{-8}$ Webers

5.4. Magnetising force H of a long, straight conductor

$= \dfrac{I}{2\pi r} = \dfrac{2000}{2 \times \pi \times 0.8}$ ampere-turns / metre

or H at conductor X, due to current in Y,

$= \dfrac{1000}{0.8 \times \pi}$ ampere-turns / metre

and B at conductor X due to current in $Y = \mu_0 \times H$

$= \dfrac{4 \times \pi \times 10^{-7} \times 1000}{0.8 \times \pi} = \dfrac{10^{-3}}{2}$ teslas

So $F = BI\ell$ newtons

$= \dfrac{10^{-3}}{2} \times 2000 \times 1 = 1$ newton/metre length

5.5. Current to give f.s.d.

$$= \frac{50 \times 10^{-3}}{10} = 5 \times 10^{-3}\,\text{A}$$

Force exerted on 1 conductor

$$= BI\ell = 0.1 \times 5 \times 10^{-3} \times 25 \times 10^{-3}$$

$$= 12.5 \times 10^{-6}\ \text{newtons}$$

Force exerted on all conductors on both sides of the coil

$$= 100 \times 2 \times 12.5 \times 10^{-6}$$

$$= 2500 \times 10^{-6}\ \text{newtons}$$

Torque exerted by coil

$$= \text{Force} \times \text{radius}$$

$$= 2500 \times 10^{-6}\, \frac{30}{2} \times 10^{-3}$$

$$= 37.5 \times 10^{-6}\ \text{Nm}$$

Therefore the controlling torque of the spring

$$= 37.5 \times 10^{-6}\text{Nm}$$

$$\text{or} = 37.5\,\mu\text{Nm}$$

5.6. Flux density B in air gap $= \dfrac{0.05}{650 \times 10^{-6}}$

$$= \frac{5 \times 10^{2}}{6.5}\ \text{teslas}$$

Also $B = \mu_0 \times H$ ∴ $H = \dfrac{B}{\mu_0} = \dfrac{5 \times 10^{2}}{6.5 \times 4 \times \pi \times 10^{-7}}$

$$= 6.12 \times 10^{7}\ \text{ampere-turns/metre}$$

Air gap $= 3\text{mm} = 3 \times 10^{-3}\,\text{m}$

∴ Required ampere-turns $= 6.12 \times 10^7 \times 3 \times 10^{-3}$

$= 183\ 600At$

5.7. $F = BI\ell$ newtons $= 0.6 \times 150 \times 1 = 90N/m$

Assuming current flows away from the observer, then the force acts right to left to move the conductor horizontally.

5.8. $F = BI\ell$ newtons $= 0.5 \times 25 \times 400 \times 10^{-3} = 5N$

5.9. Force on 1 conductor $= 0.6 \times 0.8 \times 250 \times 10^{-3}$

Force on 800 conductor $= 0.6 \times 8 \times 250 \times 10^{-3} \times 8 \times 10^2$

$$= 9.6 \times 10^2\ N$$

Torque on armature $= 9.6 \times 10^2 \times 100 \times 10^{-3} = 96Nm$

Power developed is given by $\dfrac{2\pi NT}{60}$ watts

$$= \frac{2 \times \pi \times 1000 \times 96}{60} = 10.05kW$$

5.10. Magnetising force H of a long straight conductor

$$= \frac{I}{2\pi r}\ \text{ampere-turns / metre}$$

$$= \frac{250}{2 \times \pi \times 25 \times 10^{-3}}$$

$$= \frac{10^4}{2 \times \pi}$$

Also $B = \mu_0 \times H$

or $B = 4 \times \pi \times 10^{-7} \times \dfrac{10^4}{2 \times \pi}$

$= 2 \times 10^{-3}$ teslas

Again $F = BI\ell = 2 \times 10^{-3} \times 250 \times 1$

$= 0.5$ newtons

Mutual force per metre rur $= 0.5N$

Chapter 6

6.1. (a) Total m.m.f., $F = 5 \times 500 = 2500\text{At}$

Mean circumference $= \pi d = \pi \times 300 \times 10^{-3}$

$$= 0.942\text{m}$$

So magnetising force, $H = \dfrac{F}{\ell} = \dfrac{2500}{0.942}$

$$= 2654\text{At/m}$$

(b) Since $\mu_0 = \dfrac{B}{H}$ then $B = \mu_0 H$

and $B = 4 \times \pi \times 10^{-7} \times 2654$

$$= 0.0033\text{T} = 3.3\text{mT}$$

(c) Total flux, $\Phi = BA$

$$= 0.0033 \times 1000 \times 10^{-6} \text{ weber}$$

$$= 3.3 \times 10^{-6}\text{Wb or} = 3.3\mu\text{Wb}$$

6.2. $B = \dfrac{\Phi}{A} = \dfrac{500 \times 10^{-6}}{400 \times 10^{-6}} = 1.25\text{T}$

Also since $B = \mu H = \mu_r \mu_0 H$ then

$$H = \frac{1.25}{2500 \times 4 \times \pi \times 10^{-7}} = \frac{1.25}{\pi \times 10^{-3}} \text{ ampere-turns/metre}$$

Length of iron $= 250 \times 10^{-3}$ m

So total m.m.f., $F = \dfrac{250 \times 10^{-3} \times 1.25}{\pi \times 10^{3}}$

or $F = 99.7\text{At}$

Required ampere-turns $= 99.7$, say 100.

6.3. (a) $B = \dfrac{\Phi}{A}$ then $B = \dfrac{400 \times 10^{-6}}{500 \times 10^{-6}} = \dfrac{400}{500} = 0.8T$

Also, as H is given by $\dfrac{\text{total magnetomotive force}}{\text{length}}$

Then $H = \dfrac{F}{\ell} = \dfrac{500}{1} = 500 \text{At} / \text{m}$

Also, since $B = \mu H = \mu_o \, \mu_r \, h$ then

$\mu_r = \dfrac{B}{\mu_o H} = \dfrac{0.8}{4 \times \pi \times 10^{-7} \times 500}$

or relative permeability = 1275

(b) Reluctance $= \dfrac{\text{Length}}{\mu \times \text{Area}} = \dfrac{\ell}{\mu_o \mu_r \times A}$ ampere-turns / weber

$\qquad = \dfrac{1}{4 \times \pi \times 10^{-1} \times 1275 \times 500 \times 10^{-6}} = 1.25 \text{MA} / \text{Wb}$

6.4. (a) $H =$ ampere-turns per metre $= \dfrac{F}{\ell} = \dfrac{400 \times 2.5}{1.25}$

Thus $H = \dfrac{1000}{1.25} = 800 \text{At/m}$

Also $B = \dfrac{\phi}{A} = \dfrac{0.000\ 75}{1500 \times 10^{-6}} = 0.5 \text{T}$

Again $B = \mu H$ or $\mu = \dfrac{B}{H} = \dfrac{0.5 \times 1.25}{1000}$

Also $\mu = \mu_r \, \mu_o$

$\therefore \ \mu_r = \dfrac{\mu}{\mu_o} = \dfrac{0.625}{1000 \times 4\pi \times 10^{-7}}$

Thus relative permeability = 497.5

(b) Reluctance, $S = \dfrac{\ell}{\mu A} = \dfrac{1.25 \times 10^{-3}}{0.525 \times 1500 \times 10^{-6}}$ ampere-turns / weber $= 1.33 \text{MA} / \text{Wb}$

(c) Since $F = H\ell = 800 \times 1.25 = 1000 \text{At}$

6.5. Area of air gap $= 1200 \times 10^{-6} \text{ m}^2$

$\therefore \ B_A = \dfrac{1.13 \times 13^{-3}}{12 \times 10^{-4}} = \dfrac{11.3}{12} \text{ teslas}$

Also since $B = \mu_o H$

Then H for air $= \dfrac{B}{\mu_0} = \dfrac{11.3}{12 \times 4\pi \times 10^{-7}}$

$= 75 \times 10^4$ ampere-turns/metre

The m.m.f. for the air gaps is given by: $75 \times 10^4 \times 2 \times 2 \times 10^{-3}$ ampere-turns = 3000At

The area of iron is the same as for the air gaps, and the B value of the iron is the same.

or $B_1 = \dfrac{1.13 \times 10^{-3} \times 10^4}{12} = 0.942\text{T}$

Using the graph of figure 10, for a flux density of 0.942T, the ampere-turns per metre length of the iron = 850.

Since length of iron path = 0.6m

∴ M.m.f. for iron = $0.6 \times 850 = 6 \times 85 = 510$At

Total m.m.f. required = 3000 + 510 or 3510At

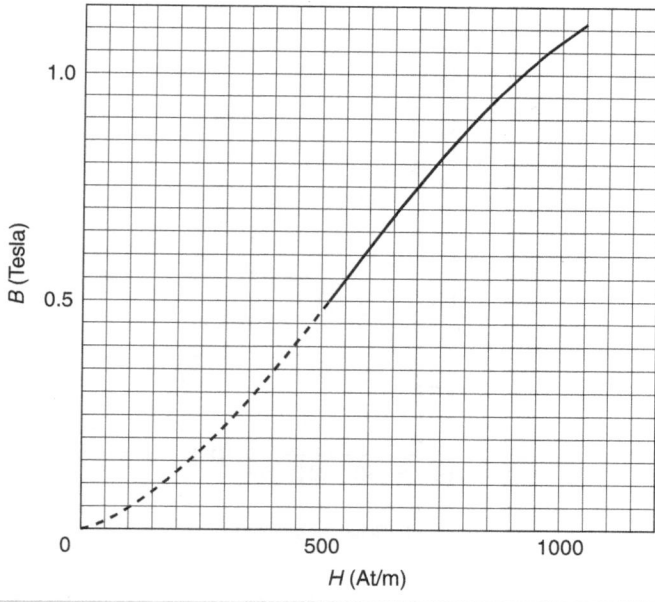

▲ **Figure 10** *B-H characteristics*

6.6. Circumference of flux path = $\pi \times 0.2 = 0.628$m

Length of air gap = $2 \times 10^{-3} = 0.002$m

Length of iron = 0.626m

This problem is best solved by trial and error thus:

Assume a flux density of 0.5T in the air gap and iron since these are of the same cross-sectional area. Then from the graph of figure 10:

M.m.f. for iron = 520 × 0.625 = 326At

Since $B = \mu_0 H$ ∴ $H = \dfrac{B}{\mu_0} = \dfrac{0.5}{4\pi \times 10^{-7}}$

$$\frac{0.5 \times 2 \times 10^{-3}}{4\pi \times 10^{-7}} = 795At$$

Total m.m.f. would be (326 + 795) = 1121At. Thus too low a flux density has been assumed.

Again, assume a flux density 0.6T, then:

M.m.f. for iron = 585 × 0.626 = 365.21At

M.m.f. for air $= \dfrac{0.6 \times 2 \times 10^{-3}}{4\pi \times 10^{-7}}$ or $\dfrac{6}{5}$ of that required for 0.5T

$$= \frac{6}{5} \times 795 = 954At$$

Total m.m.f. would be 365.2 + 954 = 1319At – still too low. Assume a flux density of 0.7T. Then:

M.m.f. for iron = 660 × 0.626 = 413.16At

M.m.f. for air $= \dfrac{7}{5} \times 795 = 1113At$

Total m.m.f. would be 413 + 1113 = 1526At

Thus for an exciting ampere-turn value of 3 × 500 = 1500, the estimated flux density in the air gap would be a little less than 0.7T.

6.7. As the *B* value in the cores is 1.2T the At/m required will be 650, as seen from the graph of figure 11. The total m.m.f. for the cores will be:

$$2 \times 160 \times 10^{-3} \times 650 = 208At.$$

In the yokes, the flux is the same as that for the cores and the flux density will therefore be different, as the areas are different.

Thus flux, $\Phi = 1.2 \times \dfrac{\pi}{4} \times 50^2 \times 10^{-6}$ weber

B in yokes $= 1.2 \times \dfrac{\pi}{4} \times \dfrac{25 \times 10^{-4}}{47 \times 47 \times 10^{-6}}$ teslas

$\therefore B = 1.066\text{T}$

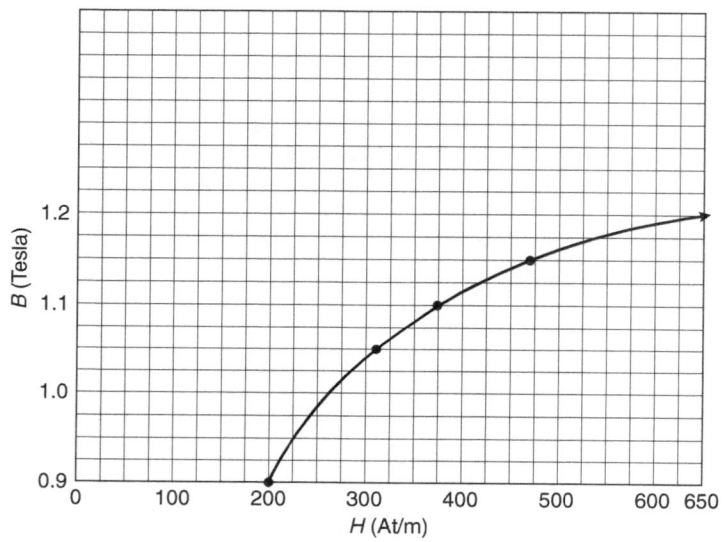

▲ **Figure 11** *B-H characteristics*

From the graph, the At/m for a density of 1.066T is 330 mean length of flux path in yokes $= (2 \times 130) + (2 \times 47) = 354\text{mm} = 0.354\text{m}$

Total m.m.f. for the yokes $= 330 \times 354 \times 10^{-3} = 116.8\text{At}$

Total m.m.f. for complete magnetic circuit $= 208 + 116.8 = 324.8$ say 325At

6.8. (a) Pull of magnet $= 196.2\text{N}$ or 98.1N per contact face

Also Pull $= \dfrac{B^2 A}{2\mu_0} = \dfrac{B^2 A}{2 \times 4\pi \times 10^{-7}}$ newtons

$\therefore 98.1 = \dfrac{B^2 A}{8\pi \times 10^{-7}}$

Whence $B^2 = \dfrac{98.1 \times 8p \times 10^{-7} \times 4}{\pi \times 15 \times 15 \times 10^{-6}}$

$= 1.396$

and $B = \sqrt{1.396} = 1.185T$

From the graph of figure 11, for a B value of 1.185T, the H value = 560At/m.

Area of one contact face $= \dfrac{\pi}{4} \times 15^2 \times 10^{-6} \text{ m}^2$

Flux, $\Phi = BA = 1.185 \times \dfrac{\pi}{4} \times 225 \times 10^{-6}$ weber

$= 2.09 \times 10^{-4}$ Wb

Since B for the horseshoe magnet = 1.185T

then, from figure 11, the H value = 560 At/m

Length of magnet path $= \pi \times \dfrac{115}{2} = 180.5\text{mm} = 0.1805\text{m}.$

Note. Mean circumference of ring

$= 2\pi \times$ (radius of ring)

$= 2\pi \times$ (50 + radius of rod)

$= 2\pi \times (50 + 7.5)$

$= \pi \times 115$ millimetres

Then m.m.f. for magnet = 550 × 0.1805 = 101.1 At

B value for armature $= \dfrac{\text{Flux}}{\text{Area}} = \dfrac{2.09 \times 10^{-4}}{15 \times 10^{-6}} = 0.932T$

and H value = 215At/m (from figure 11)

Length of armature path = (115 + 15) = 130mm

$= 130 \times 10^{-3}$ m

So m.m.f. required = 215 × 130 × 10⁻³ = 27.95At

Total m.m.f. required = 101.1 + 27.95 = 129At

Current $= \dfrac{129}{480} = 0.268A$

(b) In the air gap $B = 1.15T$

∴ Flux, $\Phi = 1.15 \times \dfrac{\pi}{4} \times 15^2 \times 10^{-6} = 2.03 \times 10^{-4}$ weber

Flux density in core $= \dfrac{2.03 \times 10^{-4}}{\dfrac{\pi \times 15^2}{4} \times 10^{-6}} = 1.15\text{T}$

and from curve, H value $= 470\text{At/m}$

Length of core path $= \pi \times 57.5 = 180.5\text{mm} = 0.1805\text{m}$

M.m.f. for core $= 0.1805 \times 470 = 84.8\text{At}$

Flux density in armature $= \dfrac{2.03 \times 10^{-4}}{15^2 \times 10^{-6}} = 0.905\text{T}$

From curve H value $= 205\text{At/m}$

Length of armature path $= 130\text{mm} = 0.13\text{m}$

M.m.f. for armature $= 0.13 \times 205 = 26.65\text{At}$

Flux density in 1 air gap $= 1.15\text{T}$, but $B = \mu_o H$

$\therefore H = \dfrac{B}{\mu_o} = \dfrac{1.15}{4\pi \times 10^7} = \dfrac{1.15 \times 10^7}{12.56}$ ampere-turns / metre

M.m.f. for 2 air gaps

$$= \dfrac{2 \times 0.5 \times 10^{-3} \times 1.15 \times 10^7}{12.56}$$

$$= 9.125 \times 10^2 \text{ ampere-turns} = 912.5\text{At}$$

Total m.m.f. for circuit $= 84.8 + 26.65 + 912.5$

$$= 1023.95 \text{ At}$$

6.9. (a) Area of air gap $= \dfrac{\pi 100^2}{4} \times 10^{-6}$

$$= \dfrac{\pi \times 10^{-2}}{4} \text{ m}^2$$

Volume of air gap $= \dfrac{\pi \times 10^{-2}}{4} \times 2.5 \times 10^{-3}$

$$= \dfrac{\pi}{16} \times 10^{-4} \text{m}^3$$

Flux density in gap $= \dfrac{0.004 \times 4}{\pi \times 100^2 \times 10^{-6}} = 0.508\text{T}$

Energy stored in joules $= \dfrac{B^2}{2\mu_o} \times \text{volume}$

$$= \frac{0.508^2}{2 \times 4\pi \times 10^{-7}} \times \frac{\pi}{16} \times 10^{-4} = 2\text{J}$$

(b) Pull (newtons) $= \dfrac{B^2 A}{2\mu_o} = \dfrac{0.508^2 \times \pi \times 10^{-2}}{2 \times 4\pi \times 10^{-7} \times 4} = 806\text{N}$

6.10. Air gap. Useful flux = 0.05Wb/pole

Flux density in air gap $= \dfrac{0.05}{60000 \times 10^{-6}} = 0.833\text{T}$

Also $B = \mu_o H$

$\therefore H$ value for air $= \dfrac{B}{\mu_o} = \dfrac{0.833}{4\pi \times 10^{-7}}$

$= 66.2 \times 10^4$ At/m

M.m.f. for air gap $= 66.2 \times 10^4 \times 5 \times 10^{-3} = 3310$At

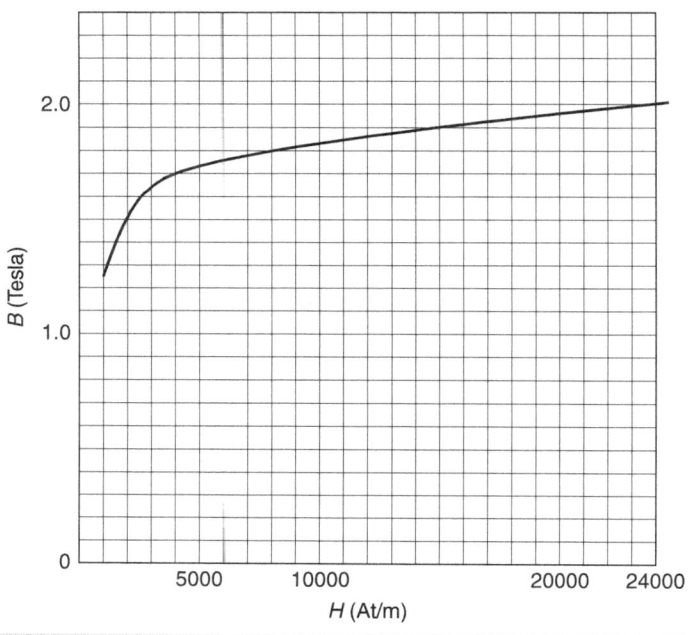

▲ **Figure 12** *B-H characteristics*

Pole

Total flux = 0.05 × 1.2 = 0.06Wb

Flux density in pole $= \dfrac{0.06}{40000 \times 10^6}$ or $B = 1.5T$

From the magnetic characteristic, plotted for figure 12, a flux density (B) of 1.5T gives an H value of 2000At/m.

Thus m.m.f. required for pole = 2000 × 250 × 10^{-3} = 500At

Teeth

Total flux = 0.05Wb (same as the gap)

Flux density in teeth $= \dfrac{0.05}{25000 \times 10^{-6}} = 2T$

From the characteristic, a flux density of 2T gives an H value of 24000At/m.

Thus m.m.f. required for teeth = 24 000 × 45 × 10^{-3}

$$= 1080At$$

Total field coil m.m.f. = 3310 + 500 + 1080 = 4890At

Chapter 7

7.1. Dynamic Induction $E = B\ell v$

$$= 40 \times 10^{-6} \times 1.4 \times \frac{100 \times 10^3}{3600} \text{ volts} = 1.55 \times 10^{-3} \text{ volts}$$

$$= 0.001\ 55V \text{ or } 1.55mV$$

7.2. In 1 revolution, the flux cut = 6 × 64 × 10^{-3} webers

The number of revolutions per second $= \dfrac{1000}{60} = \dfrac{100}{6}$

∴ Flux cut per second $= 6 \times 64 \times 10^{-3} \times \dfrac{10^2}{6}$

$$= \text{generated voltage} = 6.4V$$

The conductors in series per parallel path $= \dfrac{468}{6} = 78$

So voltage per parallel path $= 6.4 \times 78 = 499.2V$

$\qquad\qquad\qquad = $ the e.m.f. of the machine

Current per conductor $=$ current per parallel path $= 50A$

Current in 6 parallel paths $= 50 \times 6 = 300A$

So power developed $= \dfrac{499.2 \times 300}{1000} = 149.76kW$

Check $E = \dfrac{Z\Phi N}{60} \times \dfrac{P}{A}$

Note. This formula is introduced in Chapter 12. Its use is shown here.

Thus $E = \dfrac{468 \times 64 \times 10^{-3} \times 1000}{60} \times \dfrac{6}{6} = 499.2V$

7.3. Change of flux $= (30 - 2)\, 10^{-3}$ weber $= 28mWb$

Induced e.m.f. $=$ rate of change of flux-linkages

$$= \dfrac{2000 \times 28 \times 10^{-3}}{0.12} = 466.6V$$

7.4. Diameter of armature $= 0.2m$

Circumference $= \pi d = \pi \times 2 \times 10^{-1} = 0.628m$

In 1 second the armature turns $\dfrac{500}{60}$ revolutions

$\therefore$ In 1 second a coil side travels $\dfrac{500}{60} \times 0.628$ metres

or $v = 5.233m/s$

So $E = B\ell v = 1.2 \times 2 \times 0.4 \times 5.233$ volts $= 5.024V$

7.5. Let E_1, Φ_1 and N_1 be the values under the original conditions and E_2, Φ_2 and N_2 be the values under the final condition. Here N_1 and N_2 are the speed conditions.

Also for a generator $E \propto \Phi N$ or $E = k\Phi N$

We know that $\Phi \propto B$ and $N \propto v$ thus the more general form $E \propto \Phi N$ (introduced later) can be used. k is a constant.

$$\dfrac{E_2}{E_1} = \dfrac{k\Phi_2 N_2}{k\Phi_1 N_1} \quad \text{or} \quad E_2 = \dfrac{E_1\Phi_2 N_2}{\Phi_1 N_1}$$

and $E_2 = \dfrac{200 \times 19 \times 10^{-3} \times 1100}{20 \times 10^{-3} \times 1000}$ $E_2 = 209V$

7.6. Diameter of coil = 0.2m

Circumference of coil = $\pi \times 0.2 = 0.628$m

$$\text{Speed} = \frac{1200}{60} = 20\text{rev}/\text{s}$$

∴ velocity of coil side = $20 \times 0.628 = 12.56$m/s

Now $E = B\ell v$ volts

$$= 0.02 \times 0.3\ 0012.56 \text{ volts per conductor}$$

Total e.m.f. = $2 \times 3 \times 10^{-3} \times 12.56 \times 400 = 30.1$V

This is true if the conductors cut the field at right angles and so the maximum value of e.m.f. generated is 30.1V

Time for 1 revolution = $\frac{1}{20}$ seconds. In one revolution one cycle is generated, so the frequency of the generated e.m.f. = 20 cycles per second or 20 Hertz

7.7. Average e.m.f. in volts = rate of change of flux-linkages

Change of flux = $(4 - 1.5)10^{-3} = 2.5 \times 10^{-3}$ webers

Time for change = 0.04s

So rate of change of flux-linkages = $1200 \times \dfrac{2.5 \times 10^{-3}}{4 \times 10^{-2}} = 75$V

7.8. Flux per pole = $0.09 \times 0.92 = 8.28 \times 10^{-2}$ webers

In one revolution a conductor cuts $4 \times 8.28 \times 10^{-2}$

$$= 33.12 \times 10^{-2} \text{ webers}$$

Also in 1 second the armature revolves $\dfrac{600}{60} = 10$ times

So by 1 conductor, the flux cut per second

$$= 33.12 \times 10^{-2} \times 10 = 3.312\text{Wb}$$

and the induced voltage per conductor = 3.312V

The number of armature conductors is $2 \times 210 = 420$

These are arranged in 4 parallel paths.

There are thus $\dfrac{420}{4} = 105$ conductors in series per parallel path

The e.m.f. of 1 parallel path $= 105 \times 3.312 = 347.76$V

The generated e.m.f. = the e.m.f. of 1 parallel path $= 347.76$V

7.9. (a) Ampere-turns of solenoid $= 400 \times 6 = 2400$At

The magnetising force H at the centre = ampere-turns/metre

$$= \frac{2400}{1.5} = 1600 \text{At/m}$$

The flux density B at the centre of the solenoid and small coil

$$= \mu_0 H = 4 \times \pi \times 10^{-7} \times 1600 = 64 \times \pi \times 10^{-5}$$

Area of small coil $= \dfrac{\pi d^2}{4} = \dfrac{\pi \times (10 \times 10^{-3})^2}{4} = \dfrac{\pi \times 10^{-4}}{4}$ m^2

So the flux linked $= 64 \times \pi \times 10^{-5} \times \dfrac{\pi \times 10^{-4}}{4}$ weber $= 0.158\mu$Wb

(b) Average induced e.m.f. = rate of change of flux-linkages

$$= \frac{50 \times 16 \times \pi^2 \times 10^{-9}}{50 \times 10^{-3}} = 0.158 \text{mV}$$

7.10. Coil A. Associated flux $\Phi = 18 \times 10^{-3}$ webers

Associated flux-linkages during reversal = turns × flux decrease to zero and then its build-up to full value Φ in the *reversed* direction.

$$= 1000 \,[0.018 - (-0.018)] = 1000(0.018 + 0.018)$$

$$= 1000 \times 2 \times 0.018 = 36 \text{ weber-turns}$$

Time of reversal $= 0.1$s

and the induced e.m.f. = rate of change of flux-linkages

$$= \frac{36}{0.1} = 360\text{V (average value)}$$

Coil B. Only 80% flux is associated and there are 500 turns.

∴ Associated flux-linkages during reversal

$$= 500 \times 0.8 \times 2 \times 0.018 = 14.4 \text{ weber-turns}$$

Induced e.m.f. $= \dfrac{14.4}{0.1} = 144\text{V} \left(\text{average value}\right)$

Alternatively:

Proportion of e.m.f. in Coil B to e.m.f. in Coil A

$$= 360 \times \dfrac{500}{1000} = 180\text{V, if full flux is associated}$$

For only 80% flux, e.m.f. is reduced in proportion

$$= 180 \times 0.8 = 144\text{V}$$

Chapter 8

8.1. For a series combination, the equivalent capacitance is given

By C, where $\dfrac{1}{C} = \dfrac{1}{0.02} + \dfrac{1}{0.04}$ (in microfarads)

So $C = \dfrac{0.04}{3} = 0.0133 \mu\text{F}$ Also $Q = CV$

∴ $Q = 0.0133 \times 10^{-6} \times 10^2$ coulombs

Then $V_1 = \dfrac{1.33 \times 10^{-6}}{0.02 \times 10^{-6}} = 66.7\text{V}$

and $V_2 = \dfrac{1.33 \times 10^{-6}}{0.04 \times 10^{-6}} = 33.3\text{V}$

The voltage drops are respectively 66.7V and 33.3V

8.2. The final 2 parallel $5\mu F$ capacitors are equivalent to 1 unit of $10\mu F$.

The capacitance C of the branch consists of: $20\mu F$, $10\mu F$ and $20\mu F$ in series given by:

$$\frac{1}{C} = \frac{1}{20} + \frac{1}{10} + \frac{1}{20} \text{ or } C = 5\mu F$$

The series circuit is in parallel with a $5\mu F$ capacitor with an equivalent capacitance $= 10\mu F$. The final arrangement between A and B is equivalent to a $20\mu F$, $10\mu F$ and $20\mu F$ capacitor in series. The equivalent capacitance is given by:

$$\frac{1}{C} = \frac{1}{20} + \frac{1}{10} + \frac{1}{20} \text{ or } C = \frac{20}{4} = 5\mu F$$

8.3. Since $Q = CV$. $\therefore$ the quantity of electricity received *initially* is given by $Q = 1000 \times 10^{-6} \times 100 = 10^5 \times 10^{-6}$

$$= 10^{-1} \text{ coulombs.}$$

Since the plates are separated by an insulated rod there is no loss of charge and hence Q remains the same.

Under the new condition since, as before, $Q = CV$

Then $V = \dfrac{Q}{C} = \dfrac{10^{-1}}{300 \times 10^{-6}} = 333.3V$

Hence the P.D. will increase by $333.3 - 100 = 233.3V$

8.4. The capacitor is made from 10 plates in parallel, making 1 assembly, interleaved with 9 plates in parallel forming the other plate assembly. There will be 18 mica separators or 18 electric fields and the total capacitance will be 18 times the capacitance between 1 pair of plates.

Thus C of 1 pair of plates $= \dfrac{\in A}{d} = \dfrac{\epsilon_o \epsilon_r A}{d}$

or $C = \dfrac{8.85 \times 10^{-12} \times 7 \times 2580 \times 10^{-6}}{0.1 \times 10^{-3}}$

$$= 1.6 \times 10^{-9} \text{ farads}$$

or with 18 units in parallel $C = 18 \times 1.6 \times 10^{-9}$ farads

$$= 0.0288\mu F$$

8.5. Since $Q = CV$, then $Q = 3 \times 10^{-4} \times 10^{-6} \times 10 \times 10^{3}$

$$= 3 \times 10^{-6} \text{ coulombs}$$

$$\therefore \text{ Flux density, } D = \frac{Q}{A} = \frac{3 \times 10^{-6}}{10000 \times 10^{-6}}$$

$$= 3 \times 10^{-4} \text{ coulomb per m}^2$$

$$\text{Also, permittivity, } \in = \frac{\text{electricity flux density}}{\text{electric force}} = \frac{D}{E}$$

$$\text{And electric force, } E = \frac{V}{d} = \frac{10 \times 10^{3}}{1 \times 10^{-3}} 10 \times 10^{6} \text{ volts m}^{-1}$$

$$\text{Hence } \in = \frac{3 \times 10^{-4}}{10 \times 10} \text{ also } \in = \in_{o} \times \in_{r}$$

$$\text{And } \in_{r} = \frac{\in}{\in_{o}} = \frac{3 \times 10^{-4}}{10 \times 10^{6} \times 8.85 \times 10^{-12}}$$

$$\text{or } \in_{r} = 3.39$$

8.6. $C = \dfrac{\in A}{d}$ $A = 6 \times 10^{4} \times 10^{-6} \text{m}^2 = 6 \times 10^{-2} \text{ m}^2$

$$d = 3.5 \times 10^{-3} \text{ m}$$

and $\in = \in_{o} \times \in_{r} = 8.85 \times 10^{-12} \times 3$

$$\text{Hence } C = \frac{8.85 \times 10^{-12} \times 3 \times 6 \times 10^{-2}}{3.5 \times 10^{-3}} = 4.55 \times 10^{-10} \text{ F}$$

Energy, $W = \dfrac{1}{2} CV^2$ joules

$$= \frac{1}{2} \times 4.55 \times 10^{-10} \times 300^2$$

$$= 20.475 \times 10^{-6} \text{ joules} = 20.48 \mu J$$

8.7. A 10-plate capacitor is made from two 5-plate assemblies interleaved with each other and separated by the dielectric. There are thus 9 electric fields or the final capacitance is 9 times that of 1-plate arrangement.

Hence $C = \dfrac{\epsilon A}{d}$ where $A = 1500 \times 10^{-6}\,m^2$

$$d = 0.3 \times 10^{-3}\ metres$$

$$\epsilon = \epsilon_o \times \epsilon_r$$

$$C = \frac{8.85 \times 10^{-12} \times 4 \times 15 \times 10^{-4}}{3 \times 10^{-4}} = 1.77 \times 10^{-10}\ farads$$

$$= 1.77 \times 10^{-4}\ microfarads$$

Total capacitance $= 9 \times 1.77 \times 10^{-4} = 0.0016\,\mu F$

8.8. Let $C =$ capacitance of the series arrangement,

then $\dfrac{1}{C} = \dfrac{1}{20} + \dfrac{1}{30} = \dfrac{5}{60}$ or $C = 12\,\mu F$

The charge stored is given by $Q = CV = 12 \times 10^{-6} \times 600$

$$= 72 \times 10^{-4}\ coulombs.$$

P.D. across $20\,\mu F$ capacitor $A = \dfrac{7.2 \times 10^{-3}}{20 \times 10^{-6}} = 360V$

P.D. across $30\,\mu F$ capacitor $B = 600 - 360 = 240V$

If P.D. across B is 400V then P.D. across the parallel arrangement will be 200V. The equivalent capacitance must be $60\,\mu F$ (double) since the voltage is half that across B. C must be $30\,\mu F$, being in parallel with A.

Also the energy stored, $w = \dfrac{1}{2}CV^2$

$$= \frac{1}{2} \times 40 \times 10^{-6} \times 200^2\ joules.\ Thus\ w = 0.8J$$

8.9. Since

$Q = CV$ and $Q = It$ then $It = CV$

or $I = C\,\dfrac{V}{t}$ where $V =$ the voltage change then,

a. $I = 40 \times 10^{-6} \times \dfrac{100}{1 \times 10^{-3}}\ amperes = 4A$

b. $I = 40 \times 10^{-6} \times \dfrac{50}{1 \times 10^{-3}}$ amperes $= 2A$

c. $I = 40 \times 10^{-6} \times \dfrac{0}{1 \times 10^{-3}}$ amperes $= 0A$

d. $I = 40 \times 10^{-6} \times \dfrac{100}{1 \times 10^{-3}}$ amperes $= 4A$

e. $I = 40 \times 10^{6} \times \dfrac{50}{1 \times 10^{-3}}$ amperes $= 2A$

Figure 13 shows the current and voltage conditions.

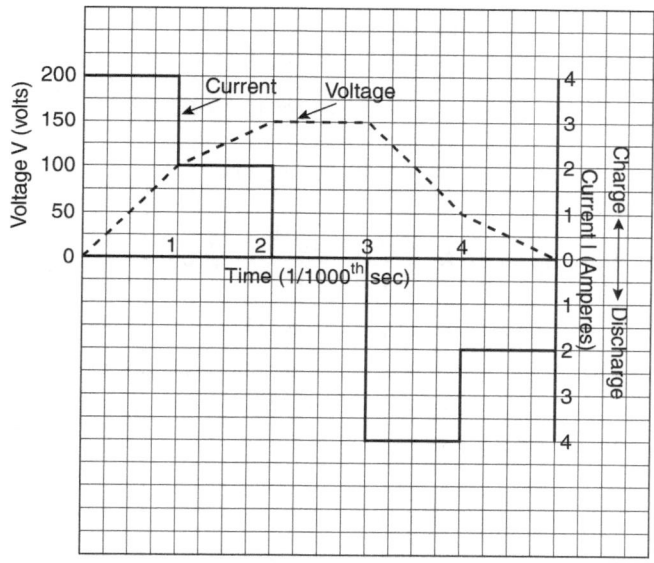

▲ **Figure 13** *Voltage and current vs time for charging and discharging circuit*

8.10. The mean diameter of the insulation $= 10 + 12 = 22mm$

The dielectric area = mean circumference × length

$$= \pi d \times 1000 \ m^2$$

$$= \pi \times 22 \times 10^{-3} \times 10^3$$

$$= 69.1 \ m^2$$

Also, from the above,

$$D = \frac{CV}{A} = \frac{0.289 \times 10^6 \times 11 \times 10^3}{69.1} \text{ coulomb m}^{-2}$$

$$= 46 \times 10^{-6} \text{ C/m}^2$$

Again $\in = \dfrac{D}{E} = \dfrac{46 \times 10^{-6}}{11 \times 10^5}$

Also $\in = \in_0 \times \in_r$

$$\therefore \in_r = \frac{46 \times 10^{-6}}{11 \times 10^{-5} \times 8.85 \times 10^{-12}}$$

$$\therefore \in_r = 4.73$$

Chapter 9

9.1. A scale of 10mm = 1A is used and I_1 is the reference phasor drawn horizontally. The diagram, drawn geometrically to scale, shows the solution with I the resultant current = 9.23A lagging I_1, by 6°.

If the above is checked mathematically

$$I_H = 4 \cos 0 + 6 \cos 30 + 2 \cos 90$$

$$= (4 \times 1) + (6 \times 0.866) + (2 \times 0)$$

$$= 4 + 5.196 = 9.196 \text{A}$$

$$I_V = 4 \sin 0 - 6 \sin 30 + 2 \sin 90$$

$$= (4 \times 0) - (6 \times 0.5) + (2 \times 1) = 0 - 3 + 2$$

$$= -1 \text{A}$$

$$I = \sqrt{(9.196 \times 9.196 + 1)} = 9.24 \text{A}$$

$$\cos \theta = \frac{9.2}{9.24} = 0.995 \text{ and } \theta = 6° \text{ (approx.)}$$

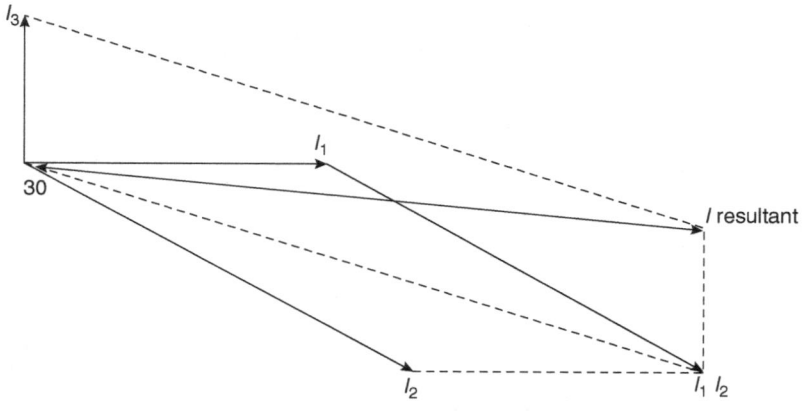

▲ **Figure 14** *Phasor diagram*

9.2. Since $v = V_m \sin (2\pi ft)$

$$\therefore \sin (2\pi ft) = \frac{200}{282.8} = 0.707$$

Now the angle whose sine is 0.707 is = 45°

$\therefore 2\pi ft = 45°$

or $t = \dfrac{45}{22 \times 180 \times 25}$

$= 0.005$ seconds $= 5$ms

(a) The first time is 5ms after zero value

(b) Time for 1 cycle $= \dfrac{1}{25} = 0.04$s

Time for $\dfrac{1}{2}$ cycle $= 0.02$s

The second time is $0.02 - 0.005 = 0.015$s or 15ms after zero value

9.3. A simple phasor diagram illustrates the problem and its mathematical solution.

Horizontal component $V_H = 100 + 80 \cos 60$

$$= 100 + 80 \times 0.5 = 140\text{V}$$

Vertical component $V_v = 0 - 80 \sin 60$

$$= 0 - 80 \times 0.866$$

$$= -69.28\text{V}$$

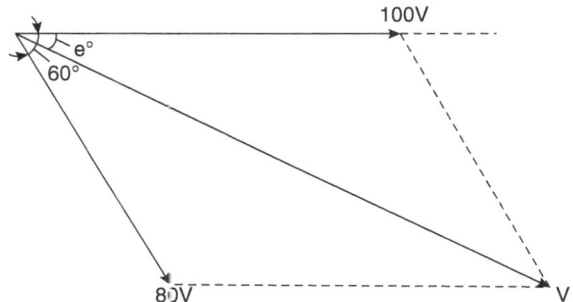

▲ **Figure 15** *Phasor diagram*

Resultant $V = \sqrt{140^2 + 69.28^2}$

$$= 156.3V$$

$\text{Cos } \theta = \dfrac{140}{156.3}\ 0.8955. \quad \therefore \quad \theta = 26°26'$

Since maximum values were used for the phasor the resultant is a maximum which lags the 100V values by 26°26'

9.4. The waveform is shown in figure 16. Erecting mid-ordinates, measuring and squaring these gives the following columns.

$i_1 = 0.22 \qquad i_1^2 = 0.05$

$i_2 = 0.60 \qquad i_2^2 = 0.36$

$i_3 = 0.92 \qquad i_3^2 = 0.85 \qquad$ Total of $i_2 = 19.17$

$i_4 = 1.25 \qquad i_4^2 = 1.56 \qquad$ Average of $i^2 = \dfrac{19.17}{10}$

$i_5 = 1.55 \qquad i_5^2 = 2.40 \qquad = 1.917$

$i_6 = 1.8 \qquad i_6^2 = 3.24 \qquad \therefore$ r.m.s. value $= \sqrt{1.917}$

$i_7 = 1.97 \qquad i_7^2 = 3.87 \qquad = 1.385A$

$i_8 = 1.92 \qquad i_8^2 = 3.68$

$i_9 = 1.56 \qquad i_9^2 = 2.44 \qquad$ Power dissipated obtained from

$i_{10} = 0.85 \qquad i_{10}^2 = 0.72 \qquad I^2R = 1.917 \times 8 = 15.34W$

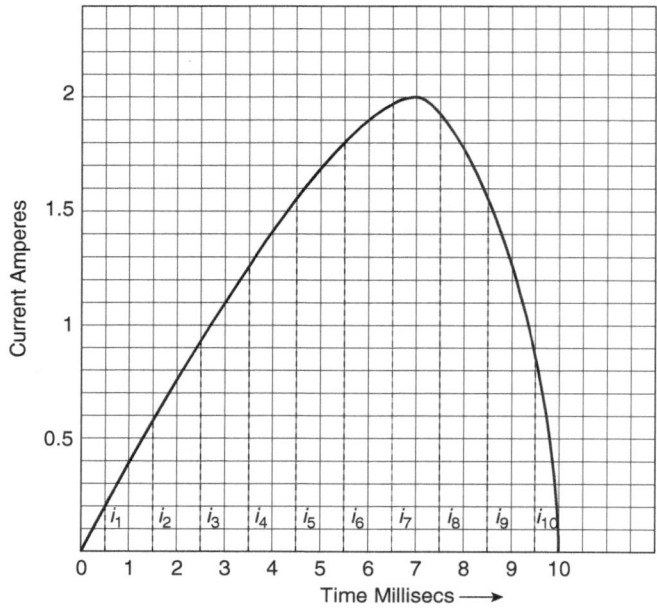

▲ **Figure 16** *Current vs time*

9.5. The phasor diagram is shown (figure 17), but not to scale with the resultant obtained mathematically. The second current is the reference along the horizontal.

Sum of horizontal components

$$I_H = 17.32 \cos 0 + 20 \cos 60 + 10 \cos 90$$
$$= (17.32 \times 1) + (20 \times 0.5) + (10 \times 0)$$
$$= 17.32 + 10 + 0 = 27.32A$$

Sum of vertical components

$$I_V = 17.32 \sin 0 - 20 \sin 60 + 10 \sin 90$$
$$= (17 + 0) - (20 \times 0.866) + (10 \times 1)$$
$$= 0 - 17.32 + 10 = -7.32A$$

$$\text{Resultant } I = \sqrt{27.32^2 + 7.32^2}$$
$$= \sqrt{799.4} = 28.32A$$

Cos $\theta = 0.965$ So $\theta = 15°12'$

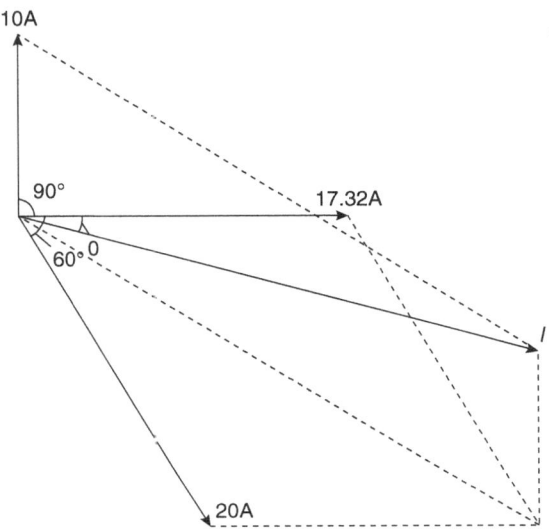

▲ **Figure 17** *Phasor diagram*

As peak or maximum values were used for the individual phasors, the maximum value of the resultant current is 28.32A lagging 105°12′ behind the 10A current.

9.6. For a sine wave voltage applied to a resistor, a sinusoidal current results, and the maximum value of this current is at the instant of maximum voltage.

$$\text{Maximum current} = \frac{\text{maximum voltage}}{\text{resistance}} = \frac{340}{24}$$

$$= 14.14\text{A}$$

R.m.s. value of current $= 0.707 \times 14.14 = 9.996\text{A}$

$$= 10\text{A}$$

9.7. The phasor diagram is drawn as shown in figure 18. For the construction a scale of 10mm = 25V is used. The phase relation between the various phasors are determined thus:

$$e_1 = 100 \sin \omega t$$

$$e_2 = 50 \cos \omega t = 50 \sin \left(\omega t + \frac{\pi}{2} \right)$$

$$= 50 \sin (\omega t + 90°)$$

$$e_3 = 75 \sin \left(\omega t + \frac{\pi}{3} \right) = 75 \sin(wt + 60°)$$

$$e_4 = 125 \cos \left(\omega t - \frac{2\pi}{3} \right)$$

$$= 125 \sin \left(\omega t - \frac{2\pi}{3} + \frac{\pi}{2} \right)$$

$$= 125 \sin \left(\omega t - \frac{\pi}{6} \right)$$

$$= 125 \sin (\omega t - 30°).$$

The diagram is drawn using the maximum value E_{1m} of the first voltage as the reference. For the diagram $E_{1m} = 100$V.

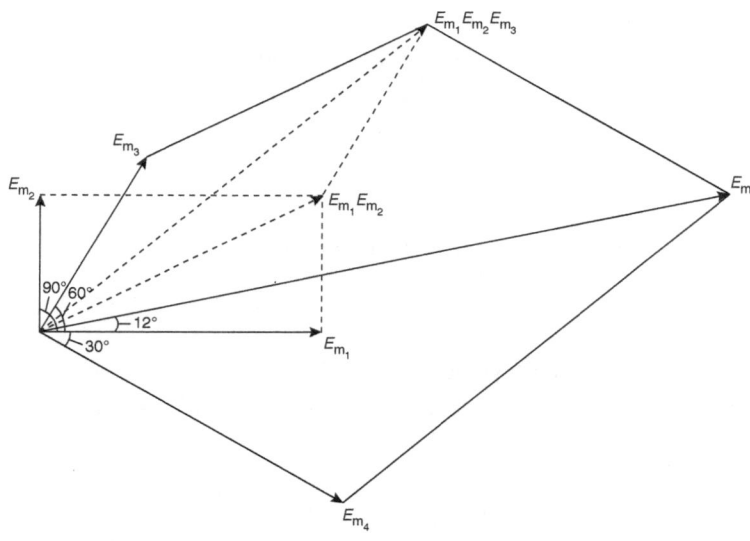

▲ **Figure 18** *Phasor diagram*

$E_{2m} = 50V$ $\qquad$ $E_{3m} = 75V$ $\qquad\qquad$ $E_{4m} = 125V$

For the resultant $E = 100.5mm = 252V$

The angle $\theta = 12°$ (approx.)

But $12° = \dfrac{180}{15}$ or $\dfrac{\pi}{15}$ radians (leading)

The required expression can be written:

$$e_1 + e_2 + e_3 + e_4 \text{ or } e = 252 \sin\left(\omega t + \frac{\pi}{15}\right)$$

9.8. (a) Alternators in step $V = 100 + 200 = 300.0V$

(b) When the phase displacement is 60°

$$V = \sqrt{100^2 + 200^2 + 2 \times 100 \times 200 \cos 60}$$

$$= \sqrt{10\,000 + 40\,000 + (40\,000 \times 0.5)}$$

$$= 264.8V$$

(c) When the phase displacement is 90°

$$V = \sqrt{100^2 + 200^2}$$

$$= 223.7V$$

(d) When the phase displacement is 120°

$$V = \sqrt{50\,000 + 40\,000 \cos 120}$$

$$= \sqrt{50\,000 + 40\,000(-\cos 60)}$$

$$= 173.2V$$

(e) When the phase displacement is 180°

$$Cos\,\theta = -1$$

$$V = \sqrt{50\,000 - 40\,000} = 100V$$

The above is obvious, i.e. since the voltages oppose, the resultant is the arithmetical difference, i.e. 100.0V

9.9. The waveform, when plotted, is a stepped shape but each half wave is regular and similar to its other half, except that it is reversed. The r.m.s. value is obtained by considering a half wave only, as the reversal mentioned, will not affect this value.

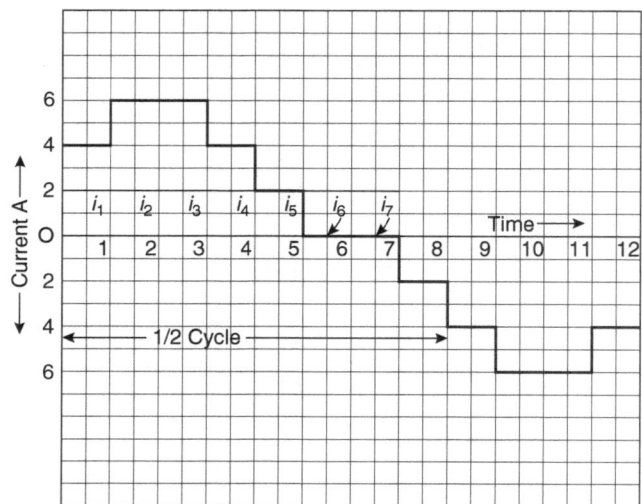

If the time interval 0–8 is considered as the base of the half wave, 8 mid-ordinates can be used, giving:

$i_1 = 4$ and $i_1^2 = 16$

$i_2 = 6$ and $i_2^2 = 36$ The sum of $i^2 = 112$

$i_3 = 6$ and $i_3^2 = 36$ The mean of $i^2 = \dfrac{112}{8} = \sqrt{14}$

$i_4 = 4$ and $i_4^2 = 16$ The r.m.s. value $= \sqrt{14}$

$i_5 = 2$ and $i_5^2 = 4 = 3.75A$

$i_6 = 0$ and $i_6^2 = 0$

$i_7 = 0$ and $i_7^2 = 0$

$i_8 = -2$ and $i_8^2 = 4$

The required value of D.C. will be 3.75A

9.10. The problem calls for a solution by phasor construction and is worked by drawing; the scale chosen being 10mm = 2V and the first e.m.f. OA is used as the reference.

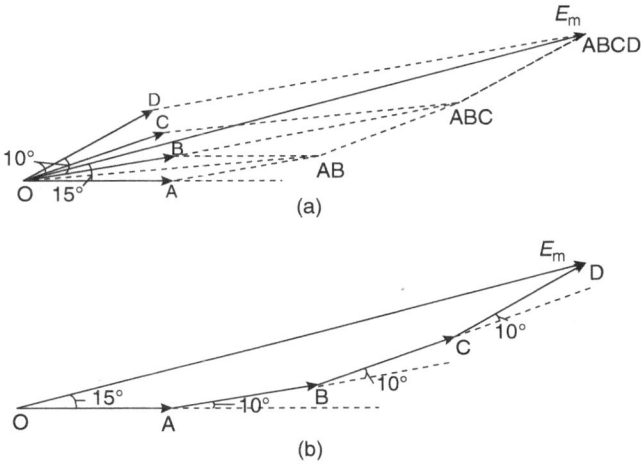

(a)

(b)

▲ **Figure 20** *Phasor diagram*

The resultant OD or E_m is measured to be 15.6V and

$$\theta = 15° \text{ or } \frac{\pi}{12} \text{radians}$$

Thus $e = 15.6 \sin\left(\omega t + \frac{\pi}{12}\right)$

The above diagrams show **alternative solutions**, thus (a) uses the parallelogram method, (b) uses the polygon method.

Chapter 10

10.1. Inductive circuit reactance $X_L = 2\pi f L$

$= 2 \times \pi \times 50 \times 0.01 = 3.14\Omega$

(a) Circuit impedance $= \sqrt{3^2 + 3.14^2} = 4.34\Omega$

(b) Power factor $= \dfrac{R}{Z} = \dfrac{3}{4.34} = 0.69$ **(lagging)**

(c) Power absorbed $= I^2R = \left(\dfrac{V}{Z}\right)^2 R = \dfrac{V^2}{Z} \times \dfrac{R}{Z} = \dfrac{V^2}{Z} \cos \phi$

$$= \dfrac{60^2}{4.34} \times 0.69 = 572.3\text{W}$$

10.2. Lamp current $= \dfrac{P}{V} = \dfrac{100}{100} = 1\text{A}$

Lamp resistance $= \dfrac{100}{1} = 100\Omega$

(a) Total resistance to give 1A with 220V applied $= \dfrac{220}{1} = 220\Omega$

∴ Series resistance $= 220 - 100 = 120\Omega$

Power absorbed by circuit $= I^2R = 1^2 \times 220 = 220\text{W}$

(b) When a coil is used for voltage dropping

Impedance of circuit $Z = \dfrac{220}{1} = 220\Omega$

Reactance of circuit $= \sqrt{220^2 - 100^2} = 196\Omega$

Also $X_L = 2\pi fL$ ∴ $L = \dfrac{196}{2 \times \pi \times 50} = 0.624\text{H}$

Power absorbed by circuit, $P = I^2R = 1^2 \times 100 = 100\text{W}$

10.3. Power absorbed, $P = VI \cos \Phi$

∴ $\cos \phi = \dfrac{P}{VI} = \dfrac{2500}{240 \times 15} = 0.694$ **(lagging)**

Current in the circuit $I = 15\text{A}$ (this data is given). Since $P = I^2R$

∴ $R = \dfrac{P}{I^2} = \dfrac{2500}{15^2} = 11.1\Omega$

$Z = \dfrac{V}{I} = \dfrac{240}{15} = 16\Omega$

$X = \sqrt{16^2 - 11.1^2} = 11.6\Omega$

In figure 21 $V_R = IR = 15 \times 11.1 = 166.5\text{V}$

$V_X = IX = 15 \times 11.6 = 174\text{V}$

$I = 15\text{A} \qquad V = 240\text{V}$

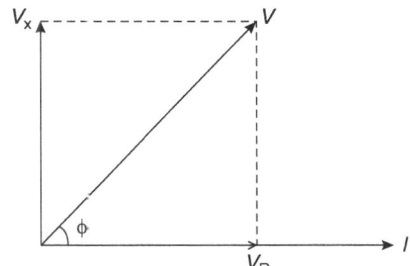

▲ **Figure 21** *Phasor diagram*

10.4. $R_A = 120\Omega$ $R_B = 100\Omega$

$X_A = 2\pi fL$ $X_B = 2\pi fL$

$= 2 \times \pi \times 50 \times 250 \times 10^{-3} = 2 \times \pi \times 50 \times 400 \times 10^{-3}$

$= 78.5\Omega$ $= 125.6\Omega$

$Z_A = \sqrt{120^2 + 78.5^2}$ $Z_B = \sqrt{100^2 + 125.6^2}$

$= 143\Omega$ $= 160.5\Omega$

Total circuit $R = 120 + 100 = 220\Omega$

$X = 78.5 + 125.6 = 204.1\Omega$

$Z = \sqrt{220^2 + 204.1^2} = 300\Omega$

(a) $I = \dfrac{230}{300} = 0.766\text{A}$

(b) $\cos\phi = \dfrac{220}{300} = 0.733$ (lagging) and $\phi = 42°46'$

(c) Voltage across A $= 0.766 \times 143 = 109.6\text{V}$

Voltage across B $= 0.766 \times 160.5 = 122.9\text{V}$

(d) $\cos\phi_A = \dfrac{120}{143} = 0.833$ (lagging) or $\phi_A = 33°7'$

$\cos\phi_B = \dfrac{100}{160.5} = 0.623$ (lagging) or $\phi_B = 51°27'$

Thus phase difference $\Phi = 51°27' - 33°7' = 18°20'$

10.5. D.C. condition $R_A = \dfrac{20}{2} = 10\Omega$ $R_B = \dfrac{30}{2} = 15\Omega$

A.C. condition $Z_A = \dfrac{140}{2} = 70\Omega$ $Z_B = \dfrac{100}{2} = 50\Omega$

$$X_A = \sqrt{70^2 - 10^2} = 69.3\Omega$$

$$X_B = \sqrt{50^2 - 15^2} = 47.7\Omega$$

Since X is proportional to frequency

Therefore at 50Hz $X_A = 69.3 \times \dfrac{5}{4} = 86.6\Omega$

$$X_B = 47.7 \times \dfrac{5}{4} = 59.7\Omega$$

For the total series circuit $R = 10 + 15 = 25\Omega$

$X = 86.6 + 59.7 = 146.3\Omega$

So $Z = \sqrt{25^2 + 146.3^2} = 148.1\Omega$

Current $I = \dfrac{230}{148.1} = 1.55A$

10.6. Current in the line $I = \dfrac{P}{V_F \cos\phi_F} = \dfrac{750 \times 1000}{3300 \times 0.8} = 284A$

Resistance voltage drop in the line, $V_R = IR = 284 \times 1 = 0.284kV$

Reactance voltage drop in the line, $V_X = IX = 284 \times 2.5 = 0.710kV$

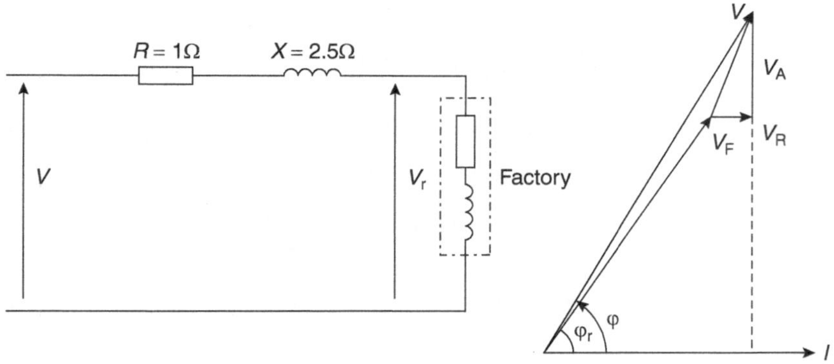

▲ **Figure 22** *Circuit and phasor diagram*

From the phasor diagram (figure 22)

$$V = \sqrt{(3.3 \times 0.8 + 0.284)^2 + (3.3 \times 0.6 + 0.71)^2}$$

$$= \sqrt{(2.64 + 0.284)^2 + (1.98 + 0.71)^2}$$

$$= \sqrt{2.924^2 + 2.69^2} = 3.98\text{kV}$$

So the voltage at the generator = 3.98kV

Generator power factor $\cos \phi = \dfrac{2.924}{3.98} = 0.73 \text{ (lagging)}$

Generator output $= \dfrac{3980 \times 284 \times 0.73}{1000} \text{ kilowatts} = 825\text{kW}$

10.7. Current in 8Ω resistor $= \dfrac{64}{8} = 8\text{A} = \text{current in circuit}$

(b) Power absorbed in resistor $= I^2R = 8^2 \times 8 = 64 \times 8 = 512\text{W}$

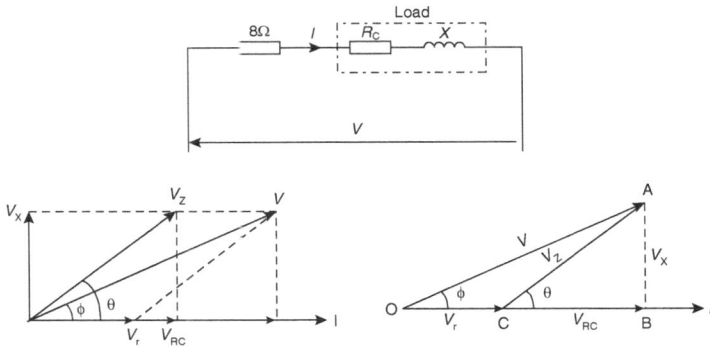

▲ **Figure 23** *Circuit and phasor diagram*

From the deduced diagrams (figure 23)

$OA^2 = OC^2 + CA^2 - 2 \times OC \times CA \times \cos(180 - \theta)$ (*Cosine formulae*)

or $100^2 = 64^2 + 48^2 + 2 \times 64 \times 48 \times \cos \theta$

$$\cos \theta = \frac{10\,000 - 4096 - 2304}{128 \times 48} = 0.586$$

(d) Power factor of load = 0.586 (lagging)

Voltage drop in resistance of inductive load $= V_{RC}$

$$= V_Z \cos \theta = 48 \times 0.586 = 28.128\text{V} = IR_C$$

$$\therefore R_c = \frac{28.128}{8} = 3.52\Omega$$

(a) Power absorbed by load $= I^2 R_c = 8^2 \times 3.52 = 225.3W$

(c) Total Power $= 512 + 225.3 = 737.3W$

(d) Power factor of circuit or $\cos \phi = \dfrac{P}{VI} = \dfrac{737.3}{100 \times 8} = 0.92$ (lagging)

10.8. Let X_1 ohms = the reactance at 40Hz

then $Z_1 = \sqrt{R_1^2 + X_1^2}$

and $Z_1 = \dfrac{200}{6.66} = 30.3\Omega$

Let X_2 ohms = the reactance at 50Hz

then $Z_2 = \sqrt{R^2 + X_2^2}$

and $Z_2 = \dfrac{200}{8} = 25\Omega$

Hence $30.3^2 = R_1^2 + X_1^2$

and $25^2 = R_2^2 + X_2^2$

but, since $R_1 = R_2$ then:

Subtracting $30.3^2 - 25^2 = X_1^2 - X_2^2$

or $(30.3 - 25)(30.3 + 25) = (X_1 - X_2)(X_1 + X_2)$

Hence $5.3 \times 55.3 = (X_1 - X_2)(X_1 + X_2)$

or $293.09 = (X_1 - X_2)(X_1 + X_2)$

Also since $X = \dfrac{1}{2\pi f C} \therefore X = \dfrac{k}{f}$ or $X \propto \dfrac{1}{f}$

Thus $X_1 = \dfrac{k}{f_1}$ and $X_2 = \dfrac{k}{f_2}$ or $\dfrac{X_1}{X_2} = \dfrac{(k/f_1)}{(k/f_2)} = \dfrac{f_2}{f_1}$

Hence $\dfrac{X_1}{X_2} = \dfrac{50}{40}$ and $X_1 = \dfrac{5}{4} X_2$ or $X_1 = 1.25 X_2$

Substituting

$$293.09 = (1.25X_2 - X_2)(1.25X_2 + X_2)$$

$$= 0.25X_2 \times 2.25X_2 = 0.5625X_2^2$$

Whence $X_2^2 = \dfrac{293.09}{0.5625}$ anc $X_2 = \sqrt{520} = 22.8W$

$$X_1 = \dfrac{5}{4} \times 22.8 = 28.5\Omega$$

Thus $R_2 = 25^2 - 22.8^2 = 105$

$$R = \sqrt{105} = 10.25\Omega$$

Also $X_C = \dfrac{10^6}{2\pi fC}$ or $22.8 = \dfrac{10^6}{2 \times \pi \times 50 \times C}$

Thus $C = \dfrac{10^6}{22.8 \times \pi \times 10^2} = 139\text{mF}$

For the diagram figure 24, as an example:

Resistance voltage drop $= IR = 6.66 \times 10.25 = 68.27\text{V}$

Reactance voltage drop $= IX_1 = 6.66 \times 28.5 = 190\text{V}$ etc.

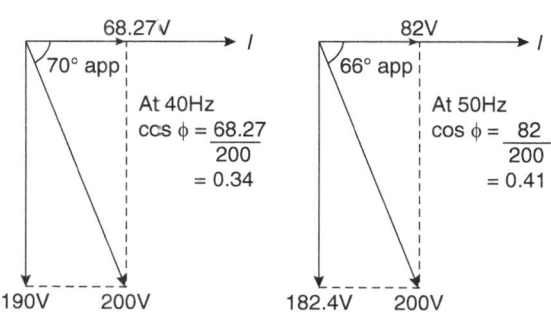

▲ **Figure 24** *Reactance diagram*

10.9.

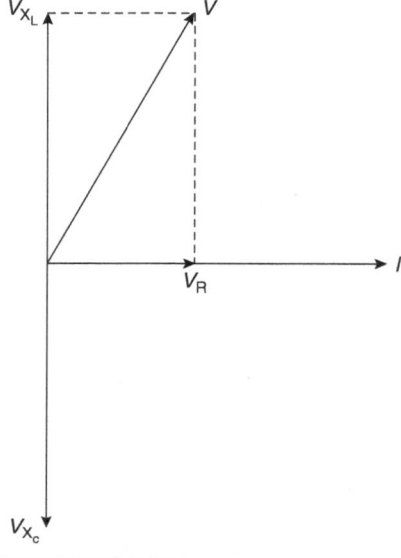

▲ **Figure 25** *Reactance diagram*

At resonance $2\pi fL = \dfrac{1}{2\pi fC}$

and $f^2 = \dfrac{1}{(2\pi)^2 LC}$ or $f = \dfrac{1}{2\pi\sqrt{LC}}$

Here I can be obtained from $I = \dfrac{V}{X_c}$

or $I = \dfrac{V}{\dfrac{1}{2\pi fC}} = V2\pi fC$

Thus $I = 100 \times 2 \times \pi \times f \times 10 \times 10^{-6} = 2\pi \times f \times 10^{-3}$

$\qquad = 2\pi \times 10^{-3} \times \dfrac{1}{2 \times \pi\sqrt{LC}}$

$\qquad = \dfrac{10^{-3}}{\sqrt{0.5\times10\times10^{-6}}} = 0.446A$

For figure 25, if actual values were required then as an example:

$$V_R = IR = 0.446 \times 60 = 26.76V$$

$$X_L = X_C = 100V.$$

10.10. At 60Hz $R = 400\Omega$ $Z = 438\Omega$

$$X_L = \sqrt{438^2 - 400^2} = 178\Omega$$

At 50Hz $X_L = 178 \times \dfrac{50}{60} = 148.3\Omega$

Circuit impedance

$$Z = \sqrt{400^2 + (X_L - X_C)^2}$$

$$= \sqrt{400^2 + \left(148.3 - \dfrac{10^6}{2 \times \pi \times 50 \times 40}\right)^2}$$

$$= \sqrt{400^2 + 68.7^2} = 405\Omega$$

$$I = \dfrac{200}{405} = 0.495A$$

$$V_C = 0.495 \times \dfrac{10^6}{2 \times \pi \times 50 \times 40} = 39.5V$$

Impedance of coil $Z_L = \sqrt{400^2 + 148.3^2} = 426.5\Omega$

$$V_{ZL} = IZ_L$$

$$= 0.495 \times 426.5\Omega = 211V$$

Chapter 11

11.1. For the D.C. circuit, $P = VI$ cr $300 = 60 \times I$

so D.C. current $= \dfrac{300}{60} = 5A$

and D.C. resistance $= \dfrac{V}{I} = \dfrac{60}{5} = 12\Omega$

For the A.C. circuit, $P = I^2R$

$\therefore 1200 = I^2 \times 12$

or $I^2 = 100$ and $I = 10A$

Since the current taken by the A.C. circuit $= 10A$

then impedance, $Z = \dfrac{V}{I} = \dfrac{130}{10} = 13\Omega$

Also $X = \sqrt{Z^2 - R^2}$ or $X = \sqrt{13^2 - 12^2}$

Thus $X = 5\Omega$ Coil reactance $= 5$ ohms

11.2. Impedance of branch A. $Z_A = \sqrt{12^2 + 3^2} = 12.4\Omega$

Current $I_A = \dfrac{V}{Z_A} = \dfrac{100}{12.4} = 8.08A$. Also $\cos \phi_A = \dfrac{12}{12.4} = 0.968$ (lagging)

and $\sin \phi_A = \dfrac{3}{12.4} = 0.242$

Impedance of branch B. $Z_B = \sqrt{8^2 + 20^2} = 21.6\Omega$

Current $I_B = \dfrac{100}{21.6} = 4.64A$ Also $\cos \phi_B = \dfrac{8}{21.6} = 0.372$ (lagging)

and $\sin \phi_B = \dfrac{20}{21.6} = 0.928$

Active components of current:

$$I_a = I_A \cos \phi_A + I_B \cos \phi_B$$

$$= (8.08 \times 0.968) + (4.64 \times 0.372) = 9.53A$$

Reactive component of current:

$$I_r = -I_A \sin \phi_A - I_B \sin \phi_B$$

$$= -(8.08 \times 0.242) - (4.64 \times 0.928) = -6.26A$$

Hence $I = \sqrt{9.53^2 + 6.26^2} = 11.4A$

11.3. Let the inductive circuit be circuit A, then:

$$X_A = 2\pi f L = 2 \times \pi \times 50 \times 0.02 = 6.28\Omega \quad R_A = 50\Omega$$

$$Z_A = \sqrt{50^2 + 6.28^2} = 50.5W$$

$$I_A = \frac{V}{Z_A} = \frac{200}{50.5} = 3.96A$$

$$\cos \phi_A = \frac{R_A}{Z_A} = \frac{50}{50.5} = 0.99 \text{ (lagging)}$$

$$\sin \phi_A = \frac{X_A}{Z_A} = \frac{6.28}{50.5} = 0124$$

Let the capacitive circuit be circuit B, then:

$$X_B = \frac{10^6}{2 \times \pi \times 50 \times 25} = 127\Omega$$

$$I_B = \frac{200}{127} = 1.575 \quad \cos \phi_B = 0 \quad \sin \phi_B = 1$$

Then $I_a = (3.96 \times 0.99) + (1\,575 \times 0) = 3.92A$

$$I_r = -(3.96 \times 0.124) + (1.575 \times 1) = 1.084A$$

or $I = \sqrt{3.92^2 + 1.084^2} = 4.1A$

$$\text{Total current} = 4.075A \quad \cos \phi = \frac{3.92}{4.075} = 0.962 \text{ (leading)}$$

Phase angle $\Phi = 15°50'$

11.4. Let the branches be A, B and C respectively. Then:

$$X_A = 2\pi f L = 2 \times \pi \times 50 \times 0.02 = 6.28\Omega$$

$$Z_A = \sqrt{8^2 + 6.28^2}$$

$$= 10.2\Omega$$

and $\cos \phi_A = \dfrac{8}{10.2} = 0.785$ (**lagging**) $\sin = \phi_A = \dfrac{6.28}{10.2} = 0.616$

$$I_A = \dfrac{100}{10.2} = 9.8A$$

$$X_B = 2 \times \pi \times 50 \times 0.05 = 15.7\Omega$$

$$Z_B = \sqrt{10^2 + 15.7^2} = 18.6\Omega$$

$$\cos \phi_B = \dfrac{10}{18.6} = 0.537 \text{ (lagging) } \sin \phi_B = \dfrac{15.7}{18.6} = 0.845$$

$$I_B = \dfrac{100}{18.6} = 5.37A$$

$$X_C = \dfrac{1}{2\pi fC} = \dfrac{10^6}{2 \times \pi \times 50 \times 80} = 39.8$$

Then $Z_C = \sqrt{20^2 + 39.8^2} = 44.54\Omega$

$$\cos \phi_C = \dfrac{20}{44.54} = 0.449 \text{ (leading) } \sin \phi_C = \dfrac{39.8}{44.54}$$

$$I_C = \dfrac{100}{44.54} = 2.24A$$

Adding the active and reactive current components.

$$I_a = I_A \cos \phi_A + I_B \cos \phi_B + I_C \cos \phi_C$$

$$= (9.8 \times 0.785) + (5.37 \times 0.537) + (2.24 \times 0.449) = 11.59A$$

$$I_r = -I_A \sin \phi_A - I_B \sin \phi_B + I_C \sin \phi_C$$

$$= (9.8 \times 0.616) - (5.37 \times 0.845) + (2.24 \times 0.894) = -8.54A$$

Then $I = \sqrt{I_a^2 + I_r^2} = \sqrt{11.59^2 + 8.54^2} = 14.38A$

$$\cos \phi = \dfrac{11.59}{14.38} = 0.805 \text{ (lagging) } \phi = 36° \text{ (approx.)}$$

11.5. Apparent power $= VI = 240 \times 50.6 \times 10^{-3}$ kilovolt amperes

$$= 12.1 \text{kVA}$$

$$\text{Power factor} = \frac{\text{true power}}{\text{apparent power}} = \frac{10}{12.144}$$

$$= 0.823 \text{ (lagging)}$$

$$\text{Efficiency} = \frac{\text{output (power)}}{\text{input (power)}} = \frac{9\text{kW}}{10\text{kW}} = 0.9$$

or $\eta = 90\%$

11.6. Output from motor $= 1.5$kW   Efficiency $= 80\%$

$$\text{Input to motor} = \frac{1500}{80} \times 100 = 1875\text{W}$$

Also power input to motor, $VI \cos \phi = 1875$W

$\therefore 1875 = 230 \times 11.6 \cos \phi$

$\therefore \cos \phi = \dfrac{1875}{230 \times 11.6} = 0.7 \text{(lagging)} \sin \phi = 0.714$

Power component of input current, $I \cos \phi = 11.6 \times 0.7 = 8.12$A

Reactive component of input current, $I \sin \phi = 11.6 \times 0.714 = 8.28$A

At the new power factor, the power component of current

$= 8.12$A $= I_1 \cos \phi_1$. Also since $I_1 \cos \phi_1 = I \cos \phi$

$\therefore I_1 \times 0.95 = 8.12$ and $I_1 = \dfrac{8.12}{0.95} = 8.55$A

Note. $\cos \phi_1 = 0.95 \sin \phi_1 = 0.327$

The reactive component cf input current, at the new power factor $= I_1 \sin \phi_1 = 8.55 \times 0.327 = 2.8$A

So reduction of reactive current = 8.28 − 2.8 = 5.48A

and capacitor current = 5.48A

Capacitor reactance $= \dfrac{230}{5.48}$ ohms $= \dfrac{10^6}{2\pi fC}$

or $C = \dfrac{10^6 \times 5.48}{2 \times \pi \times 50 \times 230} = 76\mu F$

Rating of capacitor $= 230 \times 5.48 \times 10^{-3} = 1.26VAr$

11.7. Load (a) Apparent power, $S_a = \dfrac{\text{active power}}{\text{power factor}} = \dfrac{10}{1} = 10kVA$

Reactive power, $Q_a = S_a \times \sin \Phi = 10 \times 0 = 0kVAr$

Load (b) Apparent power, $S_b = 80kVA$ at a power factor of 0.8 (lagging)

Active power, $P_b = 80 \times 0.8 = 64kW$

Reactive power, $Q_b = 80 \times 0.6 = 48kVAr$ (lagging)

Load (c) Apparent power, $S_c = 40kVA$ at a power factor of 0.7 (leading)

Active power, $P_c = 28kW$

Reactive power, $Q_c = 40 \times 0.7143 = 28.57kVAr$

Total power taken from the supply, $P = 10 + 64 + 28 = 102kW$

Total reactive power, $Q = 0 − 48 + 28.57 = −19.43kVAr$

Total apparent power from supply, $S = \sqrt{102^2 + 19.43^2} = 104kVA$

Power factor of combined load, $\dfrac{P}{S} = \dfrac{102}{104} = 0.98$ **(lagging)**

Mains current $= \dfrac{104\ 000}{250} = 416A$

11.8 (a) Phase voltage = 100V Impedance per phase of load = 10Ω

∴ Load current per phase $= \dfrac{100}{10}$ 10A

Line current = Phase current = 10A

Total power, $P = \sqrt{3}\ VI \cos \phi$

But $V = \sqrt{3} V_{ph} = 1.732 \times 100 = 173.2V$ and $I = 10A$

∴ $P = \dfrac{\sqrt{3} \times 173.2 \times 10 \times \cos 30}{1000}$ $P = 2.598kW$

(b) Line voltage $= \sqrt{3} \times 100$

$\therefore$ Voltage per phase of load $= \sqrt{3} \times 100$ volts

Current per phase of load $= \dfrac{V_{ph}}{Z_{ph}} = \dfrac{\sqrt{3} \times 100}{10} = \sqrt{3} \times 10$ amperes

Line current = phase current $= \sqrt{3} \times \sqrt{3} \times 10 = 30A$

Total power, $P = \sqrt{3}\ VI \cos \phi$

$$= \frac{\sqrt{3} \times \sqrt{3} \times 100 \times 30 \times 0.866}{1000} = 7.794W$$

(c) Line voltage = 100V

Voltage per phase of load = 100V

Current per phase of load $= \dfrac{100}{10} = 10A$

Line current $= \sqrt{3}I_{ph} = 1.732 \times 10 = 17.32A$

Total power, $P = \sqrt{3}\ VI \cos \phi$

$$= \frac{\sqrt{3}\ 1000\ \sqrt{3}\ 10 \times 0.866}{1000} = 2.598W$$

(d) Line voltage = 100V

Voltage per phase of load $= \dfrac{100}{\sqrt{3}}$

$\therefore$ Current per phase of load $= \dfrac{100}{\sqrt{3} \times 10}$ amperes

Line current = Phase current $= \dfrac{10}{\sqrt{3}} = 5.77A$

Total power, $P = \sqrt{3}\ VI \cos \phi$

$$= \frac{\sqrt{3} \times 100 \times 10 \times 0.866}{1000 \times \sqrt{3}} \text{kilowatts} = 0.866kW$$

11.9. Motor output = 45kW = 45 000W

Efficiency of motor = 88%

Motor input $= 45\,000 \times \dfrac{100}{88} = 51\,140\text{W}$

Since $P = \sqrt{3}\ VI \cos\phi$

(a) Line current, $I = \dfrac{51\,140}{\sqrt{3} \times 500 \times 0.9}$ $I = 65.6\text{A}$

(b) Alternator output = input to motor = 51.14kW

(c) Input to alternator $= \dfrac{51\,140 \times 100}{80}$ watts or motor power of prime-mover = 64kW

11.10.

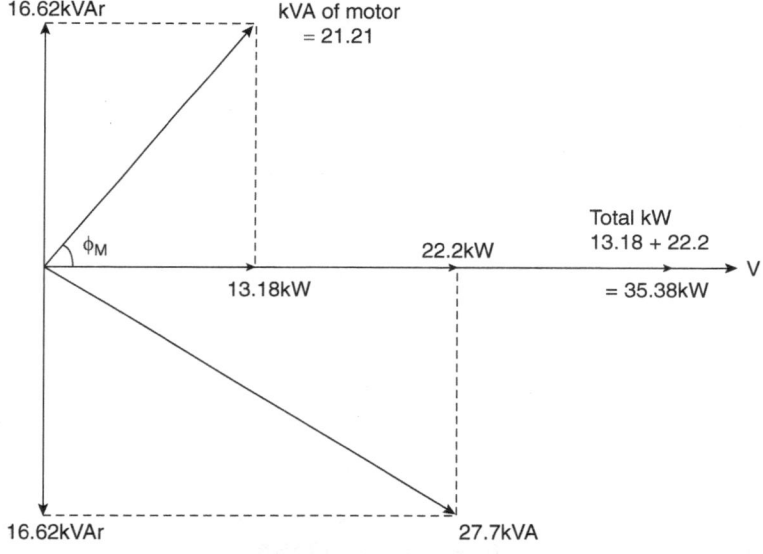

▲ **Figure 26** *kVAr diagram*

Input power to the system, $P_s = \sqrt{3}\ VI \cos\phi$

$= \dfrac{\sqrt{3} \times 400 \times 40 \times 0.8}{1000} = 22.2\text{kW}$

Power output from the motor at 91% efficiency = 12kW

∴ Power input to the motor, $P_m = \dfrac{12 \times 100}{91} = 13.18\text{kW}$

Apparent power of system, $S_s = \dfrac{\sqrt{3} \times 400 \times 40}{1000} = 27.7\text{kVA}$

Reactive power of system, $Q_s = S_s \sin \Phi = 27.7 \times 0.6 = 16.62\text{kVAr}$

To improve the power factor to unity, the motor's reactive power must equal the reactive power of the system, so the motor's reactive power $Q_m = 16.62\text{kVAr}$.

Apparent power to motor, $S_m = \sqrt{P_m^2 + Q_m^2}$

$= \sqrt{13.18^2 + 16.62^2} = 21.21\text{kVA}$

Power factor of motor, $\cos \phi_m = \dfrac{P_m}{S_m} = \dfrac{13.18}{21.21} = 0.62 \ (\textbf{leading})$

Total power taken from the mains = power supplied to the system + power supplied to the motor $= P_s + P_m = 22.2 + 13.18 = 35.38\text{kW}$

The phasor diagram shows the method of solution. Even though the problem is a 3-phase one, the diagram, as drawn, can be applied, as balanced conditions are assumed.

Chapter 12

12.1. E.m.f. generated $= \dfrac{Z\Phi N}{60} = \dfrac{P}{A}$ volts

Here Z is $144 \times 6 = 864$ $N = 600$ rev/min

$P = 4$ and $A = 4$ since this is a *lap* winding

$\therefore 216 = \dfrac{864 \times \Phi \times 600}{60} \times \dfrac{4}{4}$

or $\Phi = \dfrac{216}{864 \times 10}$ webers

If the armature is *wave* wound $A = 2$

Substituting the value of Φ

then $E = 864 \times \dfrac{216}{864 \times 10} \times \dfrac{600}{60} \times \dfrac{4}{2} = 432\text{V}$

12.2. Voltage applied to shunt field = 220V

Current through shunt field $= \dfrac{220}{176} = 1.25$A

Armature current = 250 + 1.25 = 251.25A

Voltage drop in armature and series field = 251.25(0.05 + 0.015)

$$= 251.25 \times 0.04 = 10.05V$$

Total voltage drop = 10.05 + 2V (brush voltage drop)

Induced e.m.f. E = 220 + 10.05 + 2 = 232.05V

12.3. It is noted that a change of speed is involved here and the solution cannot be affected before the O.C.C. at 900 rev/min is reached. Since $E \propto N$, the new values can be obtained by multiplying the original by a factor: $\dfrac{900}{1200} = \dfrac{3}{4}$

The table shows the adjustment for the 900 rev/min condition

Excitation current I_t (amperes)	0	0.4	0.8	1.2	1.6	2.0	2.4
E.m.f. at 1200 rev/min E (volts)	15	88	146	196	226	244	254
E.m.f. at 900 rev/min E (volts)	11.25	66	109.5	147	169.5	183	190.5

The field voltage-drop line is drawn by taking any current value and multiplying it by 90Ω. For example: 2A × 90Ω = 180V. Join this point R to the origin. The required answer 185V is obtained from the intersection point as shown.

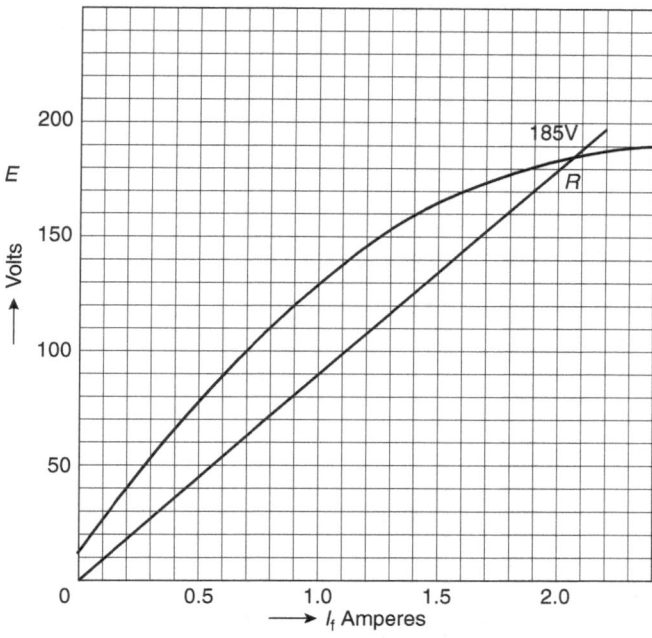

▲ **Figure 27** *E vs I diagram*

12.4. $I_f = \dfrac{220}{50} = 4.4A \; I_L = \dfrac{38\,000}{220}$

$I_a = 172.72 + 4.4 = 177.12A$

Also $E = V + I_a R_a$

$= 220 + (177.12 \times 0.1) = 237.72V$

Again $E = \dfrac{Z\Phi N}{60} \times \dfrac{P}{A}$

$\therefore 237.72 = \dfrac{700 \times \Phi \times 800 \times 4}{60 \times 2}$

or $\Phi = \dfrac{237.71 \times 3}{56 \times 10^3}$

$= 0.0128Wb$ or $12.8mWb$

12.5. O.C. e.m.f. $= 440V = k\Phi N$

So $k = \dfrac{440}{0.055 \times 620}$

Full-load current $= \dfrac{250\,000}{480} = 520A$

Total voltage drop on full load $= 520 (0.01 + 0.005 + 0.005)$

$= 520 \times 0.02 = 10.4V$

So e.m.f. generated on full load $= 480 + 10.4 = 490.4V$

$\therefore 490.4 = k\Phi_2 N_2 = \dfrac{440}{0.055 \times 620} \times \Phi_2 \times 600$

or $\Phi_2 = \dfrac{490.4 \times 0.055 \times 620}{440 \times 600} = 0.0633Wb$ or $63.3mWb$

12.6. On no load, the e.m.f. generated is caused by the shunt-field ampere-turns = 7900. These give a no-load voltage of 500V. But since the load voltage rises to 550V, the shunt-field current will rise and the shunt-field ampere-turns, on no load, increase to

$$7900 \times \dfrac{550}{500} = 8690At$$

It is found on a full-load test that 11 200At are needed and the extra (above 8690At) must therefore be supplied by the series field.

Series field must supply 11 200 – 8690 = 2510At

Now the full-load current $= \dfrac{500\ 000}{550} = 910A$

$\therefore$ The series turns required $= \dfrac{2510}{910} = 2.76$ i.e. 3 turns

12.7. Here $Z = 90 \times 6 = 540$

$$\therefore E = \frac{540 \times 0.03 \times 1500}{60} \times \frac{4}{4} = 405V$$

If $I_a = 25A$. The armature voltage drop $= 25 \times 1.0 = 25$ volts. Since the same field flux and speed are assumed, then the same e.m.f. is being generated or $V = E - I_a R_a$

$\therefore V = 405 - 25 = 380V.$

So shunt-field current $= \dfrac{380}{200} = 1.9A$

Machine output current $= 25 - 1.9 = 23.1A$

Let $I_L =$ the load current

Then $I_L \times 40 = V$ (the terminal voltage)

Also $V = E - I_a R_a$

$= 380 - I_a \times 1.0$ also $I_a = I_f + I_L$

$\therefore V = 380 - 1.0\ (I_f + I_L)$

or $40 \times I_L = 380 - I_f - I_L$

and $41 \times I_L = 380 - I_f$

also $I_f = \dfrac{V}{200} = \dfrac{40 \times I_L}{200} = \dfrac{I_L}{5}$

So the above becomes:

$41 \times I_L = 380 - \dfrac{I_L}{5}$

$205 \times I_L = 1900 - I_L$

$206 \times I_L = 1900$ and $I_L = \dfrac{1900}{206} = 9.22A$

12.8. (a) The 1000 rev/min O.C.C. is plotted and cut by the 100Ω field voltage-drop line which is plotted by drawing a straight line through any deduced point and the origin. Thus consider a I_f value of 1A, then a field voltage-drop line point would be $1 \times 100Ω = 100V$. Join this point to the origin.

The point of intersection at 108V is the answer required.

(b) The tangent is drawn to the 1000 rev/min O.C.C. The critical resistance R_c is determined by taking any voltage value on this tangent and dividing by the current.

Thus $R_c = \dfrac{145V}{0.9A} = 161.1\Omega$. The critical resistance for this speed is 161Ω (approx.)

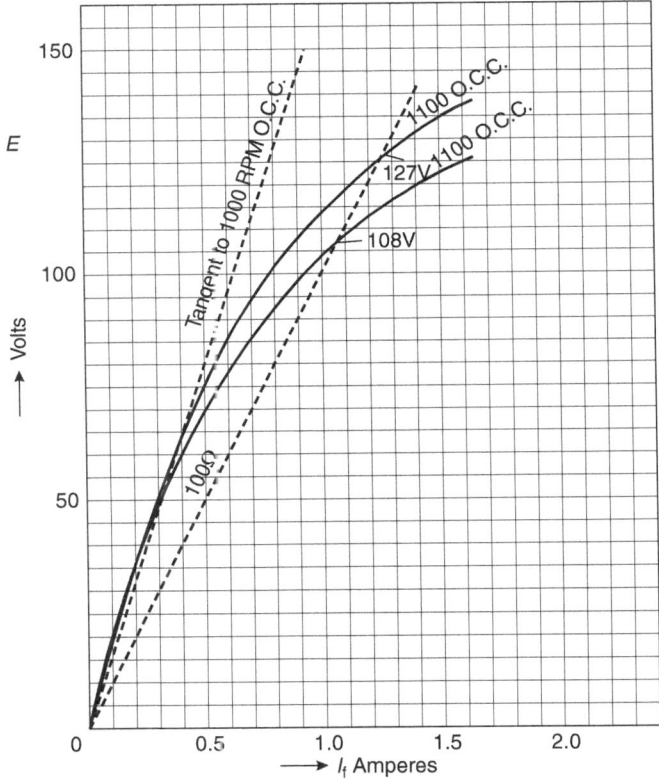

▲ **Figure 28** *E vs I diagram*

(c) The 1100 rev/min O.C.C. is obtained by multiplying the original 1000 rev/min values by a factor: $\dfrac{11}{10} = 1.1$ to give the new table:

Excitation current I_t (amperes)	0.2	0.4	0.6	0.8	1.1	1.2	1.4	1.6
E.m.f. at 1000 rev/min E (volts)	32	58	78	93	104	113	120	125
E.m.f. at 1100 rev/min E (volts)	35.2	63.8	58.8	102.3	114.4	124.3	132	137.4

This 1100 rev/min characteristic when plotted is cut by the 100Ω field voltage-drop line at 127V.

12.9. Output current $= \dfrac{50 \times 1000}{230} = 217.39\text{A}$

Shunt-field current $= \dfrac{230}{55} = 4.18\text{A}$

Armature current $= 217.39 + 4.18 = 221.57\text{A}$

Armature voltage drop $= 221.57 \times 0.034 = 7.53\text{A}$

Induced e.m.f. $= 230 + 7.53 + 2 = 239.53\text{V}$

Electrical power required to be generated

$= 239.53 \times 221.57 \text{ watts} = 53.18\text{kW}$

Total input power = electrical power input + mechanical loss

$= 53.18 + 1.6 = 54.78\text{kW}$

Thus input power $= 54.78\text{kW}$

12.10. Since the answers required are at a different speed condition, the new O.C.C. at 400 rev/min is obtained by multiplying the original values by $\dfrac{400}{200} = 2$. Thus:

Excitation current I_t (amperes)	0	1	2	3	4	5	6	7	8	9
E.m.f. at 200 rev/min E (volts)	10	38	61	78	93	106	115	123	130	135
E.m.f. at 400 rev/min E (volts)	20	76	122	156	186	212	230	246	260	270

(a) E.m.f. to which machine self-excites = 243V

This is the point of intersection between the 400 rev/min O.C.C. and the 36Ω field voltage-drop line. The latter is drawn by taking any current, say 5A, and finding the voltage drop 5 × 36 = 180V, and joining this point to the origin.

(b) Draw the tangent to 400 rev/min O.C.C. This may be difficult. Disregard the bottom part of the characteristic (due to residual magnetism) and assume that the characteristic would pass from the second point given (76V at 1A), through the origin. A straight line drawn as shown, through this point and the origin, will be sufficiently tangential to give a suitable answer.

Consider a point such as X. Then critical resistance $R_c = \dfrac{120}{1.5} = 80\Omega$

(c) For the e.m.f. to reduce to 220V, the field voltage-drop line should cut the O.C.C. at this point. The field current would then be 5.3A. So field-current resistance would be $\dfrac{220}{5.3} = 41.5\Omega$. Thus additional resistance required = 41.5 − 36 = 5.5Ω.

(d) This answer is obtained by assuming the machine slows down and is best obtained by trial and error. Multiply the O.C.C. values by various fractions of

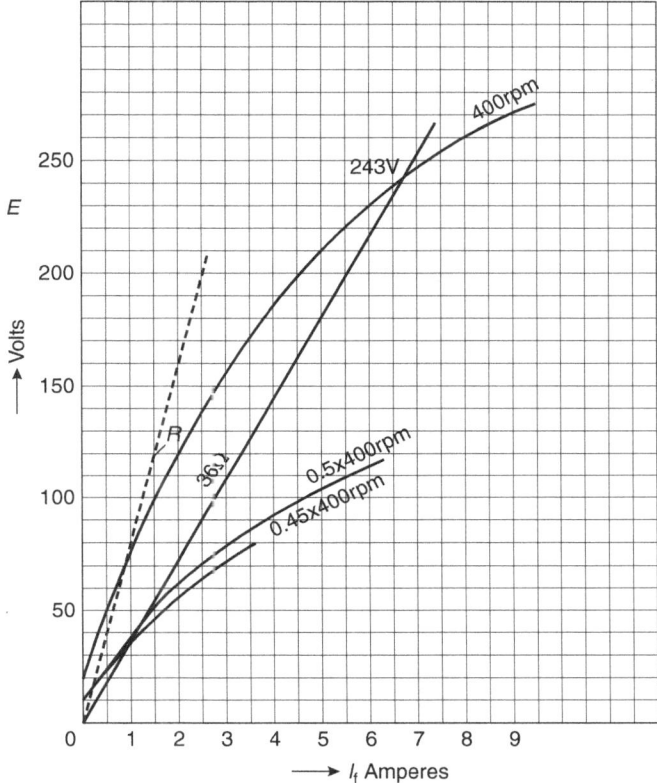

▲ **Figure 29** *E vs I diagram*

the speed to obtain a magnetisation curve that makes the 36Ω voltage-drop line a tangent. The 200 rev/min O.C.C. is a suitable starting point and is plotted. It is seen to be cut at the bottom by the 35Ω line, which therefore is not quite tangential. The required O.C.C. must be at a lower speed and 400 × 0.45 may be tried to give values of 9, 34.2, 54.9, 70.2, etc. If plotted this will be approximately correct and the answer 130 rev/min (approx.).

Chapter 13

13.1. The output is 7.5kW = 7500W.

Input is $\dfrac{7500 \times 100}{85} = 8824\text{W}$

Motor current $= \dfrac{8824}{110} = 80.2A$

$E_b = V - I_a (R_a + R_{se})$

$$= 110 - (80.2 \times 0.12) = 100.4V$$

13.2. Power input = 90 000W

Current input to motor $= \dfrac{90\,000}{500} = 180A$

Shunt-field current $I_f = \dfrac{500}{100} = 5A$

Armature current $= 180 - 5 = 175A$

Back e.m.f. $= E_b = V - I_a R_a$

$$= 500 - (175 \times 0.1) = 482.5V$$

13.3. No load. Back e.m.f. $E_{b0} = V - I_{a0} R_a$

$$= 460 - (10 \times 0.025) = 459.75V$$

Full load. Back e.m.f. $E_{b1} = V - I_{a1} R_a$

$$= 460 - (300 \times 0.025) = 452.5V$$

13.4. (a) Current in 1 parallel path $= \dfrac{40}{4} = 10A$

Current in 1 conductor = current per parallel path

So force on 1 conductor is given by $BI\ell$

$$= 1.2 \times 10 \times 0.4 = 4.8N$$

Torque of 1 conductor $\dfrac{4.8 \times 0.3}{2} = 0.72Nm$

(b) Total torque due to all conductors

$= 240 \times 0.72 = 172.8Nm$

(c) Power output $= \dfrac{2 \times \pi \times 800 \times 172.8}{60}$ joules/sec

$= \dfrac{6.28 \times 80 \times 172.8}{6}$

$= 14\,469W = 14.5W$

13.5. On full load $E_{b1} = 220 - (25 \times 0.2) = 220 - 5 = 215V$

Also, since torque is unchanged, and $T_1 = T_2$

and since $T \propto \Phi I_a$ or $T = k\Phi I_a$, we can write:

$k\Phi_1 I_{a1} = k\Phi_2 I_{a2}$

But $\Phi_2 = 0.9\Phi_1$

so $\Phi_1 I_{a1} = 0.9\Phi_1 I_{a2}$

or $I_{a2} = \dfrac{I_{a1}}{0.9} = \dfrac{25}{0.9} = 27.77A$

Thus $E_{b2} = 220 - (27.77 \times 0.2) = 214.446V$

Also since $\dfrac{E_{b2}}{E_{b1}} = \dfrac{k\Phi_2 N_2}{k\Phi_1 N_1}$

$\therefore N_2 = \dfrac{E_{b2} \times \Phi_1 \times N_1}{E_{b1} \times \Phi_2} = \dfrac{214.446 \times \Phi_1 \times 725}{215 \times 0.9\Phi_1}$

$= 804$ rev/min

13.6. *As a generator.* 50kW at 250V gives a load current of

$$i_L = \frac{50\ 000}{250} = 200A$$

Field current $I_f = \dfrac{250}{50} = 5A$

$\therefore$ Armature current $I_a = I_L + I_f = 200 + 5 = 205A$

General voltage $E = V + I_a R_a +$ brush voltage drop

or $E = 250 + (205 \times 0.02) + 2 = 256.1V$

As a motor. Input current $I_L = \dfrac{50\ 000}{250} = 200A$

Field current I_f (as before) $= \dfrac{250}{50} = 5A$

$\therefore$ Armature current $I_a = I_L - I_f = 200 - 5 = 195A$

Back e.m.f. $E_b = V - I_a R_a -$ brush voltage drop

$= 250 - (195 \times 0.02) - 2$

or $E_b = 244.1V$

Again since E and E_b are proportional to flux and speed then,

$E = k\Phi N$ and $E_b = k\Phi N$

Thus $\dfrac{E}{E_b} = \dfrac{5 \times 400}{5 \times N}$ or $\dfrac{256.1}{244.1} = \dfrac{400}{N}$

or $N = \dfrac{400 \times 244}{256.1} = 382$ rev/min

Note. Flux is assumed to be proportional to the field ampere-turns, and hence the exciting current, with the current value substituted for the flux values Φ.

13.7. *On no load.* $I_f = \dfrac{105}{90} = 1.17\text{A} \ I_{10} = 3.5\text{A}$

$I_{a0} = 3.5 - 1.17 = 2.33\text{A}$

$E_{b0} = 105 - (2.33 \times 0.25) = 104.42\text{V}$

On full load. Output $= 3 \times 1000$ watts

Input $= \dfrac{3 \times 1000 \times 100}{82} = 3660\text{W}$

Input line current $I_{L1} = \dfrac{3660}{105} = 34.86\text{A}$

and we can write $I_{a1} = 34.86 - 1.17 = 33.7\text{A}$

$E_{b1} = 105 - (33.7 \times 0.25) = 96.57\text{V}$

Now $E \propto \Phi N$ or $E = k\Phi N$

and we can write $\dfrac{E_{b0}}{E_{b1}} = \dfrac{k\Phi_0 N_0}{k\Phi_1 N_1}$

or $N_0 = \dfrac{E_{b0} \times \Phi_1 \times N_1}{E_{b1} \times \Phi_0} = \dfrac{104.42 \times 1.17 \times 1000}{96.57 \times 1.17} = 1080$ rev/min

Again, since T is constant and $T \propto \Phi I_a$, we can write

$T_2 = k\Phi_2 I_{a2}$ and $T_1 = k\Phi_1 I_{a1}$

or $\dfrac{T_2}{T_1} = \dfrac{k\Phi_2 I_{a2}}{k\Phi_1 I_{a1}}$ But $T_2 = T_1$

and $\Phi_2 = \Phi_1$

$\therefore I_{a2} = I_{a1} = 33.7\text{A}$

If R is the added resistance to reduce speed then

$E_{b2} = 105 - 33.7 (R + 0.25)$

Also since $E_b \propto \Phi$ and N, since flux is constant

then $\dfrac{E_{b2}}{E_{b1}} = \dfrac{800}{1000}$ or $E_{b2} = 96.57 \times \dfrac{800}{1000}$

Back e.m.f. (at reduced speed) = 77.26V

Thus 77.26 = 105 − 33.7 (R + 0.25)

= 105 − 33.7R − 8.43

or 33.7R = 105 − 85.69

$$R = \frac{19.31}{33.7} = 0.57\Omega$$

Note. As for Q13.6, flux being proportional to field current, I_f is substituted for Φ.

13.8. Cold condition

$$I_{f0} = \frac{230}{200} = 1.15A$$

$I_{a0} = 50 − 1.15 = 48.85A$

$E_{b0} = 230 − (48.85 \times 0.2) = 220.23V$

Hot condition

Temp rise = 60 − 15 = 45°C

$$\therefore R_{a1} = 0.2\left[1+\left(\frac{0.4}{100}\times 45\right)\right] = 0.236\Omega$$

Similarly $R_{f1} = 200\left[1+\left(\frac{0.4}{100}\times 45\right)\right]$

= 200 + (200 × 45 × 0.004) = 236Ω

So $I_{f1} = 50 − 0.975 = 49.025A$

$E_{b1} = 230 − (49.025 \times 0.235) = 218.43V$

Again since $E_b \propto \Phi$, and N and since $\Phi \propto I_f$

we write $E_{b0} = kI_{f0}N_0$

and $E_{b1} = kI_{f1}N_1$

or $\dfrac{N_1}{N_2} = \dfrac{E_{b1}}{E_{b0}} \times \dfrac{I_{f0}}{I_{f1}}$ or $N_1 = \dfrac{E_{b1} \times I_{f0} \times N_0}{E_{b0} \times I_{f1}}$

and $N_1 = \dfrac{218.43 \times 1.15 \times 1000}{220.23 \times 0.975} = 1160$ rev/min

13.9. (a) $V = E_b + I_aR_a$ or $E_b = V − I_aR_a$

= 230 − (200 × 0.35)

and $E_b = 160V$

Also since $E_b = \dfrac{Z\Phi N}{60} \times \dfrac{P}{A}$ then $N = \dfrac{E_b}{Z\Phi} \times \dfrac{60A}{P}$

or $N = \dfrac{160 \times 60 \times 2}{294 \times 0.025 \times 4} = 653\,\text{rev/min}$

(b) Again torque is given by:

$$T = 0.159 \times Z\Phi I_a \,\dfrac{P}{A}\,\text{Nm}$$

$$= 0.159 \times 294 \times 0.025 \times 200 \times \dfrac{4}{2} = 467.5\,\text{Nm}$$

13.10. $I_{f0} = \dfrac{230}{104.5} = 2.2\text{A}\quad I_{a0} = 5 - 2.2 = 2.8\text{A}$

Also since $I_{f1} = 2.2\text{A}$ then $I_{a1} = 50 - 2.2 = 47.8\text{A}$

Again $E_{b1} = 230 - (47.8 \times 0.4) - 2 = 208.88\text{V}$

And $E_{b0} = 230 - (2.8 \times 0.4) - 2 = 226.88\text{V}$

(a) Since $\dfrac{E_{b1}}{E_{b0}} = \dfrac{k\Phi_1 N_1}{k\Phi_0 N_0}$

Then $N_0 = \dfrac{E_{b0} N_1}{E_{b1}}$ assuming constant flux.

or $N_0 = \dfrac{226.88 \times 600}{208.88} = 648\,\text{rev/min}$

(b) At 600 rev/min $E_{b1} = 208.88\text{V}$

Assuming a constant flux, then for 500 rev/min

$$E_{b2} = 208.88 \times \dfrac{5}{6}$$

$\therefore E_{b2} = 174.07\text{V}$

$\therefore$ The voltage across the armature has to be reduced by:

$230 - 174.07 = 55.93\text{V}$

or since $V = E_b + I_a (R_a + R) + 2$

then $V - E_b = I_a (R_a + R) + 2$

or $55.93 = I_a R_a + I_a R + 2$

So $I_a R = 55.93 - 2 - (47.8 \times 0.4) = 34.81$

$$R = \frac{34.81}{47.8} = 0.73\Omega$$

(c) Under the new condition, Φ must be altered, hence I_f

$$\therefore \frac{E_{b3}}{E_{b1}} = \frac{k\Phi_3 N_3}{k\Phi_1 N_1} \text{ or } E_{b3} = 230 - (30 \times 0.4) - 2 = 216V$$

$$\frac{216}{208.88} = \frac{I_{f3} \times 750}{2.2 \times 600}$$

So $I_{f3} = \dfrac{216 \times 2.2 \times 600}{208.88 \times 750} = 1.82A$

Thus current – and hence the flux, is reduced to:

$$\frac{1.82}{2.2} = 0.827 = 82.7\%$$

SELECTION OF TYPICAL SECOND CLASS EXAMINATION QUESTIONS

1. A thin rectangular plate 3.5cm × 2.5cm is totally covered on both sides with Nickel 0.12mm thick in 8.25h. The current required if supplied to a voltameter causes 0.0805kg of silver to be deposited in 1h. If the Nickel E.C.E. is 304 × 10^{-9}kg/C, calculate the Nickel density in kg/m^3, if the E.C.E. of silver is 1118 × 10^{-9}kg/C.

2. A 110V D.C. lighting system comprises six 150W and forty 60W lamps. Calculate the inductance of a coil of negligible resistance that if placed in series, would operate on 230V, 50Hz mains.

3. The resistance of the armature, field coils and starter of a 220V shunt motor are 0.2, 165 and 9.8Ω respectively, the field being connected across the first stud of the starter and an armature terminal. Calculate (a) the field current at the instant of starting, (b) the field current when running and (c) the total current taken by the motor at the instant of starting, considering the armature as stationary.

4. If the instantaneous value of a current is represented by $i = 70.7 \sin 520t$, calculate the current's (a) maximum value, (b) r.m.s. value, (c) frequency and (d) instantaneous value 0.0015s after passing through zero.

5. The maximum value of a 50Hz sinusoidal current wave is 170A. Find graphically the instantaneous currents after 0.001s, 0.003s, 0.006s and 0.008s after zero and increasing positively.

6. The O.C. voltage of a cell, measured by a voltmeter of 100Ω resistance, was 1.5V, and the P.D. when supplying current to a 10Ω resistance was 1.25V, measured by the same voltmeter. Determine the cell's e.m.f. and internal resistance.

7. An alternating current series circuit consists of a coil A that has an inductance of 0.3H and negligible resistance and a resistor B of 100Ω. The supply voltage is 200V with a frequency of 50Hz. Determine (a) the circuit's impedance, (b) the current flowing and (c) the power factor.

8. What is meant by the term 'back e.m.f.' as applied to an electric motor? A 40kW, 220V shunt motor has a full-load efficiency of 90%, an armature resistance of 0.075Ω and a shunt-field resistance of 55Ω. When 'at starting', the starter handle is moved onto the first stud. It is desired to limit the current through the armature to 1.5 times the value it has when the motor is on full load. What must be the total value of the starting resistance? If, on overload, the speed falls to 90% of its normal full-load value, what would be the armature current? Neglect the effect of armature reaction.

9. State Lenz's law. An iron ring is wound with a coil of 84 turns and carries 0.015mWb of residual magnetism. When the coil is excited, the magnetic flux increases to 0.3mWb in 0.12s. Calculate the average value of the e.m.f. that will be self-induced in the exciting coil while the flux is increasing, and state the *direction* in which it will act relative to the supply voltage, giving reasons.

10. Find the impedance and power factor of an A.C. circuit consisting of 2 electronic components, A and B. A has a resistance of 2Ω and inductive reactance of 14Ω, and B has a resistance of 10Ω and a capacitive reactance of 6Ω.

11. Define the temperature coefficient of resistance of a conductor. State a conductor that has a negative temperature coefficient. When first switched on, the field winding of a 200V shunt motor takes 2A. After running for 2 hours the field current decreases to 1.7A, the supply voltage having remained constant and the shunt regulator setting not having changed. If the ambient air temperature is 15°C,

calculate the average temperature rise in the windings. Temperature coefficient of resistance of copper is 0.004 28 at 0°C.

12. A 440V single-phase motor is rated at 7.5kW and operates at a power factor of 0.8 (lagging) with an efficiency of 88%. Find the current taken from the supply.

13. A coil consumes 300W when the voltage is 60V D.C. On an A.C. circuit the consumption is 1200W when the voltage is 130V. What is the coil's reactance?

14. A motor has 4 poles, its armature is 0.36m in diameter and has 720 conductors whose effective lengths are 0.3m. The field flux density under the poles is 0.7T. Each conductor carries 30A. If the armature turns at 680 rev/min, find the torque in Newton metres and the power developed if two-thirds of the conductors are effective.

15. Define the average value and r.m.s. value of an alternating quantity. Calculate the average r.m.s. value for the stepped half wave given.

Time (ms)	0–10	10–20	20–30	30–40	40–50	50–60	60–70
Steady current (a)	2	4	6	8	6	4	2

16. Two 200V lamps are connected in series across a 400V supply. One lamp is 75W, the other is 40W. What resistance must be connected so that each lamp gives its correct illumination? What will be the power loss in the resistance?

17. Explain the meaning of the following: (a) self-inductance and (b) back e.m.f.

18. Explain the term 'power factor'. An alternator supplies 560kW at a power factor of 0.7 (lagging). What extra power would be available if the power factor is increased to 0.8 (lagging) for the same kVA output?

19. The armature winding of a 6-pole, lap-wound generator is made up from wire 250m long and 7mm² cross-sectional area. If the specific resistance of copper is 1.7 × 10^{-8}Ωm, find the armature's resistance.

20. 100V A.C. is applied to a circuit of 3Ω resistance and 4Ω reactance. Find (a) the circuit current, (b) the circuit's active e.m.f. (resistance voltage drop) and (c) the self-induced e.m.f. (reactive voltage drop).

21. A coil of 125Ω impedance has a resistance of 100Ω when connected across a 50Hz supply. Find its inductance. If the impedance falls to 120.6Ω when the frequency is varied, find the new frequency value.

22. 4.5 litres of fresh water at 17°C is heated to boiling point in 15min. If the heater is 80% efficient and the supply voltage is 220V, find the current taken from the mains and the heater's resistance. Take the density of water as 1kg/litre and the specific heat capacity as 4.2kJ/kg°C.

23. An alternating voltage of r.m.s. value 100V is applied to a circuit with negligible resistance and an inductive reactance of 25Ω. Determine the r.m.s. variation of current flowing. Show graphically the variation of current and voltage during one cycle of applied voltage. What is the value of the current when the voltage is at its *maximum* value?

24. A generator has 8 brush arms, each with 6 brushes, 30mm long and each with a bearing surface of 30mm by 20mm. The current density is 0.054A/mm² in the brushes. Find the sectional area of the cables, if the leads to the switchboard are each 9.2m and the current density must not exceed 1.0A/mm². Find the power lost in the brushes and cables. The resistivity for carbon and copper is 2550×10^{-8} and $1.7 \times 10^{-8}\Omega$m respectively.

25. An iron conductor and an aluminium conductor are connected in *parallel* to a supply. The iron conductor is 10% longer than, and half the diameter of, the aluminium conductor. Given that the ratio of the resistivities of iron to aluminium is 40 to 13, find the ratio of the currents in the 2 conductors.

26. Two currents $I_1 = 14.14$A and $I_2 = 8.5$A with a phase difference of 30° are fed into a common conductor. Find the resultant current and the heating effect in joules when it passes through a resistor of value 4Ω for a period of 2min.

27. A 6-pole D.C. generator has 498 conductors, the e.m.f. per conductor being 1.5V and the current in each 100A. Find the e.m.f. and current output of the armature, if it is (a) lap wound and (b) wave wound.

28. A circuit takes a current of 10A from 220V, 50Hz mains at a power factor of 0.866 (lagging). Find the value of the current when the voltage is (a) passing through its maximum value and (b) 0.005s later.

29. An inductance coil has a resistance of 19.5Ω and when connected to a 220V, 50Hz supply, the current passing is 10A. Find the coil's inductance.

30. State Faraday's and Lenz's laws of electromagnetic induction. A 4-pole, 250V motor has its armature removed in order to test the continuity of the field windings which

are connected in series and consist of 2000 turns each. What is the average e.m.f. induced when the current is switched off, if the flux falls from 0.026Wb to 0.001Wb in 0.2s?

31. A coil when connected across 206V A.C. mains passed a current of 10A and dissipates 500W. If it is connected in series with an impedance of 5Ω and a capacitive reactance of 4Ω, find the impedance and power factor of the complete circuit.

32. A piece of copper wire is bent to form a circle and another piece of the same wire is placed across to form a diameter, all the junctions being electrically connected. If the resistance of the straight wire is 2Ω, find the total current flowing when a P.D. of 220V is applied across the junctions.

33. A 15kW motor of efficiency 90% is supplied at 240V by a 2-wire system. The supply cables are 500m long and are of diameter 5mm. Find the current taken by the motor, the voltage at the supply point and the efficiency of the distribution system. Take the specific resistance of copper as $1.7 \times 10^{-8}\Omega$m.

34. If the impedance of a circuit is 20Ω, the resistance is 16Ω and the inductance 0.047 75H, find the frequency of the supply.

35. Differentiate clearly between the kilowatt and the kilowatt-hour. A heater with an efficiency of 85% develops 10MJ in 30min at 200V. Find the energy consumption in kilowatt-hours and the current taken. Find also the length of wire in the element if its resistance is 0.26 ohm per metre.

36. An electric heater of resistance 6.5Ω is connected in series with a coil of inductance value 0.1H. If the mains frequency is 50Hz, find the voltage to be applied to the arrangement in order to maintain 110V across the heater. If the frequency was increased by 5%, keeping the applied voltage constant, find the voltage across the heater.

37. A resistor of ohmic value 3Ω is connected in series with a coil of inductance 0.1H and resistance 1Ω. If 100V at a frequency of 50Hz is applied to the circuit, find the current flowing.

38. A 500V D.C. shunt motor has a full-load armature current of 20A. 3% of the input power is dissipated as heat in the armature. What would be the current on starting if 500V is applied across the armature? Find the value of starting resistance needed to *limit* the starting current to twice the full-load current.

39. Find the total effective reactance of a 50Hz circuit made up from a coil of inductance 100mH, in series with a capacitor of 20µF. If the coil has a resistance of 10Ω, find the circuit's impedance.

40. The armature resistance of a 200V shunt motor is 0.4Ω and the no-load armature current is 2A. When fully loaded and with an armature current of 50A, the speed is 1200 rev/min. Find the no-load speed and state the assumption made in the calculation.

SOLUTIONS TO TYPICAL SECOND CLASS EXAMINATION QUESTIONS

1. When used in the voltameter: $m = zIt$

or $I = \dfrac{0.0805}{1118 \times 10^{-9} \times 3.6 \times 10^3} = 20\text{A}$

When used for plating:

Area of coating $= 350 \times 250 \times 2 = 175\,000\text{mm}^2$

Volume of coating $= 175 \times 10^3 \times 12 \times 10^{-2}$

$$= 21\,000\text{mm}^3 = 21 \times 10^{-6}\text{m}^3$$

Also mass of nickel deposited

$= 304 \times 10^{-9} \times 20 \times 8.25 \times 3600$

$= 0.181 \text{kg}$

So density $= \dfrac{0.181}{21 \times 10^{-6}} \text{ kg/m}^3$

$= 8600 \text{kg/m}^3$

2. Total wattage $= (6 \times 150) + (40 \times 60)$

$= 3300 \text{W}$

System current $= \dfrac{3300}{110} = 30 \text{A}$

Impedance required on 230V A.C. $= \dfrac{230}{30} = 7.66\Omega$

Resistance of lamps $= \dfrac{110}{30} = 3.66\Omega$

Reactance of coil $= \sqrt{7.66^2 - 3.66^2} = 6.73\Omega$

Inductance $= \dfrac{6.73}{2 \times \pi \times 50}$

$= 0.0214 \text{H}$

3. (a) Field current at instant of starting $= \dfrac{220}{165} = 1.33 \text{A}$

(b) When running, the starter resistance is inserted into the field circuit by virtue of the position of the contact arm. Field-circuit resistance $= 165 + 9.8 = 174.8\Omega$

Field current when running $= \dfrac{220}{174.8} = 1.26 \text{A}$

(c) Armature-circuit resistance, at instant of starting $= 0.2 + 9.8 = 10\Omega$

Armature current $= \dfrac{220}{10} = 22 \text{A}$

Total current taken by motor $= 22 + 1.33 = 23.33 \text{A}$

4. (a) Maximum value of current = 70.7A

(b) The current is sinusoidal

∴ r.m.s. value = 0.707 × maximum value

$$= 0.707 \times 70.7 = 49.94A$$

(c) $i = 70.7 \sin (520 \times 0.0015)$ is in the form $i = I_m \sin \omega t$

where $\omega = 520$ radians/second. Also ω equals $2\pi f$

or $2\pi f = 520$

Thus $f = \dfrac{520}{2 \times \pi} = 82.8Hz$

(d) Again $i = 70.7 \sin 2\pi ft$, and if degrees are used for the angle, then

$i = 70.7 \sin (2 \times 180 \times 82.8 \times 0.0015) = 49.65A$

5. Plot a sine wave on graph paper with a maximum value of 170A and a base of 0.01s. A frequency of 50Hz gives the time of a half wave as 0.01s.

The wave can be plotted from a phasor of length equal to 170A or by use of tables to obtain ordinates. Thus for 30° the instantaneous value or ordinate would be 170 × sin 30° = 170 × 0.5 = 85A. So for 45° $i = 170 \times 0.707 = 120.2A$.

For 60° $i = 147.02A$. For 90° $i = 170A$, etc.

Answers from the deduced waveform:

When time $t = 0.001s$, $i = 57.5A$.

When time $t = 0.003s$, $i = 132A$.

When time $t = 0.006s$, $i = 162A$.

When time $t = 0.008s$, $i = 102A$.

6. Let E = the e.m.f. of the cell and R_i = the internal resistance.

With the voltmeter across the cell terminals only,

Current taken by voltmeter $= \dfrac{1.5}{100} = 0.015A$

Then $E = 1.5 + (0.015) \times R_i \dots$ (i)

With the voltmeter and resistor across the cell terminals,

Current taken by resistor $= \dfrac{1.25}{10} = 0.125\text{A}$

Current taken by voltmeter $= \dfrac{1.25}{100} = 0.0125\text{A}$

Current supplied by cell $= 0.125 + 0.0125 = 0.1375\text{A}$

Thus $E = 1.25 + 0.1375R_i \dots$ (ii)

Solving (i) and (ii) then $E = 1.25 + 0.1375R_i$

and $E = 1.5 + 0.015R_i$

Subtracting $0 = -0.25 + 0.1225R_i$

or $R_i = \dfrac{0.25}{0.1225} = 2.04\,\Omega$

and $E = 1.5 + (0.015 \times 2.04) = 1.5 + 0.031$

Thus cell e.m.f. $= 1.531\text{V}$

7. $R_A = 0\,\Omega$ $X_A = 2\pi fL = 2 \times \pi \times 50 \times 0.3$

$\qquad = 94.2\,\Omega$

$R_B = 100\,\Omega$ $Z = \sqrt{100^2 + 94.2^2}$

$\qquad = 137.4\,\Omega$

Current $I = \dfrac{200}{137.4} = 1.45\text{A}$

Power factor, $\cos\phi = \dfrac{R}{Z} = \dfrac{100}{137.4} = 0.73$ (lagging)

8. Output = 40 × 1000 watts

$$\text{Input} = 40 \times 10^3 \times \frac{100}{90}\text{watts} = 44\,444\text{W}$$

$$\text{Input current} = \frac{44\,444}{220} = 202.02\text{A}$$

$$\text{Shunt-field current} = \frac{220}{55} = 4\text{A}$$

Armature current = 202.02 − 4 = 198.2A

Armature starting current = 198.2 × 1.5 = 297.3A

$$\text{Resistance of armature circuit} = \frac{220}{297.3} = 0.74\Omega$$

Resistance to be added = 0.74 − 0.075 = 0.665Ω

On normal load E_b = 220 − (198.2 × 0.075) volts

$$= 205.13\text{V}$$

On 90% speed E_{b1} = 0.9 × 205.13 = 184.62V

∴ Armature voltage drop = 220 − 184.62 = 35.38V

$$\text{Armature current} = \frac{35.38}{0.075} = 471.7\text{A}$$

9. $E_{av} = \dfrac{N(\Phi_2 - \Phi_1)}{t}$ volts

$$= \frac{84(3\times10^{-4} - 1.5\times10^{-5})}{0.12}$$

$$= 0.1995\text{V or } 199.5\text{mV}$$

The induced voltage will *oppose* the applied supply voltage, thus reducing the rate of current growth.

10. Total circuit resistance $= 2 + 10 = 12\Omega$

Total circuit reactance $= 14 - 6 = 8\Omega$ (inductive)

From impedance relationship:

$$Z = \sqrt{12^2 + 8^2} = 14.4\Omega$$

The circuit impedance is 14.4Ω

$$\cos \phi = \frac{R}{Z} = \frac{12}{14.4} = 0.832$$

Thus power factor $= 0.832$ (lagging)

11. Carbon has a negative temperature coefficient.

$$\text{Resistance of field (cold)} = R_1 = \frac{200}{2} = 100\Omega$$

$$\text{Resistance of field (hot)} = R_2 = \frac{200}{1.7} = 117.64\Omega$$

Then $100 = R_0 (1 + 0.004\,28 \times 15)$

and $117.64 = R_0 (1 + 0.004\,28 \times T)$

Dividing $\dfrac{117.64}{100} = \dfrac{1 + 0.0024\,28T}{1 + 0.0642}$

or $1.1764 \times 1.0642 = 1 + 0.004\,28T$

and $1.2519 - 1 = 0.004\,28T$

$$\therefore T = \frac{0.2519}{0.00428} = 58.85°C$$

Temperature rise of winding $= 58.85 - 15°C$

$$= 43.85°C$$

12. Motor output $= 7.5 \times 1000 = 7500W$

Motor input $= 75000 \times \dfrac{100}{88} = 8523W$

But $VI \cos \phi = P \quad \therefore I = \dfrac{P}{V \cos \phi} = \dfrac{8523}{440 \times 0.8}$ amperes

or $I = \dfrac{8523}{352} = 24.21A$

Thus motor current $= 24.2A$

13. When on D.C. $P = \dfrac{V^2}{R} \quad \therefore R = \dfrac{V^2}{P} = \dfrac{60 \times 60}{300} = 12\Omega$

When on A.C. $P = 1200W$ also $P = I^2 R$

So $I^2 = \dfrac{1200}{12} = 100$ or $I = 10A$

The impedance Z of the circuit $= \dfrac{V}{I} = \dfrac{130}{10} = 13\Omega$

So reactance $X = \sqrt{13^2 - 12^2} = 5\Omega$

Thus reactance of coil $= 5\Omega$

14. Force on 1 conductor is given by $F = BI\ell$ newtons

or $F = 0.7 \times 30 \times 0.3 = 6.3N$

No of conductors in the field at any instant $= \dfrac{2}{3} \times 720 = 480$

Total force $= 480 \times 6.3 = 3024N$

Torque $=$ force $\times$ radius $= 3024 \times 0.18$ Nm

$\qquad = 544.32Nm$

So torque exerted = 544Nm

Power developed $= \dfrac{2\pi NT}{60}$

$$= \dfrac{2\times \pi \times 680 \times 544}{60}$$

$$= 38.7\text{kW}$$

15. If plotted, this waveform wi l be found to be made up of 7 rectangular blocks, the mid-ordinates of which are 2, 4, 6, etc., as given.

$\therefore$ Average value $= \dfrac{2+4+6+8+6+4+2}{7} = \dfrac{32}{7}$

$$= 4.57\text{A}$$

Also r.m.s. value $= \sqrt{\dfrac{2^2 + 4^2 + 6^2 + 8^2 + 6^2 + 4^2 + 2^2}{7}}$

$$= \sqrt{\dfrac{4 + 16 + 36 + 64 + 36 + 16 + 4}{7}}$$

$$= \sqrt{\dfrac{176}{7}} = \sqrt{25.14} = 5\text{A}$$

16. 75W lamp. $I = \dfrac{75}{200} = 0.375\text{A}$

40W lamp. $I = \dfrac{40}{100} = 0.2\text{A}$

With lamps in series 40W lamp will only pass 0.2A

$\therefore$ (0.375 – 0.2) amperes must be passed through a shunt resistor connected across the 40W lamp. This resistor is also to be suitable for 200V and its resistance value:

$$= \dfrac{200}{0.175} = 1143\Omega$$

Power loss in this resistor = 200 × 0.175 = 35W

17. Standard explanations, written in the student's own words, should be drawn from the relevant chapter to provide a satisfactory explanation of the meaning of both (a) self-inductance and (b) back e.m.f.

18. Since $P = VI \cos \phi$, then $VI = \dfrac{P}{\cos \phi}$

or kilovolt amperes $(S) = \dfrac{\text{kilowatts } (P)}{\text{power factor } (\cos \phi)}$

Thus $S = \dfrac{560}{0.7} = 800\text{kVA}$

If this kVA or S value is to be maintained

New $P = 800 \times 0.7 = 640\text{kW}$

Thus extra power available $= 640 - 560 = 80\text{kW}$

19. Here $\rho = 1.7 \times 10^{-8}\,\Omega\text{m}$. $\ell = 250\text{m}$. $A = 7 \times 10^{-6}\text{m}^2$.

Then $R = \dfrac{\rho \ell}{A} = \dfrac{1.7 \times 10^{-8} \times 250}{7 \times 10^{-6}} = 0.607\Omega$

This would be the resistance of the length of wire, but this is a lap-wound generator, with 6 parallel paths in the armature. Thus resistance of 1 parallel path $= \dfrac{0.607}{6} = 0.101\,16\Omega$.

But there are 6 paths in parallel so the equivalent resistance is one-sixth of the above

$= \dfrac{0.101\,16}{6} = 0.016\,86\Omega$

20. Impedance of circuit $= \sqrt{3^2 + 4^2}$

$= 5\Omega$

Thus $Z = 5\Omega$

(a) Current $I = \dfrac{V}{Z} = \dfrac{100}{5} = 20A$

(b) Resistive voltage drop $V_R = IR = 20 \times 3 = 60V$

(c) Reactive voltage drop $V_X = IX = 20 \times 4 = 80V$

21. Reactance X_L at 50Hz $= \sqrt{Z^2 - R^2} = 100\sqrt{1.25^2 - 1^2} = 75\Omega$

Also $X_L = 2\pi fL \quad \therefore L = \dfrac{75}{2 \times \pi \times 50} = 0.24H$

At new frequency $X_L = \sqrt{120.6^2 - 100^2} = 67.4\Omega$

So $\dfrac{\text{new frequency}}{50} = \dfrac{67.4}{75}$

or new frequency $= 67.4 \times \dfrac{50}{75}$

$= 44.93Hz.$ The new frequency value will be 45Hz

22. Mass of water $=$ volume $\times$ density $= 4.5 \times 1 = 4.5kg$

Heat received by water $= 4.5 \times 4.2\,(100 - 17)$

$= 18.9 \times 83 \text{ kilojoules} = 1569kJ$

Electrical energy supplied to heater $= 1569 \times \dfrac{100}{80}$

$= 1961kJ$

Power rating of heater $= \dfrac{\text{energy}}{\text{time}} = \dfrac{1961}{15 \times 60}$

$= 2.18kW$

Current taken from mains $= \dfrac{2180}{220} = 9.9A$

Mains current $= 10A$ (approx.)

Resistance of heater $= \dfrac{220}{9.9} = 22.2\Omega$

23. Here $R = 0\Omega$ and $Z = X = 25\Omega$

$$\therefore \text{ r.m.s. value of current } I = \frac{V}{Z} = \frac{100}{25} = 25A$$

The graphical solution consists of a sinusoidal voltage wave with a sinusoidal current wave lagging it by 90°, since the circuit is wholly inductive. When voltage is maximum, current is zero. When voltage has fallen to zero, the current has risen to its maximum value and as voltage rises to its negative maximum the current falls to zero.

When V is a maximum the current value is zero.

24. Bearing surface of 1 brush = 30×20 mm² = 600mm²

With a current density of 0.054A/mm², the current carried by one brush = $600 \times 0.054 = 32.4A$

With 8 brush arms, there are 4 positive and 4 negative brush arms. Also since there are 6 brushes per arm, the number of brushes in parallel carrying current into or out of the armature = $4 \times 6 = 24$ brushes, or total current carried by brushes = $32.4 \times 24 = 777.6A$.

Current display density in the cable is limited to 1.6A/mm².

$$\therefore \text{ The cable is able to carry } 777.6A = \frac{777.6}{1.6} = 486\text{mm}^2$$

$$\text{Resistance of cable} = \frac{\rho \ell}{A} = \frac{1.7 \times 10^{-8} \times 9.2 \times 2}{486 \times 10^{-6}} = 0.643 \times 10^{-3} \text{ ohms}$$

Voltage drop in cable = $777.6 \times 0.643 \times 10^{-3} = 0.5V$

Power loss in cable = $777.6 \times 0.5 = 388.8W$

Resistance of the brushes is given by $R = \frac{\rho \ell}{A}$ where ℓ is the length of a +ve plus a −ve brush and A is the area of half the total number of brushes.

$$\text{Thus } R = \frac{2550}{10^8} \times \frac{2 \times 30 \times 10^{-3}}{600 \times 10^{-6} \times 24} = 0.106 \times 10^{-3} \text{ ohms}$$

Power loss in the brushes is given by I^2R

$= 777.6^2 \times 0.106 \times 10^{-3} = 64\ 1$ watts

Thus power loss in brushes $= 64.1W$

25. The equations for the iron and aluminium conductors can be written as $R_i = \dfrac{\rho_i \ell_i}{A_i}$

and $R_a = \dfrac{\rho_a \ell_a}{A_a}$

$\therefore \dfrac{R_i}{R_a} = \dfrac{\rho_i \ell_i}{A_i} \Big/ \dfrac{\rho_a \ell_a}{A_a} = \dfrac{\rho_i \ell_i A_a}{\rho_a \ell_a A_i}$

But $\ell_i = 1.1\ \ell_a$

and $d_i = \tfrac{1}{2} d_a$ $\qquad \therefore A_i = \dfrac{A_a}{4}$ since area $\propto$ diameter^2.

So $\dfrac{R_i}{R_a} = \dfrac{40 \times 1.1 \times \ell_a \times A_a \times 4}{13 \times \ell_a \times A_a}$

$= \dfrac{13.54}{1}$

Since the resistance ratio of iron to aluminium is 13.54 to 1, and as the wires are in *parallel*, the currents in the wires are in the ratio iron:

aluminium $= 1 : 13.54$.

26. Resultant current $I = \sqrt{I_1^2 + I_2^2 + (2 \times I_1 \times I_2 \times \cos 30)}$

$= \sqrt{14.14^2 + 8.5^2 + (2 \times 14.14 \times 8.5 \times 0.866)}$

$= 21.92A$

Resultant current $= 21.92A$

Power dissipated $= I^2R = 21.9^2 \times 4$ watts

Energy at heat = I^2Rt joules

$$= 21.92^2 \times 4 \times 2 \times 60$$

$$= 230\ 602J = 230.602kJ$$

27. (a) *Lap wound.* A = P = 6

$$\text{Conductors in series} = \frac{\text{total conductors}}{\text{parallel paths}} = \frac{498}{6} = 83$$

E.m.f. of 1 parallel path = e.m.f. of the machine

$$= 1.5 \times 83 = 124.5V$$

Current per parallel path = current in 1 conductor = 100A

Current of 6 paths in parallel = 6 × 100 = 600A

(b) *Wave wound.* A = 2

$$\text{Conductors in series} = \frac{\text{total conductors}}{\text{parallel paths}} = \frac{498}{2} = 249$$

E.m.f. of 1 parallel path = e.m.f. of machine

$$= 1.5 \times 249 = 373.5V$$

Current per parallel path = current in 1 conductor = 100A

Current of 2 paths in parallel = 2 × 100 = 200A

28. The power factor of this circuit is 0.866 (lagging) or cos ϕ = 0.866 (lagging) and ϕ = 30°, where ϕ is the angle of lag between the voltage and the current – the latter lagging the former.

The values of 10A and 220V as given can be assumed to be r.m.s. values. So the maximum value of current is given by $I = 0.707\ I_m$ (here I_m is the maximum value). Sine-wave working is assumed.

or $I_m = \dfrac{10}{0.707} = 14.14A$

Also $V = 0.707\ V_m$ (maximum value).

or $V_m = \dfrac{220}{0.707} = 311.08V$

The voltage and current are written as:

$v = V_m \sin \omega t$

and $i = I_m \sin (\omega t - \phi)$ ϕ is in radians

At a frequency of 50Hz, time for 1 cycle $= \dfrac{1}{50}$ seconds

(a) When the voltage is at a maximum, the time is for $\dfrac{1}{4}$ cycle or

$t = \dfrac{1}{4 \times 50} = 0.005s$

Current at this instant is given by substituting in

$i = 14.14 \sin (2\pi 50 \times 0.005 - \phi)$

or $i = 14.14 \sin (2 \times 180 \times 50 \times 0.005 - 30)$. π and ϕ in degrees

$= 14.14 \sin (90 - 30)$

$= 14.14 \sin 60° = 12.25A$

(b) At an instant 0.005s later t would be 0.01s

$\therefore i = 14.14 \sin (2 \times 180 \times 50 \times 0.01 - 30)$

or $i = 14.14 \sin (180 - 30) = 14.14 \sin 150°$

$i = 14.14 \sin 30° = 7.07A$

29. From the information given, the impedance Z of the circuit is $= \dfrac{V}{I}$ or

$Z = \dfrac{220}{10} = 22\Omega$

The resistance R is 19.5Ω. Therefore the reactance X is obtained from $X = \sqrt{Z^2 - R^2} = \sqrt{22^2 - 19.5^2} = 10.15\Omega$

Also $X = 2\pi fL$ $\therefore L = \dfrac{10.15}{2\times\pi\times 50} = 0.032H$

30. From Faraday's law.

$$E_{av} = \frac{N(\Phi_1 - \Phi_2)}{t}\ volts$$

$$= \frac{200\times 4(2.6\times 10^{-2} - 0.1\times 10^{-2})}{0.2}$$

Induced e.m.f. $= 1000V$ or $1kV$

31. Z of coil $= \dfrac{206}{10} = 20.6\Omega$

Also since $P = I^2R$, then R of coil $= \dfrac{500}{10^2} = 5\Omega$

Thus reactance X of coil $= \sqrt{Z^2 - R^2}$

or $X = \sqrt{20.6^2 - 5^2}$

$\qquad = 19.98\Omega$

Z of additional circuit $= 5\Omega$

X of additional circuit $= 4\Omega$ (capacitive)

$\therefore R$ of additional circuit $= \sqrt{5^2 - 4^2} = 3\Omega$

Total resistance of circuit $= 5 + 3 = 8\Omega$

Total reactance of circuit $= 19.98 - 4 = 15.98\Omega$

Note. The inductive and capacitive reactances have been subtracted.

Total circuit impedance $= \sqrt{8^2 + 15.98^2} = 17.88\Omega$

Since $\cos\phi = \dfrac{8}{17.88} = 0.44\,(lagging)$

Thus power factor = 0.44 (lagging), since circuit is net inductive

32. Since the circuit is built up from wire of the same material and cross-sectional area, then the resistances of various parts of the circuit are proportional to length.

The resistance of the diameter = 2 ohms

∴ The resistance of the circumference = $\pi d = 2\pi$ ohms.

The resistance of 1/2 circumference = π ohms

The circuit is made up of a diameter and two $\dfrac{1}{2}$ circumferences in parallel. ∴ if R is

the circuit resistance $\dfrac{1}{R} = \dfrac{1}{2} + \dfrac{1}{\pi} + \dfrac{1}{\pi} = \dfrac{\pi + 2 + 2}{2\pi} = \dfrac{\frac{22}{7} + 4}{2 \times \frac{22}{7}}$

or $\dfrac{1}{R} = \dfrac{\dfrac{22 + 28}{7}}{\dfrac{44}{7}} = \dfrac{50}{44} = \dfrac{25}{22}$

or $R = \dfrac{22}{25} = 0.88\Omega$

With 220V applied across R, the current would be $= \dfrac{220}{0.88} = 250\text{A}$

33. Output of motor = 15kW

Input to motor $= \dfrac{15}{0.9}$ kilowatts $= 16\,666\text{W}$

Input current or current in cables $= \dfrac{16\,666}{240} = 69.44\text{A}$

Resistance of cable is given by $\dfrac{\rho \ell}{A}$

or $R = \dfrac{1.7 \times 10^{-8} \times 500 \times 2}{\dfrac{\pi}{4} \times (5 \times 10^{-3})^2} = 0.866\Omega$

Voltage drop in cable $= 69.44 \times 0.866 = 60.14\text{V}$

Input voltage at supply cables $= 240 + 60.14 = 300.14\text{V}$

$$\text{Efficiency of distribution} = \frac{\text{Power output from cables}}{\text{Power input to cables}}$$

$$= \frac{240 \times 69.44}{300.14 \times 69.44}$$

$$= 0.799 = 79.9\%$$

34. Here $Z = 20\Omega$ and $R = 16\Omega$ $\therefore X = \sqrt{20^2 - 16^2} = 12\Omega$

Also $X = 2\pi fL = 2 \times \pi \times f \times 0.047\,75 = 12$

Thus $f = \dfrac{12}{2 \times \pi \times 0.047\,75}$

$\qquad = 40\text{Hz}$

The frequency of the supply is 40Hz.

35. Output of heater $= 10\text{MJ}$

also $1\text{kWh} = 3600 \times 1000 = 36 \times 10^5$ joules

Now energy output of heater $= \dfrac{10 \times 10^6}{36 \times 10^5}$

$$= 2.78\text{kWh}$$

$$\text{Energy input} = \frac{\text{output}}{\text{efficiency}} = \frac{2.78}{0.85} = 3.27\text{kWh}$$

$$\text{Power input} = \frac{\text{energy}}{\text{time}} = \frac{3.27}{0.5} = 6.54\text{kW}$$

Also, since $P = I^2R = I \times R \times I = V \times \dfrac{V}{R} = \dfrac{V^2}{R}$

$\therefore R = \dfrac{V^2}{P}$

Thus $R = \dfrac{200^2}{6540} = 6.12\Omega$

Length of element at 0.26 Ω/m $= \dfrac{\text{resistance}}{\text{ohms per metre}} = \dfrac{6.12}{0.26}$

$$= 23.54$$

Current taken $= \dfrac{P}{V} = \dfrac{6540}{200} = 32.7\text{A}$

36. If 110V is maintained across the heater, the current will be $\dfrac{110}{6.5} = 16.92\text{A}$

At 50Hz reactance of coil $= 2\pi fL$

$$= 2 \times \pi \times 50 \times 0.1 = 31.4\Omega$$

Impedance Z of complete circuit $= \sqrt{6.5^2 + 31.4^2} = 32.07\Omega$

Applied voltage for $16.92\text{A} = 16.92 \times 32.07 = 542.6\text{V}$

If frequency rises 5%, reactance rises 5%.

New reactance $= 3.14 \times 1.05 = 32.97\Omega$

New impedance $= \sqrt{6.5^2 + 32.97^2} = 33.6\Omega$

Circuit current $= \dfrac{542.6}{33.6} = 16.15\text{A}$

Voltage across heater $= 16.15 \times 6.5 = 104.97 = 105\text{V (approx.)}$

37. Total resistance of circuit $= 3 + 1 = 4\Omega$

Reactance of circuit is given by $X = 2\pi f L$

or $X = 2 \times \pi \times 50 \times 0.1 = 31.4\Omega$

Circuit impedance $Z = \sqrt{4^2 + 31.4^2} = 31.65\Omega$

Current flowing $= \dfrac{100}{31.65} = 3.16A$

38. Neglecting the field current, as it will be small

Input power $= VI$

$$= 500 \times 20 = 10\,000W$$

3% of the input power $= \dfrac{3}{100} \times 10\,000 = 300W$

This is dissipated as heat in the armature, i.e. it is a copper or I^2R loss.

$\therefore I^2R_a = 300$ or $20^2 R_a = 300$

and $R_a = \dfrac{300}{400} = 0.75\Omega$

The starting current with only armature resistance to limit the armature current:

$I_{as} = \dfrac{500}{0.75} = 666.66A$

Twice full-load current $= 20 \times 2 = 40A$

$\therefore$ Total resistance required in the armature circuit to limit starting current to

$40A = \dfrac{500}{40} = 12.5\Omega$

Since $R_a = 0.75\Omega$

Series resistance will be $12.5 - 0.75 = 11.75\Omega$

39. Inductive reactance $X_L = 2\pi fL$

$$= 2 \times \pi\,50 \times 100 \times 10^{-3}$$

or $X_L = 31.4\Omega$

Capacitive reactance $X_C = \dfrac{10^6}{2\pi fC} = \dfrac{10^6}{2\times\pi\times50\times20} = 159.23\Omega$

Total effective reactance $X = 159.23 + 31.4$

or $X = 127.83\Omega$ (capacitive)

Impedance of circuit $Z = \sqrt{R^2 + X^2}$

$$= \sqrt{10^2 + 127.83^2} = 128.3\Omega$$

40. On no load: $E_{b0} = V - I_a R_a = 200 - (2 \times 0.4)$

$$= 200 - 0.8 = 199.2V$$

On full load: $E_{b1} = 200 - (50 \times 0.4) = 200 - 20 = 180V$

As this is a shunt motor, constant field current and therefore the same constant flux is assumed for the no-load and full-load conditions.

Since $E_{b0} = k\Phi_0 N_0$ and $E_{b1} = k\Phi_1 N_1$ and $\Phi_0 = \Phi_1$

then $\dfrac{E_{b0}}{E_{b1}} = \dfrac{k\Phi_0 N_0}{k\Phi_1 N_1}$ or $N_0 = \dfrac{N_1 E_{b0}}{E_{b1}}$

This gives $N_0 = 1200 \times \dfrac{199.2}{180} = 1328$ rev/min

SELECTION OF TYPICAL FIRST CLASS EXAMINATION QUESTIONS

1. The rudder motor of a steering gear (Ward-Leonard type) is a compound-wound machine, details of which are given below. From this information find: (a) armature current, (b) the torque developed by the motor and (c) output power of motor. Armature – volts 90, resistance 0.0288Ω, rev/min 370. IR drop over armature is taken as 8.5V. Shunt field – separately excited from 110V supply, resistance 65Ω, turns per pole 1000. Series field – separately excited by a line current of 325A, 11 turns per pole. Torque Nm = (armature current (A) × flux (kWb) per pole × 10^{-4} × 6.8) − 15.

M.m.f. per pole (At)	3000	3500	4000	4500	5000	5500	6000
Flux per pole (kWb)	2500	2700	2860	2990	3100	3190	3275

2. A 500V installation consists of a synchronous motor taking 50kW working in parallel with a load of 90kW having a power factor of 0.6 (lagging). If the power factor of the combined load is 0.8 (lagging), find the motor's power factor and reactive kVA.

3. A 175kVA, 6600/440V, single-phase transformer has an iron loss of 2.75kW. The primary and secondary windings have resistances of 0.4Ω and 0.0015Ω respectively. Calculate the efficiency on full load when the power factor is 0.9.

4. A shunt-wound generator has the following O.C. characteristic. If the actual field-resistance value is half that of the critical field resistance, above which the machine fails to excite, find the O.C. voltage. The e.m.f. when the generator is operating at a load of 200A falls to 135V. Find the terminal voltage and the armature resistance.

Field current (A)	0.5	1.0	2.0	3.0	4.0	5.0
O.C. voltage (V)	55	90	133	160	179	193

5. What are the units of length, mass and force in the SI system of units?

A current of 12A produces a magnetic flux of 0.4mWb in a coil of 60 turns. What e.m.f. is induced in the coil if the current of 12A was reversed in 25ms? What is the inductance of the coil in henries?

6. A 2-wire ring main, 2km long, is supplied at a point 'X' with 220V. At a point 'Y' situated 400m from 'X' there is a load of 110A. The resistance of 1km of single, main conductor is 0.032Ω. Calculate the current in each section of the main and the voltage at the load.

7. In a supply, the voltage and current vary sinusoidally at a frequency of 50Hz, the r.m.s. values being 311.2V and 70.7A respectively, at a power factor of 0.866 (lagging). Plot the graphs, for a single cycle, of the voltage and current in their correct relative positions. From them, derive the instantaneous values of voltage and current at 3, 6, 11 and ˙8ms from the time that the voltage passed through zero, in the course of it increasing positively.

8. The moving coil of a permanent-magnet voltmeter is made of copper and has a resistance of 5Ω when at 20°C. The instrument is connected in series with a resistance of 995Ω at 20°C. Calculate the percentage error, high or low, of the reading at 50°C, if the series resistance is made of (a) copper and (b) manganin. Take the temperature coefficient of copper as 0.004 28/°C at 0°C and that of manganin as zero.

9. The field windings of a motor consist of 8 coils connected in series, each coil having 1200 turns. The flux linked with each coil is 0.05Wb when the current is 5A. Calculate the circuit's inductance and the value of the average e.m.f. induced, if the current was cut off in 50ms.

10. A 550kVA, 50Hz, single-phase transformer has 1875 and 75 turns in the primary and secondary windings respectively. If the secondary voltage is 220V, calculate (a) primary voltage, (b) primary and secondary currents and (c) maximum flux value.

11. The armature and field resistances of a 220V shunt motor are 0.25 and 110 ohms respectively and, when running on no load, the motor takes 6A. Calculate the losses attributable to iron, friction and windage and, assuming this value to remain constant on all loads, determine the efficiency when the current supplied is 62A.

12. Explain the principles underlying the necessity for the introduction of the term 'power factor' when considering A.C. machinery. An alternator is supplying a load of 560kW at a power factor of 0.7 (lagging). If apparatus is installed that raises the power factor to 0.8 (lagging), calculate the increase in power available for the same kVA loading.

13. A heater unit of negligible inductance has a resistance of 6.5Ω and is intended for use with 100V mains. For what 50Hz voltage would it be suitable when placed in series with an external device, of negligible resistance, having an inductance of 0.01H? If the frequency rises by 5% and the voltage remains constant, what would be the resulting change of voltage at the heater terminals?

14. The magnetic field in the air gap of a 2-pole motor has a flux density of 0.8T. The armature is wound with 246 conductors, each of 400mm effective length, mounted at 150mm effective radius, and at full load each conductor carries a current of 20A. Assuming that the actual torque produced is equivalent to that due to two-thirds of the number of conductors cutting the lines of force at right angles, find (a) the torque in newton metres and (b) the shaft power developed at 500 rev/min.

15. State briefly the meaning of the expressions 'star-connected' and 'delta-connected' as applied to 3-phase A.C. practice. What is the ratio of the maximum line voltage to the maximum phase voltage in each case.

Determine the line current taken by a 440V, 3-phase, star-connected motor having an output of 45kW at 0.88 (lagging) power factor and an efficiency of 93%.

16. A battery is to consist of a number of cells connected in series. Each cell has an e.m.f. of 1.5V and an internal resistance of 0.5Ω. The external load has a resistance of 100Ω and requires approximately 2W for satisfactory operation. Determine how many cells will be required.

17. A battery comprises 6 cells each of e.m.f. 2.2V and internal resistance 0.1Ω. Determine the value of the load resistance connected when the battery delivers maximum power and evaluate this maximum power.

18. If an alternator supplies the following loads: (a) 200kW lighting load at unity power factor, (b) 400kW induction-motor load at 0.8 (lagging) power factor, (c) 200kW synchronous-motor load, find the power factor of the synchronous-motor load, to give an overall power factor of 0.97 (lagging).

19. A D.C. shunt-wound machine is run as a motor, being supplied with 55kW at 220V when its speed is 500 rev/min. Find the speed at which this machine must be driven to generate an output of 55kW with a terminal P.D. of 220V. The resistance of the armature is 0.02Ω and that of the field, which is the same for each case, is 110Ω.

20. A single-phase, 50Hz transformer has a core with a square cross-section, each side being 270mm. The transformation ratio is 3500/440V and the maximum flux density in the core is not to exceed 1.4T. Find the number of turns of the windings required if the frequency is 50Hz.

21. A 500V, 3-phase alternator supplies a balanced delta-connected load in parallel with a balanced star-connected load. The delta load is 30kW at a power factor of 0.92 (leading) and the star load is 40kW at a power factor of 0.85 (lagging). Calculate the line current and the power factor of the supply.

22. A 440V shunt motor takes an armature current of 30A at 700 rev/min. The armature resistance is 0.7Ω. If the flux is suddenly reduced by 20%, to what value will the armature current rise momentarily? Assuming unchanged resisting torque to motion, what will be the new steady values of speed and armature current? Sketch graphs showing armature current and speed as functions of time during the transition from the initial to the final steady state condition.

23. A load takes 250A at 240V and a power factor of 0.8 (lagging) from a 50Hz supply. The supply cable is then operating at its full rating. Find graphically, the additional power that might be supplied, without the cable exceeding its rating, when an 800μF capacitor is connected across the load.

24. Two shunt generators work in parallel. Each has an armature resistance of 0.015Ω and a shunt-field resistance of 85Ω. Machine A is excited so that its e.m.f. is 600V, while the other machine B is excited so that its e.m.f. is 620V. What is the output of each machine when they jointly supply a load of 2500A and what is the bus bar voltage?

25. A 230V motor, which develops 10kW at 1000 rev/min with an efficiency of 85%, is to be used as a generator. The armature resistance is 0.15Ω and the shunt-field resistance is 220Ω. If it is driven at 1080 rev/min and the field current adjusted to 1.1A, with a shunt regulator, what output in kW may be expected as a generator if the armature copper loss was kept down to that when running as a motor?

26. Two coils are connected in parallel across a 220V, 60Hz supply. At the supply frequency, their impedances are 16Ω and 25Ω respectively, and their resistances are 3Ω and 7Ω respectively. Find the current in each coil, the total current and the total power. Draw a complete phasor diagram for the system.

27. A shunt-wound generator has a magnetisation curve given in the table below. The total resistance in the field circuit is 20Ω and the armature resistance is 0.02Ω. With the machine on load, estimate the e.m.f. generated and the armature current when the terminal voltage of the machine is 140V.

Field current (I_f) – amperes	1.2	2.8	5.0	7.0	7.7	9.0	11.0
Generated e.m.f. (E) – volts	46	88	126	149	154	162	168

28. A 12-pole, 3-phase, delta-connected alternator runs at 600 rev/min and supplies a balanced star-connected load. Each phase of the load is a coil of resistance 35Ω and inductive reactance 25Ω. The line terminal voltage of the alternator is 440V. Determine (a) frequency of supply, (b) current in each coil, (c) current in each phase of the alternator and (d) total power supplied to the load.

29. A coil of 100Ω resistance and 0.1H inductance is connected in series with a 0.1μF capacitor to a 230V variable frequency A.C. supply. Calculate the resonant frequency and the P.D. across the capacitor at resonance.

30. Three D.C. generators connected in parallel each supply a load of 640A to a set of 220V bus bars. The e.m.f. of one generator is raised from 230V to 235V. If the load and the resistances are constant, determine the current supplied from each generator and the voltage at the bus bars.

31. A balanced delta-connected load and a balanced star-connected load are connected in parallel to a 220V, 3-phase supply. The delta-connected load takes a total power of 50kW at a power factor of 0.75 (lagging), and the star-connected load, 40kW at a power factor of 0.62 (leading). Calculate the power, volt amperes and power factor of the supply.

32. A section of supply cable AB 1km long has a fault to earth such that, when end B is disconnected, the resistance measurement from end A to earth is 5Ω. When end A is disconnected, the resistance reading from end B to earth is 3Ω. The length of the cable AB has a resistance of 4Ω when intact. Find the distance of the fault from end A.

33. A series-connected D.C. motor has a field and armature resistance of 0.1Ω and runs at 600 rev/min when taking a full-load current of 100A from a 210V supply. Calculate the speed of the motor when the torque is reduced 75%.

34. Each phase of a star-connected load consists of a resistor of 14Ω in parallel with a 400µF capacitor. Calculate the line current, power and power factor when the above load is connected to a 440V, 60Hz, 3-phase supply. What power would be dissipated in the load, if it is reconnected in delta?

35. A ring main, 900m long, is supplied at a point A at a P.D. of 220V. At a point R, 240m from A, a load of 45A is drawn from the main, and at a point C, 580m from A, measured in the same d rection, a load of 78A is taken from the main. If the resistance of the main (lead and return) is 0.25Ω per kilometre, calculate the current that will flow in each direction round the main from the supply point A and the P.D. across the main, at the load where it is lowest.

36. A non-inductive coil of 6Ω resistance is connected in parallel with an inductive coil of 3Ω resistance and 9Ω impedance at 50Hz. If a P.D. of 110V is applied to the terminals, find the current in each coil and in the mains. If a capacitor of 600µF is connected in parallel with these coils, calculate the total current.

37. A 3-phase transformer has 560 turns on the primary and 42 turns on the secondary. The primary windings are connected to a line voltage of 6.6kV. Calculate the secondary line voltage when the transformer is connected (a) star-delta or (b) delta-star.

38. Two batteries, of e.m.f.s 220V and 225V and internal resistances of 0.2Ω and 0.3Ω respectively, are connected in parallel to supply a load resistor of 10Ω. Find the current supplied by each battery and the terminal voltage.

39. A kettle, when connected to a 220V D.C. supply, boils 1 litre of water initially at 11°C in 3.5min. Calculate the percentage time difference when the water is boiled by connecting the kettle to a 220V, 50Hz A.C. supply. The inductance of the element is 0.05H. One litre of water has a mass of 1kg and its specific heat capacity is 4.2kJ/kg°C.

40. In a shunt motor the 4 field coils are connected in *series*. Each coil is wound to give 750 ampere-turns, the length of each turn being 450mm. At the safe working temperature, there are 45W dissipated at each coil. If the supply voltage is 220V, find (a) the field current, (b) the diameter of the wire and (c) the length of wire in each coil. Take the resistivity of copper as $2.0 \times 19^{-8}\Omega$m.

SOLUTIONS TO TYPICAL FIRST CLASS EXAMINATION QUESTIONS

1. This problem requires some understanding of the electrical connections of a Ward-Leonard (1861–1915) motor control system, which enables a D.C. generator to drive at constant speed, ideal for lift systems and one that is still in use today.

   Armature-resistance voltage drop 8.5V $R_a = 0.0288\Omega$

   $$\therefore I_a = \frac{8.5}{0.0288} = 295.15A$$

   Shunt-filed current $I_{sh} = \frac{110}{65} = 1.692A$

   Shunt ampere-turns/pole = $1.692 \times 1000 = 1692$

Series ampere-turns/pole = 325 × 11 = 3575

Total ampere-turns/pole = 1692 + 3575 = 5267

Plot the graph. From this Φ = 3150 kilo webers per pole.

From the given expression:

$T = 295.15 \times 3150 \times 10^{-4} \times 6.8 - 15 = 617.2\text{Nm}$

Output power $= \dfrac{2\pi NT}{60} = \dfrac{2 \times \pi \times 370 \times 617.2}{60} = 24\text{kW}$

2. Total active power of combined load, P = 50 + 90 = 140kW

Total apparent power of combined load, $S = \dfrac{140}{\cos\phi} = \dfrac{140}{0.8} = 175\text{kVA}$

Total reactive power of combined load, Q = 175 × 0.6 = 105kVAr

Active power of original load, P_1 = 90kW

Apparent power of original load, $S_1 = \dfrac{90}{\cos\phi} = \dfrac{90}{0.6} = 150\text{kVA}$

Reactive power of original load, Q_1 = 150 × 0.8 = 120kVAr

Thus the reactive power is reduced by 120 – 105 = 15 kVAr and this kVAr figure is that of the motor, Q_2 operating at a leading power factor. Thus reactive power Q_2 of motor = 15kVAr

Also apparent power of motor, $S_2 = \sqrt{50^2 + 15^2} = 52.2\text{kVA}$

Power factor of motor $= \dfrac{P_2}{S_2} = \dfrac{50}{52.2} = 0.96$ (leading)

3. Although the transformer has not been dealt with in any detail in this book, this problem can be worked from first principles.

Thus efficiency $(\eta) = \dfrac{\text{output}}{\text{input}} = \dfrac{\text{output}}{\text{output} + \text{losses}}$

In a transformer there are no rotational losses and the expression can be written:

$$\eta = \frac{\text{output (kW)}}{\text{output (kW)} + \text{copper loss (kW)} + \text{iron loss (kW)}}$$

Thus $\eta = \dfrac{\text{kVA} \cos \phi}{\text{kVA} \cos \phi + P_c + P_{Fe}}$

Here $P_{Fe} = 2.75\text{kW}$

Now primary current $= \dfrac{175\,000}{6600} = 26.51\text{A}$

Primary copper loss $= \dfrac{26.5^{-2} \times 0.4}{1000} = 0.2812\text{kW}$

Similarly, secondary current $= \dfrac{175\,000}{440} = 397.7\text{A}$

Secondary copper loss $= \dfrac{397.7^2 \times 0.0015}{1000} = 0.237\text{kW}$

Thus $\eta = \dfrac{175 \times 0.9}{175 \times 0.9 + (0.28^{\cdot}2 + 0.237) + 2.75}$

$= 0.98$

Efficiency on full load = 98%.

4. The O.C. characteristic is plotted as shown and a tangent drawn. The resistance value obtained from this tangent is the critical resistance. Consider a field current of 1A. The voltage value for this current, reading off the critical resistance line, is 120V. The critical resistance is $= \dfrac{120}{1} = 120\Omega$.

According to the given data, the field resistance is $\dfrac{120}{2} = 60\Omega$ and the required

O.C. voltage is obtained by taking any field-current value, say 2A, and finding the voltage value for this current. Thus $2 \times 60 = 120$V. Plot this value and join the point to the origin to obtain the field voltage-drop line. The intersection with the O.C.C. gives the required voltage = 146V.

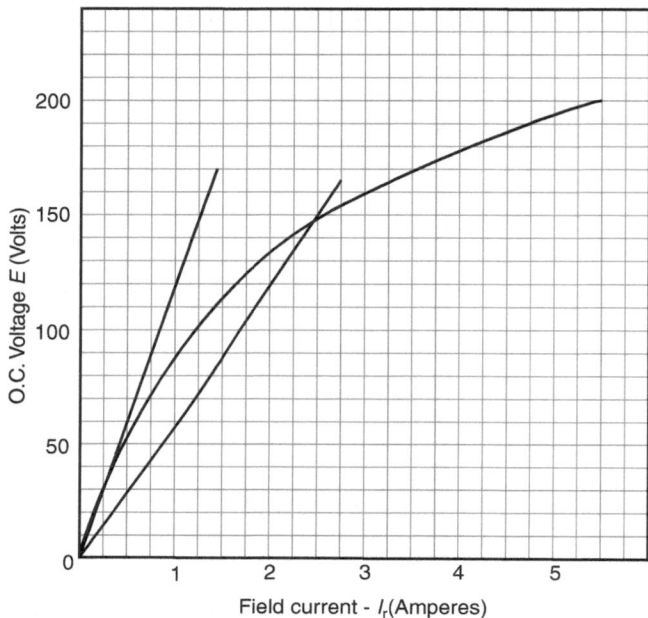

▲ **Figure 1** *O.C. voltage vs field current*

Again, from the O.C.C., an e.m.f. of 135V requires a field current of 2.1A. It must be remembered that the field resistance has *not been altered* and is 60Ω. The voltage necessary to maintain 2.1A through this field resistance is the terminal voltage.

Terminal voltage = 2.1 × 60 = 126V

Armature voltage drop = 135 − 126 = 9V

Also $9 = I_a R_a$

$$\therefore R_a = \frac{9}{200 + 2.1} = 0.045\Omega$$

Note. I_a = line current + shunt-field current.

5. Using the expression $E_{av} = \dfrac{N(\Phi_1 - \Phi_1)}{t}$

We have $E_{av} = \dfrac{60\,[(0.4 \times 10^{-3}) - (-0.4 \times 10^{-3})]}{25 \times 10^{-3}}$ volts or $E_{av} = 1.92$V.

Also $E_{av} = L \times$ average rate of change of current

or $1.92 = L \times \dfrac{I}{t}$ Thus $1.92 = L \times \dfrac{12}{12.5 \times 10^{-3}}$

Note. The time taken for the current to fall to zero has been taken.

so, $L = \dfrac{1.92 \times 12.5 \times 10^{-3}}{12} = 0.002$H

The inductance of the coil is 2mH

L can also be found from first principles, as the inductance value is determined from the flux-linkages per ampere

Thus $L = \dfrac{N\Phi}{I} = \dfrac{60 \times 0.4 \times 10^{-3}}{12} = 0.002$H

6. Let I be the current in the short section of the ring, i.e. in the 400m length. Therefore $(110 - I)$ is the current in the $(2000 - 400) = 1600$m length.

Resistance of 400m of cable (double conductor)

$= \dfrac{0.032 \times 800}{1000} = 0.0256\Omega$

Resistance of 1600m of cable

$= 0.0256 \times 4 = 0.1024\Omega$

Since points X and Y are connected by both sections of the ring, it follows that the voltage drop in the shorter section = the voltage drop in the longer section

or $I \times 0.0256 = (110 - I) \times 0.1024$.

and $0.0256I = 0.1024 \times 110 - 0.1024I$

or $0.128I = 0.1024 \times 110$

$\therefore I = 88$A

Current in shorter section $= 88$A

Current in longer section $= 110 - 88 = 22A$

Voltage at load $= 220 - (88 \times 0.0256) = 217.75V$

7. Cos $\phi = 0.866$. $\therefore$ $\phi = 30°C$. Thus current lags voltage by 30°.

Maximum value of voltage $= \dfrac{311.2}{0.707} = 440V$

Maximum value of current $= \dfrac{70.7}{0.77} = 100A$

Time for 1 cycle $= \dfrac{1}{50}$ second $= 0.02s$

Time for $\dfrac{1}{2}$ cycle $= 0.01s$

Since voltage is sinusoidal, then $v = 440 \sin (2 \times 180 \times 50 \times t)$

When $t = 0.001s$, $v = 440 \sin (2 \times 180 \times 50 \times 0.001)$

$= 440 \sin 18° = 136V$

When $t = 0.002s$, $v = 440 \sin 36° = 259V$

Since voltage is sinusoidal, then t $= 0.003s$, $v = 440 \sin 54° = 356V$

When $t = 0.004s$, $v = 440 \sin 72° = 418V$

When $t = 0.005s$, $v = 440 \sin 90° = 440V$

When $t = 0.006s$, $v = 418V$, etc.

The voltage wave is plotted to a time base t, as shown.

Similarly for the current wave.

Since $i = 100 \sin (2 \times 180 \times 50 \times t)$

When $t = 0.001s$, $i = 100 \sin 18° = 30.9A$

When $t = 0.002s$, $i = 100 \sin 36° = 58.8A$

When $t = 0.003s$, $i = 100 \sin 54° = 80.9A$

When $t = 0.004s$, $i = 100 \sin 72° = 95.1A$

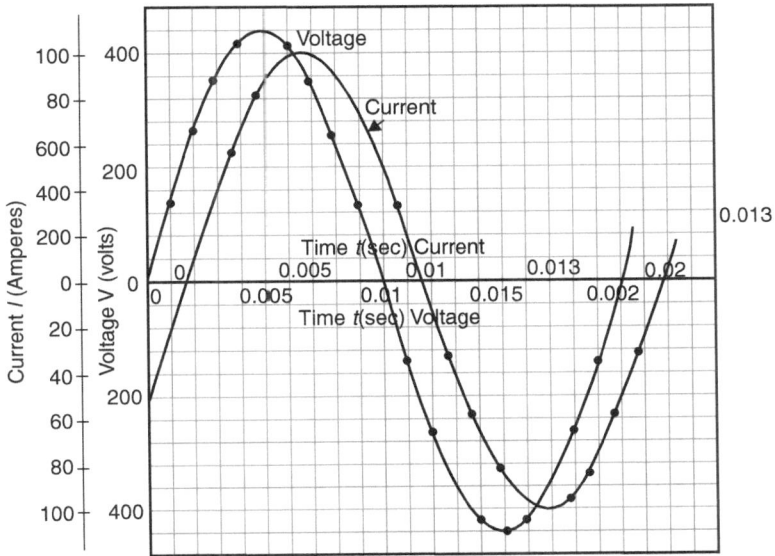

▲ **Figure 2** *Current and voltage vs time*

When $t = 0.005$s, $i = 100 \sin 90° = 100$A

When $t = 0.006$s, $i = 95.1$A, etc.

Draw the current wave to a new scale with its zero displaced from that of the voltage-time scale by 0.00166s. Use the ordinates derived above for this new time scale.

Note. If a time of $90° = 0.005$s, $30° = 0.00166$s.

Read off the required answers from the original voltage-time scale on both the voltage and current waveforms.

Thus when

$t = 3 \times 10^{-3} = 0.003$s, $v = 355$V and $i = 38$A

$t = 0.006$s, $v = 415$V and $i = 96$A

$t = 0.011$s, $v = -140$V and $i = 25$A

$t = 0.018$s, $v = -260$V and $i = 93$A

8. (a) Assume the meter indicates correctly at 20°C when the resistance is 1000Ω. Let I_1 be the meter current at this temperature. Also let R_2 and T_2 be the resistance and temperature at 50°C and let R_1 and T_1 be the resistance and temperature at 20°C.

Then from the resistance to temperature relation,

$$\frac{R_2}{R_1} = \frac{R_0(1+\alpha T_2)}{R_0(1+\alpha T_1)} \quad \therefore \quad R_2 = \frac{1000(1+\alpha 50)}{(1+\alpha 20)} \text{ ohms}$$

or $R_2 = \dfrac{1000(1+0.004\ 28 \times 50)}{1+0.004\ 28 \times 20}$

and $R_2 = 1118.2\Omega$

Meter current I_2 at 50°C $= \dfrac{V}{118.2}$ where $I_1 = \dfrac{V}{1000}$

The voltage across the meter can be assumed constant.

$$\therefore \quad \frac{I_2}{I_1} = \frac{1000}{1118.2} \text{ or } I_2 = 0.894I_1 \text{ amperes}$$

Now meter deflection is proportional to current

$$\therefore \quad \text{Percentage error} = \frac{\text{difference in readings}}{\text{true reading}} \times 100$$

or $\dfrac{I_1 - I_2}{I_1} \times 100 = \dfrac{I_1 - 0.894I_1}{I_1} \times 100$

Thus percentage error $= \dfrac{0.106I_1}{I_1} \times 100 = 10.6\% \ \text{(low)}$

(b) If the series resistor is made from manganin, its value at 50°C does not vary and the resistance of the 5Ω copper coil is given by R_2

$$\therefore \quad \frac{R_2}{R_1} = \frac{R_0(1+0.004\ 28 \times 50)}{R_0(1+0.004\ 28 \times 20)}$$

or $R_2 = 5.58\Omega$

The resistance of the meter at t_2 is now $995 + 5.58$ ohms $= 1000.58\Omega$

New current $I_2 = \dfrac{V}{1000.58}$ or as before $I_1 = \dfrac{V}{1000}$ A

$\therefore \dfrac{I_2}{I_1} = \dfrac{1000}{1000.58}$ So new current $I_2 = 0.9994I_1$

$\therefore$ Percentage error, as deduced before,

$= \dfrac{I_1 - 0.9994I_1}{I_1} \times 100 = \dfrac{1 - 0.9994}{1} \times 100$

or percentage error $= 0.0006 \times 100 = 0.06\%$ (low)

9. Using the expression:

   Inductance (in henries) = flux-linkages per ampere

   Then $L = \dfrac{N\Phi}{I} = \dfrac{8 \times 1200 \times 50 \times 10^{-3}}{5} = 96H$

   Also $E_{av} = L \times$ rate of change of current

   $= 96 \times \dfrac{5}{50 \times 10^{-3}} = 9600V$

   Thus the value of induced e.m.f. is 9.6kV.

10. (a) Although no work has been done in this book on the transformer, it can be stated that the voltages induced in the primary and secondary windings are in direct proportion to their turns.

    Thus $\dfrac{V_1}{V_2} = \dfrac{N_1}{N_2}$ or $V_1 = V_2 \times \dfrac{N_1}{N_2}$

    Hence $V_1 = 220 \times \dfrac{1875}{75} = 220 \times 25$

    and primary voltage = 5500V or 5.5kV

    (b) The kVA rating applies equally to the primary and secondary sides.

Thus primary current $= \dfrac{\text{kVA rating} \times 1000}{V_1}$

$$= \dfrac{550 \times 1000}{5500} = 100\text{A}$$

Secondary current $= \dfrac{\text{kVA rating} \times 1000}{V_2}$

$$= \dfrac{550 \times 1000}{220} = 2500\text{A}$$

(c) This part of the question is answered by a knowledge of the e.m.f. formula for the transformer. This is developed in Volume 7 but can be memorised.

Thus $V_1 = 4.44\Phi_m f N_1$ volts.

or $5500 = 4.44 \times \Phi_m \times 50 \times 1875$

Thus $\Phi_m = \dfrac{5500}{2.22 \times 100 \times 1875}$

$\quad = 13.2 \times 10^{-3}$ webers

Maximum value of flux $= 13.2\text{mWb}$

11. Since $I_f = \dfrac{220}{110} = 2\text{A}$, then $I_a = I_L - I_f = 6 - 2 = 4\text{A}$

Input power $= 220 \times 6 = 1320\text{W}$

Copper loss $\left(\text{armature}\right)$, $I_a^2 R_a = 4^2 \times 0.25 = 16\times$

Copper loss (field), $I_f^2 R_f = 2^2 \times 110 = 440\text{W}$

Total copper loss $= 444\text{W}$

Rotational loss $= 1320 - 444 = 876\text{W}$. These are the losses attributable to iron, friction and windage.

When the current is 62A, the input is $220 \times 62 = 13\,640$W

The output = input – losses (all values in watts)

$$= 13\,640 - \text{copper losses} - \text{rotational losses}$$

$$= 13\,640 - (60^2 \times 0.25 + 2^2 \times 110) - 876$$

$$= 13\,640 - (900 + 440) - 876$$

$$= 13\,640 - 2216 = 11\,424\text{W}$$

So efficiency $= \dfrac{11\,424}{13\,640} = 0.837$ or 83.7%

12. A basic definition of power factor is that it is the ratio of the 'active power' to the 'apparent power' being expended in a circuit.

Thus power factor, $\cos \phi = \dfrac{\text{active power}}{\text{apparent power}} = \dfrac{P}{S}$

or active power = apparent power × power factor

For the problem:

$P = S \cos \phi$ or $\text{kW} = \text{kVA} \times \cos \phi$

Hence $560 = \text{kVA} \times 0.7$

Thus apparent power supplied $= \dfrac{560}{0.7} = 800\text{kVA}$

With the power factor increased to 0.8 and the apparent power kept constant, the new 'active power' $= 800 \times 0.8 = 640\text{kW}$.

The increase in active power would be:

$640 - 560 = 80\text{kW}$

13. Resistance of heater $= 6.5\Omega$ Working voltage 100V

$\therefore$ Rated current $= \dfrac{100}{6.5} = 15.38\text{A}$

Reactance of coil, $X_L = 2\pi f L$

or $X_L = 2 \times \pi \times 50 \times 0.01 = 3.14\Omega$

Impedance of circuit, $Z = \sqrt{6.5^2 + 3.14^2}$

or $Z = 7.22\Omega$

Circuit voltage would be $= 15.38 \times 7.22 = 111.04V$

Applied voltage should be 111.04V to give 100V on heater.

If frequency rose to $50 + \left(\dfrac{5}{100} \times 50\right) = 52.5Hz$, X_L would rise in proportion.

$\therefore$ New reactance $X_{L_1} = 3.14 \times \dfrac{52.5}{50} = 3.297\Omega$

New impedance $Z_1 = \sqrt{6.5^2 + 3.297^2}$

or $Z_1 = 7.288\Omega$

New heater current $= \dfrac{V}{Z_1} = \dfrac{111.04}{7.288} = 15.23A$

New voltage across heater terminals $= 15.23 \times 6.5 = 98.995V$

Change of voltage $= 100 \times 98.995 = 1.005V$

14. Force on 1 conductor:

$F = BI\ell$ newtons

$= 0.8 \times 20 \times 400 \times 10^{-3}$ newtons $= 6.4N$

Force due to all *active* conductors

$= \dfrac{2}{3} \times 246 \times 6.4 = 164 \times 6.4 = 1050N$

(a) Torque produced $= 1050 \times 150 \times 10^{-3} = 157.5Nm$

(b) Shaft power developed $= \dfrac{2\pi NT}{60}$ Watts

$\dfrac{2 \times \pi \times 500 \times 157.5}{60} = 8.24kW$

15. Output from motor = 45kW

$$\text{Input to motor} = \frac{45 \times 1000 \times 100}{93} \text{ watts} = 48.4\text{kW}$$

But 3-phase power (watts) $= \sqrt{3}VI \cos \phi$

$$\therefore 48.4 \times 10^3 = \sqrt{3} \times 440 \times I \times 0.88$$

$$I = \frac{48.4 \times 10^3}{1.732 \times 440 \times 0.88} = 72\text{A}$$

16. Let n = the no. of cells in series.

Battery e.m.f. $= n \times 1.5$ volts

Battery internal resistance $= n \times 0.5$ ohms

Also since $P = I^2R$, then for the load: $2 = I^2 \times 100$

$$\text{or } I = \frac{\sqrt{2}}{10} = 0.1414\text{A}$$

Also for the circuit, current I is given by:

$$I = V/R = \frac{1.5n}{100 + 0.5n} \text{ or } \frac{1.5n}{100 + 0.5n} = 0.1414$$

$$\therefore 1.5n = 14.14 + 0.0707n$$

$$\text{or } 1.4293n = 14.14$$

giving $n = 10$ (approx.)

$$\text{As a check: } I = \frac{1.5 \times 10}{100 + (0.5 \times 10)} = 0.143\text{A}$$

So power dissipated $= 0.143^2 \times 100$

$$= 2.045\text{W}$$

or the given rating, 2W (approx.)

17. Total internal resistance = $6 \times 0.1\Omega = 0.6\Omega$. For maximum power; external load resistance and internal battery resistance are equal.

Thus load resistance = 0.6Ω

Load current $I = \dfrac{E}{R + R_i} = \dfrac{6 \times 2.2}{0.6 + 0.6}$

$$= 11\text{A}$$

$\therefore$ Maximum power $P = I^2 R$

$$= 11^2 \times 0.6$$

$$= 72.6\text{W}$$

An **alternative solution** is possible using calculus.

$$I = \dfrac{E}{R + R_i} \text{ and } P = I^2 R$$

$$\therefore P = \left(\dfrac{E}{R + R_i}\right)^2 R = \dfrac{E^2 R}{R^2 + 2RR_i + R_i^2}$$

$$= \dfrac{E^2}{R + 2R_i + R_i^2 R^{-1}}$$

$$= \dfrac{13.2^2}{R + 1.2 + 0.6^2 R^{-1}}$$

$$\therefore P = \dfrac{174.24}{R + 1.2 + 0.36 R^{-1}}$$

P is maximum when the denominator is minimum. Differentiate denominator with respect to R.

Thus $\dfrac{d(R + 1.2 + 0.36 R^{-1})}{dR} = 1 + 0 - 0.36 R^{-2}$

$$= 0 \text{ for min. value}$$

$1 + 0 - 0.36R^{-2} = 0$

$\therefore R = 0.6\Omega$

The second differential determines whether R is maximum or minimum value.

$$\frac{d^2(R + 1.2 + 0.36R^{-1})}{dR^2} = \frac{d(1 + 0 - 0.36R^{-2})}{dR}$$

$$= 0.72R^{-3}$$

Since the second differential is positive, R is a minimum value.

Thus when $R = 0.6\Omega$ P is a maximum value.

Calculate power as shown previously.

18. (a) Apparent power $S_a = 200\text{kVA}$ Reactive power $Q_a = S_a \sin \phi_a = 0\text{kVAr}$

(b) Active power $P_b = 400\text{kW} \cos \phi_b = 0.8 \sin \phi_b = 0.6$

$$S_b = P_b/\cos \phi_b = \frac{400}{0.8} = 500\text{kVA}$$

$Q_b = S_b \sin \phi_b = 500 \times 0.6 = 300\text{kVAr}$

(c) $P_c = 200\text{kW}$

Total active power of loads, $P = 200 + 400 + 200 = 800\text{kW}$

Total apparent power of all loads, $S = \dfrac{P}{\cos \phi}$

$$= \frac{800}{0.97} = 824.74\text{kVA}$$

Also $\cos \phi = 0.97$. $\therefore \phi = 14°4'$ and $\sin \phi = 0.234$

So reactive power of all loads, $Q = S \sin \phi = 824.74 \times 0.243$

$= 200.4\text{kVAr}$

Thus lagging kVAr value is reduced by $300 - 200.4 = 99.6$. This must therefore be the leading reactive power Q_c of the synchronous motor. Apparent power rating of motor

$$S_c = \sqrt{200^2 - 99.6^2}$$

$$= 223.3\text{kVA}$$

Power factor of motor, $\cos \phi_c = \dfrac{200}{223.3} = 0.89$ (leading)

19. Motor condition

Input power $= 55\text{kW}$ Line current $= \dfrac{55\ 000}{220} = 250\text{A}$

Field current $= \dfrac{220}{110} = 2\text{A}$

Armature current $= 250 - 2 = 248\text{A}$

Back e.m.f. $= 220 - (248 \times 0.02) = 220 - 4.96$

$$= 215.04\text{V}$$

Generator condition

Output power $= 55\text{kW}$ Line current $= \dfrac{55\ 000}{220} = 250\text{A}$

Field current $= \dfrac{220}{110} = 2\text{A}$

Armature current $= 250 + 2$

$$= 252\text{A}$$

Generated e.m.f. $= 220 + (252 \times 0.02)$

$$= 225.04\text{V}$$

As e.m.f. is proportional to speed, then $\dfrac{225.04}{215.04} = \dfrac{N_2}{500}$

or $N_2 = \dfrac{225.04}{215.04} \times 500$

Thus generator speed $= 1.0464 \times 500$

$$= 523.3 \text{ rev/min}$$

20. If $B_m = 1.4T$ and area of core $= 0.27 \times 0.27 = 0.0729 m^2$

Then $\Phi_m = 1.4 \times 0.0729 = 0.102 Wb$

Substituting in the formula:

$$V_1 = 4.44 \Phi_m f N_1 \text{ or } N_1 = \frac{3500}{4.44 \times 0.102 \times 50} \text{ turns}$$

$$\therefore N_1 = \frac{3500}{2.22 \times 10.2} = 154.7 \text{ turns}$$

or primary turns = 155 (approx.)

$$\text{Secondary turns} = \frac{440}{3500} = \frac{N_2}{155}$$

$$\text{or } N_2 = 155 \times \frac{44}{350} = 19.5 \text{ turns}$$

Thus secondary turns = 20 (approx.)

21. Delta-connected load

Active power, $P_1 = 30kW$ at a power factor or 0.92 (leading)

$$\therefore \text{ Apparent power, } S_1 = \frac{30}{0.92} = 32.61 kVA$$

$\cos \phi = 0.92 \therefore \phi = 22°56' \sin \phi = 0.3896$

So the reactive power, $Q_1 = 32.61 \times 0.3896 = 12.7 kVAr$

Star-connected load

Active power, $P_2 = 40kW$ at a power factor of 0.85 (lagging)

$$\text{Apparent power, } S_2 = \frac{40}{0.85} = 47.1 kVA$$

$\cos \phi = 0.85 \therefore \phi = 31°37' \sin \phi = 0.5242$

Thus reactive power, $Q_2 = 47.1 \times 0.5242 = -24.7\text{kVAr}$

Note. −ve sign given to the lagging reactive power value to distinguish it from the leading reactive power value of the other load.

Total active power on alternator, $P = P_1 + P_2$

$$= 30 + 40 = 70\text{kW}$$

Total reactive power loading, $Q = Q_1 + Q_2$

$$= 12.7 - 24.7$$

$$= -12\text{kVAr}$$

So apparent power, loading on alternator is:

$$S = \sqrt{70^2 + 12^2}$$

$$= 71.06\text{kVA}$$

Again 3-phase kilovolt amperes is given by $\dfrac{\sqrt{3}VI}{1000}$

Hence $\dfrac{1.732 \times 500 \times I}{1000} = 71.06$

or $I = \dfrac{71.06 \times 2}{1.732} = 82.1\text{A}$

Line current $= 82.1\text{A}$

Supply power factor $= \dfrac{P}{S} = \dfrac{70}{71.06} = 0.98$ (lagging)

The lagging condition is determined from the resultant −ve sign of the total reactive power value.

22. Original conditions:

$E_{b1} = 440 - (30 \times 0.7) = 419\text{V}$

Original flux condition Φ_1

Final flux condition: $\Phi_2 = 0.8\Phi_1$.

Assume no speed change then since generated e.m.f. is proportional to flux:

new E_b value $= 0.8 \times 419 = 335.2$V

and momentary current is given by $\dfrac{V - E_b}{R_a}$ amperes

or $I_a = \dfrac{440 - 335.2}{0.7} = 149.7$A

If the final torque condition $T_2 =$ original torque T_1 and since torque is proportional to flux *and* armature current, then

$T = k\Phi I_a$

Thus we can write $T_1 = T_2$ or $k\Phi_1 I_{a1} = k\Phi_2 I_{a2}$

and $I_a = \dfrac{\Phi_1 I_{a1}}{\Phi_2} = \dfrac{\Phi_1 I_{a1}}{0.8\Phi_1} = \dfrac{3.0}{0.8} = 37.5$A

New armature current will be 37.5A

New back e.m.f. $= 440 - 37.5 \times 0.7$

$$= 413.75\text{V}$$

Also since $E_b \propto \Phi N$

We can write $\dfrac{E_{b2}}{E_{b1}} = \dfrac{k\Phi_2 N_2}{k\Phi_1 N_1}$ or $\dfrac{413.75}{419} = \dfrac{0.8\Phi_1 N_2}{\Phi_1 700}$

So $N_2 = \dfrac{413.75 \times 70}{419 \times 0.8} = 864.4$ rev/min

23. Original power transmitted by supply cable is given by:

$P_1 = VI \cos \phi = \dfrac{240 \times 250 \times 0.8}{1000} = 48\text{kW}$

Current taken by an 800μF capacitor is given by:

$$I_c = \frac{V}{X_c} \text{ where } X_c = \frac{1}{2\pi fC} = \frac{10^6}{2\times\pi\times50\times800}$$

Thus $X_c = \dfrac{10^2}{25.12} = 3.98 = 4\Omega$ (approx.)

and capacitor current $= \dfrac{240}{4} = 60A$

There are now 2 currents, $I_L = 250A$, *lagging* the voltage by 36°44′ (Note cos 36°44′ = 0.8) and $I_c = 60A$, *leading* the voltage by 90°. The problem calls for a graphical solution. Draw a voltage ordinate horizontally. Choose a suitable current scale and, from the origin, i.e. left-hand point of the voltage ordinate, draw I_c, to scale, vertically up. Next from the origin draw I_L to scale below the voltage ordinate by 36°44′. Complete the parallelogram for the current vectors, draw and measure the resultant current for magnitude and phase. This part of the problem is worked out here mathematically.

Thus $I = \sqrt{I_L^2 - I_c^2 - 2I_L I_c \cos\theta}$

$\qquad = \sqrt{250^2 + 60^2 - (2\times250\times60\times\cos53°16′)}$

$\qquad = 219.4A$

Assuming the same power of 48kW was being supplied, the new power factor of the supply would be obtained from:

$$P_2 = VI\cos\phi_2 \text{ or } \cos\phi_2 = \frac{480\,000}{240\times219.4}$$

$\qquad = 0.91$ (lagging)

The question is not clear as to the final loading conditions but it is assumed that this power-factor condition of 0.91 (lagging) is maintained. The maximum current would be limited to 250A and the power transmitted would be P_2.

Thus $P_2 = \dfrac{240\times250\times0.91}{1000}$ kilowatts

$\qquad = 54.6kW$

Additional power, $P_2 - P_1 = 54.6 - 48 = 6.6\text{kW}$

24. Although the parallel working of generators has not been considered in this book, this more testing problem can be solved by a direct application of Kirchhoff's laws. Thus, let I_A be the current output from machine A and I_B the current output from machine B. Let V = the common bus bar or terminal voltage.

Then $I_A + I_B = 2500\text{A}$ and 2 voltage equations can be built up as follows, since

$$I_{aA} = I_A + \frac{V}{85} \text{ and } I_{aB} = I_B + \frac{V}{85}$$

Thus $V = 600 - 0.015I_{aA}$

and $V = 620 - 0.015I_{aB}$

or $V = 600 - 0.015\left(I_A + \frac{V}{85}\right)$

or $V = 620 - 0.015\left(I_B + \frac{V}{85}\right)$

giving $V = 600 - 0.015I_A - \frac{0.015V}{85} \quad \dots \text{(i)}$

$V = 620 - 0.015I_B - \frac{0.015V}{85} \quad \dots \text{(ii)}$

Subtracting (i) from (ii),

$0 = 20 - 0.015I_B + 0.015I_A$ or $0 = 20 - 0.015(I_B - I_A)$

and $0 = 20 - 0.015\,[I_B - (2500 - I_B)]$

$= 20 - 0.015\,[I_B - 2500 + I_B]$

$= 20 - 0.03I_B + 37.5$

Thus $I_B = \frac{57.5}{0.03} = 1916.66\text{A}$

$I_A = 2500 - 1916.66 = 583.34\text{A}$

From (i) $V - 600 - (583.34 \times 0.015) - \dfrac{0.015\,V}{85}$

$= 600 - (5.8334 \times 1.5) - \dfrac{0.015\,V}{85}$

or $V = 600 - 8.75 - \dfrac{0.015\,V}{85}$

Whence $V + \dfrac{0.015V}{85} = 591.25$

giving $\dfrac{85V + 0.015V}{85} = 591.25$

or $85.015V = 591.25 \times 85$

and $V = \dfrac{591.25 \times 85}{85.015}$

Thus $V = 590.75$V

Output of machine A $= \dfrac{583.34 \times 590.75}{1000} = 344.6$kW

Output of machine B $= \dfrac{1916.66 \times 590.75}{1000} = 11.32$kW

Note. Although the kW rating of B is correct for the values given in the question, it is high for the accepted sizes of D.C. machine. The bus bar is a common term used in electrical power distribution. It is a strip or bar of metal (usually aluminium, brass or copper) that conducts electricity in switchboards, substations, distribution boards, battery banks or other such electrical systems. Its main purpose is to conduct high current electricity, in some cases tens of thousands of amperes. Their flat or hollow tube shape allows heat to dissipate more efficiently due to their high surface area. Their one purpose is to conduct electricity, and they do not function as structural members.

25. Motor output = 10kW

Motor input $= 10 \times \dfrac{100}{85} = 11.76$kW $= 11\,760$W

Current taken from supply $= \dfrac{11\,760}{230} = 51.31\text{A}$

Shunt-field current, $I_f = \dfrac{230}{220} = 1.045\text{A}$

Armature current, $I_a = 51.13 - 1.045 = 50.085\text{A}$

Back e.m.f., $E_b = 230 - (50.1 \times 0.15) = 222.5\text{V}$

As a generator, speed is increased and flux is increased in proportion to the shunt-field current.

$\therefore E = 222.5 \times \dfrac{1080}{1000} \times \dfrac{1.1}{1.045} = 252.98\text{V}$

Also since armature copper loss is the same as for the motor, armature current must be the same = 50.1A.

$\therefore$ Terminal voltage, $V = 252.98 - (50.1 \times 0.15) = 245.47\text{V}$

Output current = 50.1 − 1.1 = 49A

So output $= \dfrac{245.47 \times 49}{1000} = 12.03\text{kW}$

26. Let I_A = the current in the first coil, then $I_A = \dfrac{220}{16} = 13.75\text{A}$

$\cos \phi_A = \dfrac{3}{16} = 0.187 \quad \phi_A = 79°13' \quad \sin \phi_A = 0.9824$

I_B = the current in the second coil, then $I_B = \dfrac{220}{25} = 8.8\text{A}$

$\cos \phi_A = \dfrac{7}{25} = 0.28 \quad \phi_B = 73°44' \quad \sin \phi_B = 0.96$

Total active components, $I_a = (13.75 \times 0.187) + (8.8 \times 0.28)$ amperes

$= 2.57 + 2.464 = 5.034\text{A}$

Total reactive components, $I_r = (-13.75 \times 0.9824)$

$$= -(8.8 \times 0.96) \text{ amperes}$$

$$= -13.51 - 8.448 = -21.96A$$

Resultant current, $I = \sqrt{5.034^2 + 21.96^2}$ amperes

$$= 22.53A$$

$$\cos \phi = \frac{5.034}{22.53} = 0.223 \text{ (lagging)}$$

Power, $P = \dfrac{220 \times 22.53 \times 0.233}{1000}$ kilowatts $= 1.105kW$

The phasor diagram for this problem can be considered as one of the basics for a parallel circuit and has already been illustrated several times.

27. The magnitude of the e.m.f. generated on O.C. is determined from the point where the field voltage-drop line intersects the O.C.C.

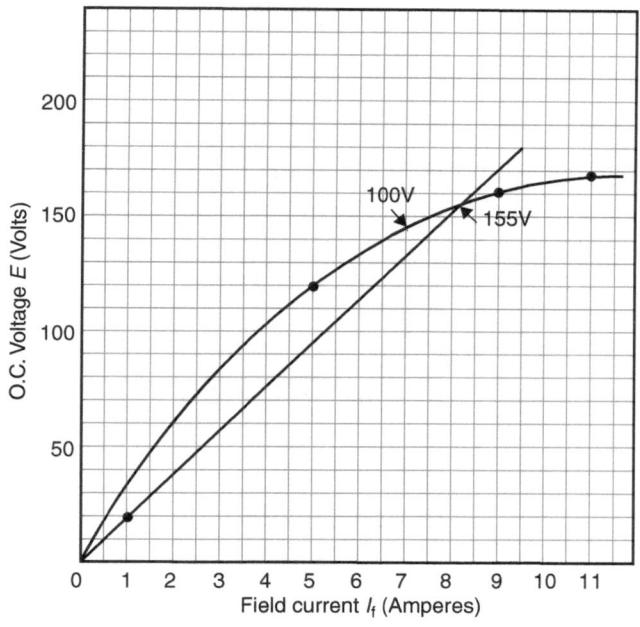

▲ **Figure 3** *O.C. Voltage vs field current*

Plot the O.C.C. and the field voltage-drop line by obtaining a typical value. Thus, for a field current of 4A, the voltage drop would be 4 × 20 = 80V. Plot this point. Draw a line from the origin through this point and produce the line to cut the O.C.C. The induced e.m.f. is then 155V.

The voltage drop across the field is also the terminal voltage of the generator and from a graph a terminal voltage of 140V is obtained when the field current is 7A, i.e. read horizontally for 140V on the voltage scale to determine the corresponding field-current value on the field voltage line. This value is 7A. For this value of field current, however, the e.m.f. generated is 149V.

Also $V = E - I_a R_a$ ∴ $140 = 149 - (I_a \times 0.02)$ or $0.02 I_a = 149 - 140$

Giving an armature current, $I_a \; = \; \dfrac{9}{0.02} = 450\text{A}$

28. A fundamental formula for the alternator and an A.C. supply is:

$f \; = \; \dfrac{PN}{120}$ where P = the number of poles.

(a) Therefore $f \; = \; \dfrac{12 \times 600}{120} = 60\text{Hz}$

(b) Since the load is balanced, the voltage across each phase, $V_{ph} \; = \; \dfrac{440}{\sqrt{3}}$ volts

Impedance of 1 phase of load, $Z_{ph} \; = \; \sqrt{35^2 + 25^2}$ ohms $= 43.01\text{W}$

For a star-connected load, current in 1 phase of load,

I_{ph} = Line current, I

or $I_{ph} \; = \; \dfrac{440}{\sqrt{3} \times 43.01} = 5.91\text{A}$

Thus current in a coil = 5.91A

(c) Current in each phase of alternator,

or $I_{ph} \; = \; \dfrac{I}{\sqrt{3}} = \dfrac{5.91}{\sqrt{3}} = 3.41\text{A}$

(d) Power factor of load $= \cos \phi = \dfrac{R}{Z} = \dfrac{35}{43.01}$

$$= 0.81 \text{ (lagging)}.$$

and total power of load $= \sqrt{3} VI \cos \phi$

$$= \frac{\sqrt{3} \times 440 \times 5.91 \times 0.81}{1000} \text{ kilowatts} = 3.65 \text{kW}$$

29. At resonance $2\pi fL = \dfrac{1}{2\pi fC}$

$$\therefore f = \frac{1}{2\sqrt{LC}}$$

$$= \frac{1}{2\pi\sqrt{0.1 \times 0.1 \times 10^{-6}}}$$

Resonant frequency $= 1592$Hz. At resonance there is *no* resultant reactance, i.e. $R = Z$

$$\therefore I = \frac{V}{R} = \frac{230}{100}$$

$$= 2.3\text{A}$$

$$X_c = \frac{1}{2\pi fC} = \frac{1}{2\pi \times 1592 \times 0.1 \times 10^{-6}} = 1000\Omega$$

P.D. across capacitor $= IX_c = 2.3 \times 1000 = 2300$V.

30. Before adjusting the e.m.f. of one of the machines, the e.m.f. of each machine is considered to be 230V. The bus bar voltage is 220V. Voltage drop due to resistance R of one machine and cables up to bus bars $= 230 - 220 = 10$V.

Current $= 640$A.

$$\therefore \text{Resistance } R \text{ of machine and cables} = \frac{10}{640} = \frac{1}{64} \text{ ohms}$$

Let V = the bus bar voltage under the new condition.

Then current supplied by generator No 1 $= \dfrac{235 - V}{\dfrac{1}{64}}$ A

$$= 64\,(235 - V)$$

Current supplied by generator No. 2 = $(230 - V)$ 64 A

Current supplied by generator No 3 = $(230 - V)$ 64 A

Power supplied by the 3 machines $= V\,(235 - V)\,64 + V\,(230 - V)\,64 + V\,(230 - V)\,64$ W

$$= 64\,(235V - V^2) + 2 \times 64(230V - V^2)$$

Now the original power supplied by 3 machines

$= 3 \times 640 \times 220$ watts $= 422\,400$W or 422.4kW

$\therefore\ 3 \times 640 \times 220 = 64\,(235V - V^2) + 2 \times 64\,(230V - V^2)$

and $6600 = 235V - V^2 + 460V - 2V^2$

or $-3V^2 + 695V = 6600$

Thus $V^2 - 231.66V + 2200 = 0$

Solving for V using the quadratic formula:

$$V = \frac{231.66 \pm \sqrt{231.66^2 - 4 \times 2200}}{2}$$

$$= \frac{231.66 \pm 211.9}{2} = \frac{443.56}{2} = 221.78V$$

Thus bus bar voltage will be 221.8V

Current of machine No. 1 $= \dfrac{231 - 221.78}{\dfrac{1}{64}}$

$$= 13.22 \times 64 = 846A$$

$$\text{Current of machine No. 2} = \frac{230 - 221.78}{\dfrac{1}{64}}$$

$$= 8.22 \times 64 = 526A$$

$$\text{Current of machine No. 3} = \frac{230 - 221.78}{\dfrac{1}{64}}$$

$$= 8.22 \times 64 = 526A$$

Check. Power supplied $= (846 + 526 + 526)\ 221.78$

$$= 1898 \times 221.78\ \text{W}$$

$$= 421\text{kW (approx.).}$$

Using the other root of the quadratic equation we have:

$$V = \frac{231.66 - 211.9}{2} = \frac{19.76}{2} = 9.88V$$

Current of machine No. 1 $= (235 - 9.88) \times 64$

$$= 225.12 \times 64 = 14\,408A$$

Current of machine No. 2 $= (230 - 9.88) \times 64$

$$= 220.12 \times 64 = 14\,088A$$

Current of machine No. 3 as for No. 2.

The above condition, though theoretical, would relate to a bus bar voltage of 9.88V and could be imagined as the result of a 'short-circuit' at the bus bars, where the power of 422kW could be assumed to be dissipated.

31. Delta load

Active power, $P_1 = 50\text{kW}\ \cos \phi = 0.75$

Hence $\phi = 41°25'$ and $\sin \phi = 0.6615$

Apparent power, $S_1 = \dfrac{50}{0.75}$

$$= 66.66\text{kVA}$$

Reactive power, $Q_1 = 66.66 \times 0.6615$

$$= -44.1 \text{kVAr (lagging)}$$

Star load

Active power, $P_2 = 40 \text{kW}$

$\cos \phi = 0.62 \; \phi = 51°41'$ and $\sin \phi = 0.7846$

Apparent power, $S_2 = \dfrac{40}{0.62} = 64.5 \text{kVA}$

Reactive power, $Q_2 = 64.5 \times 0.7846 = 50.6 \text{kVAr (leading)}$

Total power, $P = P_1 + P_2 = 50 + 40 = 90 \text{kW}$

Total reactive power, $Q = Q_1 + Q_2 = -44.1 + 50.6 = 6.5 \text{kVAr (leading)}$

Apparent power, $S = \sqrt{90^2 + 6.5^2} = 90.23 \text{kVA}$

Power factor $= \dfrac{P}{S} = \dfrac{90}{90.23} = 0.988 \text{ (leading)}$

32. Let the resistance of the fault be R ohms

Since the cable resistance itself is 4 ohms, then (Resistance of end A to earth − R) + (Resistance of end B to earth − R) = 4

Or $5 - R + 3 - R = 4$ or $8 - 2R = 4$

giving $2R = 4$ or $R = 2\Omega$

It can be assumed that for a cable, the resistance is proportional to length.

Then from end A to fault = 5Ω and the fault resistance is 2Ω

∴ cable length must have a resistance of 3Ω.

Thus fault must be $\dfrac{3}{4} \times 1000 = 750 \text{m}$ from end A. Practically, this is a useful technique to know!

33. The torque of a motor $T \propto \Phi I_a$ and since, for a series motor, $\Phi \propto I_f$ and $I_f = I_a = I_L$ then

$T \propto I_L^2$ or $T = kI_L^2$

On full load

$E_{bi} = V - I_a R = 210 - (100 \times 0.1) = 200V$

When load torque is reduced to 25% of full-load value, then $T_2 = T_1 \times \frac{1}{4}$ where, T_2

is the new torque and T_1 the original torque.

Then $\dfrac{T_2}{T_1} = \dfrac{kI_{L2}^2}{kI_{L1}^2}$ or $\dfrac{I_{L2}^2}{I_{L1}^2} = \dfrac{1}{4}$

$\therefore I_{L2}^2 = \dfrac{I_{L1}^2}{4}$ or $I_{L2} = \dfrac{I_{L1}}{2} = \dfrac{100}{2} = 50A$

On new load condition

$E_{b2} = 210 - (50 \times 0.1) = 205V$

As $E_b \propto \Phi N$ then $= \dfrac{E_{b2}}{E_{b1}} = \dfrac{k\Phi_2 N_2}{k\Phi_1 N_1}$ Thus $\dfrac{205}{200} = \dfrac{50 \times N_2}{100 \times 600}$.

Note that field current $I_f = I_a = I_L$ is now substituted for flux, since $\Phi \propto I_f$.

Thus $N_2 = \dfrac{205}{200} \times 600 \times 2 = 1230$ rev/min.

Speed of motor on 75% load torque = 1230 rev/min

34. Reactance X_c of capacitor $= \dfrac{1}{2\pi f C} = \dfrac{10^6}{2 \times \pi \times 60 \times 400} = 6.63\Omega$

Since the load is balanced:

The voltage across a phase, $V_{ph} = \dfrac{440}{\sqrt{3}} = 254.5V$

Current in the resistor $I_R = \dfrac{254.5}{14}$

$= 18.18\text{A, in-phase with } V_{ph}$

Current in the capacitor $I_C = \dfrac{254.5}{6.63}$

$= 38.38\text{A, leading } V_{ph} \text{ by } 90°$

Let $I_{ph} =$ the resultant of 18.ˉ8A and 38.38A, which are in quadrature

$\therefore I_{ph} = \sqrt{18.18^2 + 38.38^2}$

= 42.45A. This is also the line current since the load is star connected $\therefore I = 42.45\text{A}$

Power factor of load $= \cos o = \dfrac{I_R}{I_{ph}} = \dfrac{18.18}{42.45}$

$= 0.43 \text{ (leading)}$

Power of load, $P = \sqrt{3}\,VI \cos \phi$

$= \sqrt{3} \times 440 \times 42.45 \times 0.43 \text{ watts } = 13.9\text{kW}$

If the load is in delta, the current per phase would rise in the proportion of

$\dfrac{440}{254.5}$ or $= \sqrt{3} \times \text{original } I_{ph}$

$= \sqrt{3} \times 42.45 \text{ amperes}$

The line current would be $\sqrt{3}$ times this new phase current.

$\therefore \text{ New } I = \sqrt{3} \times \sqrt{3} \times 42.45 = 3 \times 42.45 = 127.35\text{A}$

The power factor of the load will remain the same

So new power, $P = \sqrt{3}\,VI \cos\phi = \sqrt{3}\times440 \times 127.35 \times 0.43$ watts $= 41.7$kW

35. Total load $= 45 + 78 = 123$A. Length BC is $(580 - 240) = 340$m. Let the current in the remaining 320m section AC be I amperes:

Then current in section AC $= I$ amperes

Then current in section AB $= (123 - I)$ amperes

Then current in section BC $= (123 - I - 45)$ amperes

$$= (78 - I) \text{ amperes}$$

Resistance of section AC $= \dfrac{0.25}{1000}\times320 = 0.08\Omega$

Resistance of section AB $= \dfrac{0.25}{1000}\times240 = 0.06\Omega$

Resistance of section BC $= \dfrac{0.25\times340}{1000} = 0.085\Omega$

By Kirchhoff's laws, the voltage drops in either section of the main feeding the load at C are equal.

$\therefore I \times 0.08 = (123 - I)\,0.06 + (78 - I)\,0.085$

or $8I = 6\,(123 - I) + 8.5\,(78 - I)$

giving $8I = 738 - 6I + 663 - 8.5I$

or $22.5I = 738 + 663 = 1401$

Thus current I in section AC $= \dfrac{1401}{22.5} = 62.27$A

Current in section AB $= 123 - 62.27 = 60.73$A

Current in section BC $= 78 - 62.27 = 15.73$A

P.D. at point C $= 220 - (62.27 \times 0.08)$

$$= 215\text{V}$$

P.D. at point B = 220 − (60.73 × 0.06)

$$= 216.36V$$

P.D. at load C is lowest = 215V

36. Let non-inductive coil of 6Ω be designated branch A.

Then

$$I_A \ = \ \frac{110}{6} = 18.33A \ \cos \phi_A \ = \ 1 \ \sin \phi_A \ = \ 0$$

Let inductive coil of impedance 9Ω be designated branch B

Then $I_B \ = \ \dfrac{110}{9} = 12.22A$

$\cos \phi_B \ = \ \dfrac{3}{9} = 0.33 \ (\text{lagging}) \ \sin \phi_B \ = 0.943$

Resolving in active and reactive components

$I_a = I_A \cos \phi_A + I_B \cos \phi_B$

$\quad = 18.33 \times 1 + 12.22 \times 0.33 = 18.33 + 4.033 = 22.363A$

and $I_r = - I_A \sin \phi_A - I_B \sin \phi_B$

$$= - 18.33 \times 0 - 12.22 \times 0.943 = -0 - 11.52 = -11.52A$$

$$\therefore I \ = \ \sqrt{22.36^2 + 11.52^2} = 25.2A$$

Thus current taken from mains is 25.2A

With capacitor of 600µF connected in parallel:

$$X_C \ = \ \frac{1}{2\pi fC} = \frac{10^6}{2 \times \pi \times 50 \times 600} = 5.3\Omega$$

Current $I_C \ = \ \dfrac{110}{5.3} = 20.75A$

Resolving as before, active component I_a remains the same but the reactive component $I_r = 20.75 - 11.52 = 9.23$A and is now vertically upwards, i.e. leading V by 90°.

Resultant current $I = \sqrt{22.36^2 + 9.23^2} = 24.2$A

37. The turns ratio per phase of the transformer are 560 to 42 or $\dfrac{560}{42} = \dfrac{13.33}{1}$. The voltages per phase will be in the same proportion.

(a) With the transformer connected, primary in star and secondary in delta, then:

Primary voltage per phase $= \dfrac{6600}{\sqrt{3}}$ volts

Secondary voltage per phase $= \dfrac{6600}{\sqrt{3} \times 13.33} = 286$V

But for a delta connection, line voltage = phase voltage

or $V = V_{ph} = 286$V

(b) With the transformer connected, primary in delta and secondary in star, then:

Primary voltage per phase = 6600V

Secondary voltage per phrase $= \dfrac{6600}{13.3} = 495.4$V

But for a star connection, line voltage $= \sqrt{3}$ phase voltage or $V = \sqrt{3}V_{ph} = \sqrt{3} \times 495.4 = 858$V

38. Let V = the terminal voltage, I_1, the current given by the 220V battery and I_2 the current given by the 225V battery. The following equations can then be set down.

$220 - 0.2I_1 = V$ ...   (i)

$225 - 0.3I_2 = V$ ...   (ii)

and $10(I_1 + I_2) = V$ ... (iii)

Subtracting (i) from (ii) we have:

$225 - 220 - 0.3I_2 + 0.2I_1 = 0$

Thus $5 = 0.3I_2 - 0.2I_1$, or $I_2 = \dfrac{5 + 0.2I_1}{0.3}$

Using (ii) and (iii) we can write:

$225 - 0.3I_2 = 10I_1 + 10I_2$ and substituting for I_2

$225 - 0.3\left(\dfrac{5 + 0.2I_1}{0.3}\right) = 10I_1 + 10\left(\dfrac{5 + 0.2I_1}{0.3}\right)$

Thus $225 - 5 - 0.2I_1 = 10I_1 + \dfrac{50 + 2I_1}{0.3}$

and $(220 \times 0.3) - 0.06I_1 = 3I_1 + 50 + 2I_1$

or $66 - 50 = 3I_1 + 2I_1 + 0.06I_1$

Hence $16 = 5.06I_1$ $\quad I_1 = \dfrac{16}{5.06} = 3.16A$

Also $I_2 = \dfrac{5 + 0.2I_1}{0.3} = \dfrac{5 + 0.632}{0.3} = 18.77A$

and $V = 10(I_1 + I_2) = 10(3.16 + 18.77) = 219.3V$

39. Heat received by the water = mass × temperature rise × specific heat capacity

= $1 \times (100 - 11) \times 4.2$ kilojoules = 373.8kJ

∴ Energy put into water = 373.8kJ and assuming an efficiency of 100% for the kettle!

Then the energy taken from mains = 373 800J

With a D.C. supply, the current taken would be $\dfrac{373\ 800}{220 \times 3.5 \times 60}$

or $I = 8.09A$

and the resistance of the kettle element $= \dfrac{220}{8.09} = 27.19\Omega$

On A.C., the reactance X of the kettle becomes effective.

$\therefore X = 2\pi f L = 2 \times \pi \times 50 \times 0.05 = 15.7\Omega$

Hence $Z = \sqrt{27.19^2 + 15.7^2} = 31.4\Omega$

Thus with an A.C. supply, current taken will be $\dfrac{220}{31.4}$

or $I = \dfrac{22}{3.14} = 7.06A$

The input power will be given by $P = I^2R$

or $P = 7.06^2 \times 27.19 = 1355W$

The time taken to produce 373 800J is given by:

$t = \dfrac{373\ 800}{1355} = 275.7s$

The percentage time difference between heating with A.C. instead of D.C. would be:

$\dfrac{276 - 210}{210} = 0.314 = 31.4\%$

40. (a) Total power dissipation of coils $= 4 \times 45 = 180W$

If applied voltage $= 220V$, then field current $t = \dfrac{180}{220}$

or $I_f = 0.818A$

(c) No. of turns per field coil $= \dfrac{750}{0.818} = 916.86$

Length of turn $= 450mm$

∴ Length of wire = 450 × 916.86 = 412 587mm

$$= 412.6m$$

(b) Since $R = \dfrac{\rho \ell}{A}$ and $R = \dfrac{220}{4} \times \dfrac{11}{9}$

$$= 67.22\Omega$$

Then $A = \dfrac{\rho \ell}{R} = \dfrac{2 \times 10^{-8} \times 412.6}{67.22} = 12.3 \times 10^{-8}\ m^2$

or $12.3 \times 10^{-8} = \dfrac{\pi d^2}{4}$ where d = diameter of wire

Then $d^2 = \dfrac{12.3 \times 10^{-8} \times 4}{\pi}$

giving $d^2 = \dfrac{15.62}{10^8}$ or $d = \dfrac{\sqrt{15.62}}{10^4} = \dfrac{3.95}{10^4}$ metres

or diameter of wire = $3.95 \times 10^{-4} \times 10^3$mm

$$= 0.395mm$$

INDEX